Jan-Bernd Lohmöller

Latent Variable Path Modeling with Partial Least Squares

Physica-Verlag Heidelberg

Dr. Jan-Bernd Lohmöller
Free University Berlin
Dept. of Political Science
(Otto-Suhr-Institute)
Ihnestraße 21
D-1000 Berlin 33, FRG

Min Öllern toudacht

ISBN 3-7908-0437-1 Physica-Verlag Heidelberg
ISBN 0-387-91363-7 Springer-Verlag New York

CIP-Titelaufnahme der Deutschen Bibliothek

Lohmöller, Jan-Bernd:
Latent variable path modeling with partial least squares / Jan-Bernd Lohmöller. - Heidelberg : Physica-Verl., 1989
ISBN 3-7908-0437-1

TEX typeset by J. B. L.
Printing: Zechnersche Buchdruckerei, Speyer
Bookbinding: Schäffer GmbH u. Co. KG., Grünstadt
7120/7130-543210

1/23/91

Preface

I am greatly indepted to my teachers – and collegues – Professors Heinz Mandl, Rolf Oerter, Rainer B. Pelka, and Jürgen W. Falter, who gave me unceasing encouragement in my work through their usefull questioning and their confronting me with many problems of a challenging nature over a period of many years.

Since 1978 when I first heard a lecture by Professor Herman Wold, I have been excited and captivated by the potentialities proferred by his PLS modeling method. In the ensuing years of collaboration I not only was introduced into and motivated by his scholarly life-style but also personally moved by his attentive and generous attitude. Thanks to his constant inspiration, criticism and encouragement and his "Advice to a Young Scientist" he became what I am proud to call my *Doktorvater*.

Experience with a particular statistical method derives from various sources, namely (a) mathematical reasoning, (b) programming, (c) application and (d) teaching. All four types of experience leave their marks on the language, style, presentation, and type of reasoning in this dissertation: (a) An attempt has been made to make the mathematical reasoning the core of this monograph. (b) The program PLS 1.8 (Lohmöller 1981a) documents how I understand PLS and is understandable to every computer but not, unfortunately, to every reader. So it has to be explained avoiding computer jargon. (c) With every new application new insight is obtained as to what the method is able to do and what not, which may, or may not, require changes in the program and the mathematical treatment. (d) Teaching proves to be the most demanding test for a method. Personally, I don't mind if this monograph is all of "too" mathematical, computerized, applied and didactic at the same time.

Summary

Model construction may be seen as the very essence of scientific work. Within the frame of reference of a model theoretical and empirical content are kept apart in order to prove that they match one another (Sc.1.1). Because of their simple algebraic tractability linear-relation models are often favoured. They are called path models, owing to their graphical representation as path or arrow diagrams. Path models with latent variables (LVP models) are applicable to a broad range of subject matters (Sc.7).

Relations in LVP models may or may not be subject to a causal interpretation (Sc.1.3). For operative use in most applications it is sufficient to specify predictors (Sc.3.1), which leads on to estimation by Wold's Partial Least Squares (PLS) method. The introductory chapter (Sc.1) is followed by a description of the estimation method, Wold's basic method (Sc.2.1), as well as of the extended method (Sc.2.2, Sc.2.3).

The core of this monograph (Sc.3) is devoted to the statistical foundation of the PLS method, in Sc.3.1 by a recapitulation of Wold's predictor specification for one relation; in Sc.3.2 and Sc.3.3 by a reconstruction of the psychometric factor score problem from the PLS theory; in Sc.3.4 by a unifying view on several traditional two-block models like Canonical analysis; in Sc.3.5 by the introduction of the Split Principal Component theorem as a basis for reflections on properties of multi-block models. The most recent developments by Manfred Glang (1988) are not integrated into this monograph.

The extension of PLS modeling to contingency tables (Sc.4) and three-mode data cubes (Sc.6) raises the question (Sc.1.3) as to whether individual scores (case values) are essential to the model (applied model, data structure model), or essential only to the estimation an merely incidential to the theory (abstract model, covariance structure model). PLS is data-structure oriented, while the ML estimation method of LISREL is covariance-structure oriented (Sc.5). The distinction of individual vs. general, of applied vs. abstract, of data vs. covariance modeling gives a rationale for the choice between PLS and ML estimation method. The ensuing difference between PLS and ML estimates are presented and explained in Sc.5.5.

Contents

List of Tables

Chapter 1

Basic Principles of Model Building

1.1 Empirical and Theoretical Concepts

Scientific knowledge of reality comes in form of models. As distinguished from everyday knowledge, scientific knowledge is characterized by the clear distinction between theoretical and empirical aspects. Because the distinction between the two cannot be part of the theoretical or the empirical, the frame of reference has to be introduced as a third element of the model for scientific knowledge. The frame of reference, formulated (more or less) in everyday language, contains a mixture of theoretical ("T") and empirical ("E") contents, as shown in Eq.1.1 which is taken from Wold (1969b):

$$\text{Scientific Knowledge} = \boxed{\boxed{T \Longleftrightarrow E}}, \tag{1.1}$$

where the rectangle marks the frame of reference.

Within the frame of reference, T and E are kept apart in order to prove that they match. Matching (denoted by the two-sided arrow) means that conclusions can be drawn from T about E (deduction) and from E about T (induction). The process of matching E and T is the model validation and can be understood as a general description of the job of science (Cronbach & Meehl 1955, Wold 1969a, 1969b, Bentler 1978, Falter 1977). Matching E and T may involve a reduction of the range of observations E explained by the theory, which is contrary to the basic esthetic qualities of theories: A theory should be as simple, elegant, consistent and general as possible.

1.1.1 Levels of Theory and Data

The process of matching becomes more transparent and controllable when the rigid distinction between E, T, and the frame of reference is relaxed in favour of more levels. Five levels T_1, T_2, E_3, E_2, E_1 may be distinguished, with a matching process between each pair of neighboring levels, as shown in Table 1.1. Each level

Table 1.1: Matching between levels of theoretical and empirical knowledge

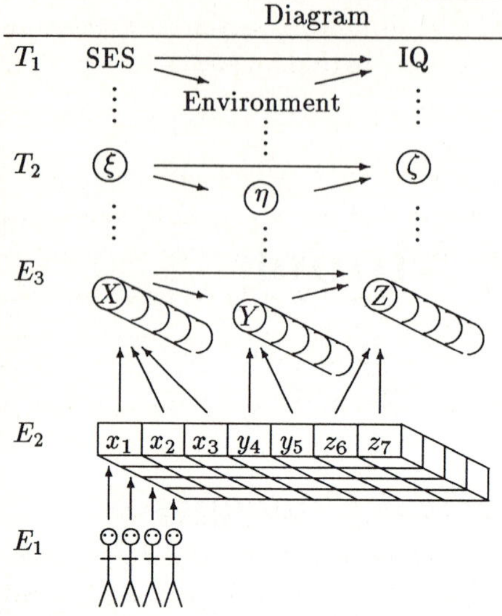

	Level	Elements	Statements
T_1	Substantive theory	Concepts, hypotheses	Causal, functional
T_2	Mathematical model	Random variables	Functional, correlational
E_3	Aggregated data	Compound variables (numbers)	Correlational (functional)
E_2	Data	Numbers	Correlational
E_1	Observations	Observational units	Verbal

has its own language, with its own primitive elements, syntax and semantics. (The example used in Tb.1.1 will be discussed in more detail in Sc.2.5.)

Level of theory. T_1 is the level of substantive theory. The elements of the theory are concepts (constructs) (MacCorquodale & Meehl 1948). The syntax which decides which elements are put together to a correct statement is the syntax of a natural language like German or English. The semantics of a statement is determined within the frame of reference of the whole model, in which the concepts, also, are defined and limited to a meaning which may be different from everyday meaning. What is depicted in the upper part of Tb.1.1 is the theoretical statement that the socioeconomic status (SES) of the parents has a direct influence on the intellectual abilities (IQ) of the children, as well as an indirect influence, which is mediated by the learning environment at home.

Level of model. T_2 is the level of a mathematical-statistical model. The elements are random variables, here denoted by ξ, η, ζ. The syntax is the algebra of expectations and linear algebra. The meaning of model T_2 is given by the correspondence to the substantive theory T_1. When limited to path models with latent variables, the elements can be inner and outer variables or inner and outer residuals, while the syntax controls the formulation of correct equations.

Level of observation. E_1 is the level of observations. The elements are observational units such as individuals, countries or time points. The statements on this level are natural-language statements about attributes (behavior, characteristics) of the units. E_2 is the level of data. The elements are real numbers. The mapping of attributes onto numbers is called measurement. E_3 is the level of functions of the data. The elements are real numbers, again, but they are formed as functions of the measured data. This can include compound variables and estimation functions for parameters.

Matching. The five levels (the number five is somewhat arbitrary) form a chain in which each member has to link up with and to adjust to its neighbor. The mathematical model T_2 must represent as completely as possible the substantive theory T_1. The statistical functions on the level E_3 must be determined from the data on E_2 as the best estimators of the model variables on T_2. The data E_2 must be gathered from the level E_1 so as to be informative for a comprehensive estimation (on E_3) of the latent variables (on T_2) which stand for the constructs (on level T_1).

Residuals. The matching of E and T is never perfect, and there remain unexplained parts on both sides. The deductive specialization of Eq.1.1 shows the empirical content as being partly a function of the theoretical content and partly unexplainable by the theory. The formal notation is:

$$E = \mathrm{E}(T) + \epsilon, \tag{1.2}$$

and the notation in plain words:

data = systematic part + residual part

data = fit + rest.

The residual ϵ may be interpreted as measurement error, prediction error, sample fluctuation, or "systematic" variation which is left out from the systematic part because it accounts only for small and specific parts of the observation (unique variation, specification error).

The inductive specialization of Eq.1.1 shows the theoretical content as being partly a function of the observation and partly unobserved:

$$T = \mathrm{T}(E) + \delta, \tag{1.3}$$

The residual δ may be interpreted as the theoretical surplus or the empirical lack of the model. If, for example, T is the theoretical concept of intelligence, and E is an intelligence test, than $\mathrm{T}(E)$ is the IQ, and δ is that what the test fails to measure, which can be understood as the lack of validity of the test.

In the process of model building the researcher tries to minimize ϵ and δ, to extend E, and to simplify T.

The notion of model is found to be used with different ranges. Some scientists set the concept of model identical to what here is called T, or $\mathrm{E}(T)$; some call the right-hand side of Eq.1.2 the model which then is contrasted to the data; some call Eq.1.2 together with Eq.1.3 the model. Wold's notion of model (our Eq.1.1) goes beyond this, in that he includes the frame of reference into the notion of a model.

Model specification. With respect to the process of model specification we will focus our attention on three dichotomies.

(i) The purpose of a model can be to provide either a prediction (stress on reconstruction or forecasting of present or future observations) or a structural explanation (stress on the hidden structure).

(ii) The data under explanation can be either raw data or summarized data (means, covariances). In the latter case the observational units (cases, people) are not present anymore.

(iii) After the variables, the parameters and the relations of a model are defined, the specifications leading to the estimation of the unknowns can be either statistical assumptions on elements of the theoretical model ($\mathrm{E}(y_n|x_n) = \alpha + \beta x_n$ or $\boldsymbol{\epsilon} \sim \mathrm{N}(\mathbf{0}, \boldsymbol{\Theta})$, e.g.) or specifications on their estimates ($\sum_n (y_n - a - bx_n)^2 \stackrel{!}{=} \min$, e.g.).

1.2 Causation and Prediction

Time, space and causality are the basic categories for our understanding of the world. We "understand" an event if we can locate it at a certain place on the world, at a certain time point, and as being framed in a flow of causes and consequences. We cannot communicate a fact, a rule, an event without anchoring it in time, space and causality. Everyday language is deeply rooted in these categories. Can science do without?

Physicists talk of bent space, but geographers use the concept of space as three-dimensional, straight and continuous, nevertheless. Even if theoretical physicists talk about time as running unsteadily, jerkily, or even backwards, historians will not change their notion of time.

Causality — an unscientific concept? Only time and space are assumed to exist really in the objective world, whereas cause is simply an interpretation, existing only in the theoretical representation of the world. This has led to the idea that causality is not a scientific concept and, as Travers (1981), with special emphasis to educational research, pointed out, belongs to medieval, pre-Newtonian science.

The model for knowledge Eq.1.1 clearly distinguishes between theoretical and empirical contents, and the concept of causality belongs to the theoretical contents. In the matching process evidence is collected in such a way that the empirical content either contradicts or does not contradict a causal theory. Whether a causal interpretation for a given substantive area is theoretically be justifiable, will be discussed later.

When the notion of causality is expelled through the front door, it slips in again through the back door. In a reply to Travers, Ennis (1982) cites from Travers' "Introduction to Educational Research" (1969) sentences and arguments containing words like "result", "hamper", "impact", "contribute to", "interfere with", "influence", "facilitate", "effect", "maximize". Ennis' point is that "to use these words in contexts in which Travers used them is to use causal terminology.

We are only fooling ourselves if we expunge the word 'cause' and use such words instead. We have not thereby stopped using causal concepts" (Ennis 1982:25, cf. the debate on causality in the *Educational Researcher* with Coleman 1981, Fearing 1981, Ericson 1982, Smith 1981).

What Travers is right to avoid is the mechanistic concept of causality which was rejected by philosophers and scientists from Hume onwards to Russell and which clearly is inappropriate to the social sciences. However, the practical need for a concept of causality led to intensive discussions about causality in analytical philosophy (Karl Popper, Mario Bunge, Wolfgang Stegmüller; cf. Falter & Ulbricht 1981:29-40) and from the viewpoint of practical research (Herman Wold, Herbert Simon, Patrick Suppes; cf. Blalock 1961, 1969; Bernert 1983).

Analytical definition of causation. Contemporary analytical philosophy understands a causal explanation as a case of deductive-nomological explanation (Hempel-Oppenheim scheme). The event Y is explained by certain side conditions X and some general rules (the nomological part), say, $Y = f(X)$. To be called causal the relation $Y = f(X)$ must satisfy three conditions, namely

(i) have concomitant variation of X and Y,
(ii) have temporal succession of Y after X, and
(iii) other causes than X must be excluded (the *ceteribus paribus* clause).

Stegmüller requires the nomological part to include at least one deterministic relation. This requirement is criticized by Falter and Ulbricht (1981:38), because under the Stegmüller concept of causality social and human sciences would never be able to formulate causal explanations.

Experimental definition of cause-effect relationship. That the cause produces an effect is most visible when it is possible to manipulate the cause and thus to produce different effects. This is the situation of the classical controlled experiment which Wold (1969b) chose as the paradigm for a causal relationship. By his definition, the relation between stimulus (X) and response (Y) in a controlled experiment is a cause-effect relationship,

$$Y = f(X) + \epsilon, \tag{1.4}$$

where the residual variable ϵ sums up the influence upon Y of uncontrolled factors that are not kept constant in the experiment and are assumed to "vary at random".

Nonexperimental definition of cause-effect relationship. Wold extends the notion of cause-effect relationship from the prototypic use in controlled experiments to nonexperimental relationships. "Let [Eq.1.4] ... be a hypothetical relation included in a nonexperimental model. The model may or may not include the hypothesis that if X could be directed in the sense of a stimulus-response experiment, then Y would present a corresponding variation, as specified in [Eq.1.4]. If the model includes this hypothesis, then [Eq.1.4] is called a cause-effect relation, with X for cause variables, and Y for effect variables" (Wold 1969b:449).

The definition of cause-effect relations thus covers both experimental and non-experimental situations. Cause-effect relationships are stimulus-response relations in both genuine and fictitious, hypothetical, experiments. The notion of causality is purely theoretical, whereas the notion of cause-effect relationship corresponds to the cognitive model Eq.1.1, which distinguishes between T, the hypothesis that Eq.1.4 is a cause-effect relationship, and E, the results of a stimulus-response experiment as the only way of verification (by repeated unsuccessful falsification) of the cause-effect hypothesis. The notion of cause-effect relations suffices to justify the causal terminology used by Travers (1969), and also suffices to be able to talk about "causal analysis", which in full would have to be called "analysis of cause-effect relationships", and to refer to "causal models" instead of "models including cause-effect relationships". The notion of causation is left to philosophy.

A reversible experiment. While it is always clear in a real experiment what is the cause and what is the effect, this can be less clear in fictitious experiments. Even some real experiments can be reversed, at least in theory, if they include a deterministic cause-effect relationship. The extreme case of an elementary chemistry experiment will be used to introduce the difference between causal and predictive modeling.

Hydrogen and oxygen burn and result in water, setting free energy in the form of heat,

$$\text{Oxidation:} \qquad (\text{water,energy}) = f(\text{hydrogen,oxygen}). \tag{1.5}$$

Water is dissolved by electrolysis, and energy is consumed in form of electricity to dissolve the water molecules,

$$\text{Electrolysis:} \qquad (\text{hydrogen,oxygen}) = g(\text{water,energy}). \tag{1.6}$$

However, when it comes to the measurement of the quantities of H_4, O_2, H_2O and energy involved in the oxidation or electrolysis experiment, the deterministic relations are subject to "superimposed" measurement error. The experimenter now has the choice between the "causal" model and the "predictive" model.

Causal model. In a causal model the experimenter is interested in establishing the "true" relation, which we know (leaving aside energy) is

$$H_4 + O_2 \;\rightarrow\; 2H_2O, \tag{1.7}$$

i.e. one hydrogen molecule and one oxygen molecule form two water molecules in an oxidation experiment. The same quantitative relation holds for the electrolysis experiment. Once it is established that the amount of energy set free by oxidation and that consumed by electrolysis are equal, one may speak of the reversibility of the experiment and interchangeability of the input and output (the cause and the effect) of the two experiments. Hence one relates Eq.1.5 and Eq.1.6 by

$$g(\cdot) = f^{-1}(\cdot). \tag{1.8}$$

Predictive model. In a predictive model the experimenter is interested in, for example, determining the quantity of hydrogen needed to produce a predetermined quantity of energy. Because no system $f(\cdot)$ is perfectly closed and because something of the substances and the energy is always lost, the experimenter will need "a little bit of extra gas" compared with the causal model, in order to acquire the desired result, so that a predictive relation could look more like

$$H_4 + O_2 \;\rightarrow\; 1.98(H_2O) + error. \tag{1.9}$$

The odd coefficient 1.98 tells us that 1% of the input substances escape the control of the experimenter.

Predictive vs. causal model. Were the experimenter really interested in determining the quantitative output, he would be unwise to use the "true" theoretical relation Eq.1.7 and the true (structural, causal) parameter "2" instead of Eq.1.8, where the coefficient "1.98" is biased and confounded with the experimental error but nevertheless gives the best estimate of the parameter for the operative use of the oxidation process. If, however, his interest is directed toward the theoretical parameters, as an illustration of Gay-Lussac's Law, for example, which states that gases combine in a proportion of small whole numbers, it may be best to do the experiment both ways, i.e. with oxidation and electrolysis, in order to get the best estimate of the true parameter. However, this helps only if the relation in question is deterministic in nature.

Predictive and causal relations can always be reversed formally, i.e. algebraically ($Y = f(X)$ and $X = f^{-1}(Y)$) and inferentially (by asking which cause could have brought these effects to the fore?). A deterministic relation may or may not be causally (i.e. experimentally, by manipulation) reversible (Wold 1973:793). However, deterministic relations as in the oxidation-electrolysis experiment are never found in social sciences. This example serves to demonstrate that, even in reversible situations, the distinction between structural ("causal") and predictive models is of importance. For the discussion cause-effect relations in social sciences, the following definition of stochastic cause-effect relations is indispensable.

Stochastic cause-effect relations. The statistical notion of causality, introduced by Poisson in 1837, formulates causality as a special case of conditional probability. The probability that one of the events y_i will take place is varying depending on the condition that one of the events x_j or x_k has taken place or not:

$$\Pr(Y = y_i|X = x_j) \neq \Pr(Y = y_i|X = x_k), \quad \exists\, i, j, k, j \neq k\ . \tag{1.10}$$

If one is not concerned with discrete exclusive events but with states measured by continuous variables X and Y, Eq.1.10 must be generalized to the conditional cumulative probabilities,

$$\Pr(Y \leq y_i|X \leq x_j) \neq \Pr(Y \leq y_i|X \leq x_k), \quad \exists\, i, j, k, j \neq k\ . \tag{1.11}$$

This definition can be narrowed down, in two drastic steps of specialization, first, by simplifying Eq.1.11 in terms of the conditional expectation:

$$E(Y|X = x_j) \neq E(Y|X = x_k), \quad \exists\, j, k, j \neq k, \tag{1.12}$$

and secondly, by assuming the conditional expectation to be linear:

$$E(Y|X) = \alpha + \beta X, \qquad \beta \neq 0, \tag{1.13}$$

which is equivalent to Wold's (1963) definition of predictor specification. The asymmetry of conditional probabilities and conditional expectations can be used as a language to express the asymmetry of effect and cause, of predictor and predictand. On the stochastic concept of causality cf. Steyer (1983, 1985).

Micro- vs. macro-events. A stochastic causal explanation does not furnish a rigid explanation (or prediction) of the single observation but only suggests a law applicable to a population of events (and a regularity within a certain sample of events). Taking the aggregation of a large number of single events (micro-events) as one observation (macro-event), this aggregated event can behave "as if" it were deterministic.

One may consider, as a first example, the random movements of the molecules of a gas (the micro-events), where the aggregate of gas follows the classical "deterministic" Boyle-Mariotte laws on pressure, volume and temperature of gases (macro-events). Had he to work only with a few molecules, the experimenter would have difficulty establishing the quasi-deterministic character of the gas laws. Are these laws in an infinitely large sample of gas truly deterministic, or should they be called only "apparently deterministic"?

As a second example, we may take the decay of radioactive substances. "One of the best-known physical examples of a probabilistic phenomenon for which no one pretends to have a deeper underlying deterministic theory is that of the radioactive decay" (Suppes 1973:524). There are deeper theories about why the half-life periods of transuranic substances are shorter than for lighter substances, but no explanations as to why one C_{14} atom disintegrates earlier than another C_{14} atom. The disintegration of a single atom occurs stochastically and is a direct consequence of the stochastic behaviour of the subatomic particles. The disintegration of a large sample of the substance seems to follow a deterministic law. The measured half-life period of the substance is used to estimate the median of the probability distribution of the decay of a single atom. Thus, the "deterministic" law for the macro-events is back-transferred to a stochastic "law" for micro-events. Eq.1.13 can now be interpreted as the stochastic causal model, were Y denotes the number of ticks of a Geiger counter in a given time interval and X the amount of a radioactive substance in an experimental setting.

For a third example, passing from atomic events to human life, one may take the life expectation (the median) of 74 years for Swedish men. On the population level this means that 50% of the Swedes born in 1908 were dead by 1982 and on an individual level that a boy born in 1908 had a 50% chance of reaching 1982. So far, the statement on the half-life period of Swedish men looks like a statement on the half-life period of C_{14} (leaving aside that the probability of "dying" is constant for atoms throughout their "life", but is higher for older than for younger people). Whereas no doubt is possible about the population of C_{14} atoms and its stochastic interpretation, questions arise regarding the population of Swedish men. Does it comprise the men born, living, working or dying in Sweden or those having Swedish citizenship by birth or by nationalization, or not at all? The Swedish census bureau has rules on how to count. Any redefinition of the population can change the estimated life expectation, because the population is by any definition not as homogeneous as the population of C_{14} atoms. Non-homogeneity means that additional characteristics for the individual may be found which are as regards his personal life expectation causal (and hence predictive), and a prediction model which takes into account only the general Swedish life expectancy and no individual characteristics would be subject to specification error.

C_{14} atoms have no individual differences, but people have. The model has to account for it.

Conclusions. Analytical definitions of causality, which are fruitless, anyway, for the social sciences, are superseded by the experimental definition which covers both deterministic and stochastic relations between stimulus and response, between cause and effect. If the relation is deterministic, it is formally reversible, and it may be even causally reversible.

Stochastic causality for a single event turns into "deterministic" causality for an infinitely large aggregate of single events, which all stem from the same population. The assumption of an infinitely large population is not feasible for all applications. The notion of causality should be left open to apply also to small-sized finite populations. Hence the notion of stochastic causality has its own place in the theory of modeling.

Empirical knowledge that is used to estimate the parameters of a stochastic model is usually superimposed by measurement and specification error, i.e. by the insufficient measurement and specification of all possible causes. This makes it likely that the following three sources of scatter of the observation around the expectation are empirically inseparable, namely the scatter due to the stochastic nature of the relationship, the scatter due to insufficient specification and the scatter due to imprecise measurement.

If a model is designed for operative use, the causal interpretation is less important, compared to the predictive power (cf. Eq.1.8). A predictive interpretation of a stochastic model also avoids the notion of an infinitely large homogeneous population of observations. Instead, the single observations are included in the model and represented by case values. This theme will be pursued in the next section and in more detail in Sc.5.

For more information on causal model building one may refer to the textbooks by Blalock (1969, 1972), Heise (1975), Davis (1985), Asher (1983), Berry (1984), Kenny (1979), Glymour, Scheines, Spirtes and Kelly (1987), on the the practical aspects structural model building cf. the textbooks by Pfeifer and Schmidt (1987), Saris and Stronkhorst (1984), Weede (1977), Loehlin (1987), Long (1983), Hayduk (1987) and the readers by Fornell (1982), Goldberger and Duncan (1973), Jöreskog and Sörbom (1979) and Hildebrandt, Rudinger and Schmidt (in press).

1.3 Data Modeling vs. Covariance Modeling

Principal Component models and Common Factor models are the two most fundamental devices in two streams of statistical modeling. The difference between these models will be shown with reference to the epistemological background. The first domain of statistical modeling will be referred to as data structure modeling, the other as covariance structure modeling.

In both domains, the basic and initial question is whether the data at hand are to be taken as being all those observations in which one is interested, or whether the observations are to be regarded as only a sample of the population on which one wants to generalize. For the sake of simplicity, let us take a rectangular matrix of psychological data, with scores in several attributes (e.g. intelligence, achievement, attitudes) of several different people (e.g. Anton, Berta, Caesar, Dora). The question now is whether we are really interested in Berta's scores with regard to intelligence and achievement or whether we could replace Berta's data with Frieda's data, without detracting our intention of making a statement about the association between intelligence and achievement.

Data structure. In the first case, we are going into data structure modeling. A model for the data matrix has to include two sets of parameters, one for the case aspect of the data matrix and one for the variables aspect. Principal component analysis does this by providing scores for the cases and loadings for the variables. The Rasch model does it by means of the person parameters and the item parameters.

Covariance structure. In the second case only the variables aspect is of interest to us. In order to facilitate the analysis, we may use a condensed form of the data matrix that preserves all relevant information; for a linear model the means and the covariances are sufficient. In this case, the theoretical statements include the words intelligence, achievement, attitude and their relationships; Anton, Berta, Caesar and Dora are not mentioned. The statistical reformulation of the theory contains (the passage from T_1 to T_2, Tb.1.1) variables with names like X, Y, Z and parameters for the relationships of the variables only; for a linear model these may be factor loadings, factor correlations, factor variances and path coefficients. The focus is on the theoretical entities and on the deduction of the observational entities from the theoretical ones. The observation used in the model, however, is not Berta's scores on intelligence but the correlation between intelligence and achievement. Berta and Frieda are simply replications of the abstract "population" person.

General vs. differential psychology. Psychometric measurements can be taken under two different goals, that of general and of differential psychology. General psychology tries to establish laws describing and explaining human behavior and governing as general laws all individual differences. One well established 'law' is the correlation of intelligence with achievement. Differential psychology

psychology differential describes the position of an individual human being in the framework established by general psychology and makes inferences about unobserved and probable future individual behavior. In this it uses general 'laws'. If Berta's intelligence score is higher than Dora's, it is likely that Berta's achievement score is higher than that of Dora, due to the general correlation stated before. General psychology is interested only in the variables side of the data matrix and belongs in the domain of covariance structure analysis. Differential psychology is interested in both sides of the data matrix, in that it uses the relationships between the variables for statements about differences and similarities between persons.

Total-survey vs. sample-survey data. Research on parliamentary elections often uses, besides survey data, official results aggregated on some geographical level like the county or constituency. Usually *all* constituencies are included in the data matrix and thus all elements of the population of interest are sampled. Here it would be natural to choose a data structure approach. Similarly, in international politics research, the population comprises all nations of interest and data sets on international relations usually include all nations (Adelman & Morris 1967, 1973; Rossa 1982; Weede 1977) but there is no interest in generalizing on more nations. Clearly, classical testing of statistical hypotheses is nonsensical because sample and population are identical. (Applicable is a test on the homogeneity of the data matrix with respect to the assumed model, which is tested by the Stone-Geisser test, cf. Sc.2.1.3). Along the same lines of argumentation, one should examine the question as to whether it makes sense to establish a covariance structure model which can be applied to more cases than those in the data matrix.

Concluding remarks. Continuing the argumentation from Sc.1.2 where the distinction between causal models concerned with the abstract relationships between attributes and predictive models concerned with the reconstruction of the data at hand was introduced, here in this section the distinction between covariance structure models and data structure models is established. Covariance structure models abstract from the single event and take it only as an exchangeable "observation" from an infinitely large population; the single event is not essential to the theoretical part of the model and serves only for the estimation of model parameters. Data structure models take the single event as a "case" which is represented in the model by case values. The distinction between these two types of modeling will be explored in more depth in Sc.5.

1.4 Notation

Most of the symbols used in this book are displayed in Table 1.2. They are designed to follow mathematical conventions: Theoretical concepts are denoted by Greek letters, empirical concepts and estimates by italic letters. Thus, the Greek letters correspond to concepts of level T_2, and italics to those of level E_3 in Tb. 1.1. Notice that $\boldsymbol{B}$ = Be and $\mathbf{B}$ = Beta, $\boldsymbol{P}$ = Pe and $\mathbf{P}$ = Rho, $\boldsymbol{T}$ = Te and

Table 1.2: Notation for the Latent Variable Path Model

		Indices
	$i,j = 1\dots J$	for blocks of MVs and LVs
	$g,h = 1\dots H$	for latent variables (LVs)
	$\ell,k = 1\dots K$	for manifest variables (MVs)
	$i,n = 1\dots N$	for observational units (cases)
	$a = 1\dots A$	for dimensions of LVs
	$k_j = 1\dots K_j$	for MVs counted within block j
	$h_j = 1\dots H_j$	for LVs counted within block j
	$k_h = 1\dots K_h$	for MVs related to LV η_h
	$h_k = 1\dots H_k$	for LVs related to MV y_k
theoretical	*estimated*	*Variables*
	$\boldsymbol{y}$, $y_k, y_{k_j}, x_\ell, y_{kn}, y_{k_jn}$	manifest variables, data
$\boldsymbol{\eta}$, $\eta_h, \eta_{h_j}, \eta_{hn}, \eta_{h_jn}$	$\boldsymbol{Y}$, $Y_h, Y_{h_j}, Y_{hn}, Y_{h_jn}$	latent variables
$\boldsymbol{\delta}$, δ_h, δ_{h_j}	$\boldsymbol{d}$, d_h, d_{h_j}	validity residuals
$\boldsymbol{v}$, v_h, v_{h_j}	$\boldsymbol{u}$, u_h, u_{h_j}	inner residuals
$\boldsymbol{\epsilon}$, $\epsilon_k, \epsilon_{k_j}, \delta_\ell$	$\boldsymbol{e}$, e_k, e_{k_j}, d_ℓ	outer residuals
		Parameters
$\mathrm{B} = [\beta_{hg}]$	$\boldsymbol{B} = [b_{hg}]$	Path coefficients
$\Pi = [\pi_{kh}] = [\pi_{k_j}]$	$\boldsymbol{P} = [p_{kh}] = [p_{k_j}]$	Loading pattern coefficients
$\Omega = [\omega_{kh}] = [\omega_{k_j}]$	$\boldsymbol{W} = [w_{kh}] = [w_{k_j}]$	Weight coefficients
$\boldsymbol{\beta}_0 = [\beta_{h0}] = [\beta_{j0}]$	$\boldsymbol{b}_0 = [b_{h0}] = [b_{j0}]$	Location parameters for LVs
$\boldsymbol{\pi}_0 = [\pi_{k0}]$	$\boldsymbol{p}_0 = [p_{k0}]$	Location parameters for MVs
		Means and covariances
$\boldsymbol{\mu} = [\mu_k] = \mathrm{E}(\boldsymbol{y})$	$\boldsymbol{m} = [m_k]$	Means of the MVs
$\boldsymbol{\nu} = [\nu_h] = \mathrm{E}(\boldsymbol{\eta})$	$\boldsymbol{n} = [n_h]$	Means of the LVs
$\Psi = [\psi_{hg}] = \mathrm{cov}(\boldsymbol{v})$	$\boldsymbol{Q} = [q_{hg}] = \mathrm{cov}(\boldsymbol{u})$	Inner residual covariance
$\Theta = [\theta_{k\ell}] = \mathrm{cov}(\boldsymbol{\epsilon})$	$\boldsymbol{C} = [c_{k\ell}] = \mathrm{cov}(\boldsymbol{e})$	Outer residual covariance
$\Gamma = [\gamma_{kh}] = \mathrm{cov}(\boldsymbol{\epsilon}; \boldsymbol{v})$	$\boldsymbol{G} = [g_{kh} = \mathrm{cov}(\boldsymbol{e}; \boldsymbol{u})$	Cov of inner and outer residuals
$\mathrm{P} = [\rho_{hg}] = \mathrm{cov}(\boldsymbol{\eta})$	$\boldsymbol{R} = [r_{hg}] = \mathrm{cov}(\boldsymbol{Y})$	LV correlation
$\Sigma = [\sigma_{k\ell}] = \mathrm{cov}(\boldsymbol{y})$	$\boldsymbol{S} = [s_{k\ell}] = \mathrm{cov}(\boldsymbol{y})$	MV covariance
$\mathrm{T} = [\tau_{kh}] = \mathrm{cov}(\boldsymbol{y}; \boldsymbol{\eta})$	$\boldsymbol{T} = [t_{kh}] = \mathrm{cov}(\boldsymbol{y}; \boldsymbol{Y})$	Loadings structure coefficients
$\mathrm{M} = [\mu_{k\ell}] = \mathrm{E}(\boldsymbol{y}\boldsymbol{y}')$	$\boldsymbol{M} = [m_{k\ell}] = \frac{1}{N}\boldsymbol{y}\boldsymbol{y}'$	Raw product matrix (Sc.4)
	$\boldsymbol{N}, \boldsymbol{Q}, \boldsymbol{D}$	Used only in Sc.4

T = Tau. I have resisted using the Swedish Beta (Ꞵ, cf. Wold 1981c) instead of the Greek Beta.

Subscripts. The subscripts are used in a systematic way and help to understand the algebraic variable they specify. Subscripts for manifest variables are k and ℓ (sometimes j), for latent variables j, i, h, g, for observational units n. The range and the limits of the subscripts are denoted by $j = 1\dots J$. Even if the range of the subscript doesn't start with 1, say $j = i+1, i+2, ..., i+J$, the cardinality of

Table 1.3: Some mathematical symbols and operators

Special mathematical operators		
$\boldsymbol{A} * \boldsymbol{B}$	$= [a_{ij} b_{ij}]$	Hadamard product (element wise product)
$\mathbf{I} * \boldsymbol{B}$	$= Diag(\boldsymbol{B})$	Hadamard product: selects the diagonal of $\boldsymbol{B}$
$\boldsymbol{A} \otimes \boldsymbol{B}$	$= [a_{ij} \boldsymbol{B}]$	Kronecker product
$\boldsymbol{A} \oplus \boldsymbol{B}$	$= \begin{bmatrix} \boldsymbol{A} & 0 \\ 0 & \boldsymbol{B} \end{bmatrix}$	Direct sum of $\boldsymbol{A}$ and $\boldsymbol{B}$, creates block-diagonal matrix
$\bigoplus \boldsymbol{A}_j$	$= [\boldsymbol{A}_1 \oplus \boldsymbol{A}_2 \oplus \cdots \oplus \boldsymbol{A}_j \oplus \cdots \oplus \boldsymbol{A}_J]$	
A^-		Generalized Inverse, $A^- A = \mathbf{I}$
$\mathrm{E}(y\|\eta)$		Conditional expectation
$R^2_{j\|1,2,3}$		Squared multiple correlation of x_j given x_1, x_2, x_3
$\boldsymbol{S}$	$= \mathrm{cov}(\boldsymbol{y})$	Symmetric covariance matrix
$\boldsymbol{T}$	$= \mathrm{cov}(\boldsymbol{y}; \boldsymbol{Y})$	Rectangular covariance matrix
Some relational operators		
x	$\equiv yz$	x is identical to yz (introduces symbol x)
i	$:= i + 1$	i is redefined by $i + 1$
$x \Longrightarrow y$		from x follows y
$x \Longleftarrow y$		from y follows x
$x \Longleftrightarrow y$		x if and only if y
x	$\hat{=}\ y$	x corresponds to y; type of correspondence clear from the context (e.g. estimation)

the range of the (lower case) subscript j is denoted by the (upper case) J. Hence in most cases it is trivial to spell out the cardinality and the range of j.

Variables. Manifest variables are denoted by x, y, z, latent variables by ξ, η, ζ, and the estimated LVs by $\hat{\xi}, \hat{\eta}, \hat{\zeta}$ or X, Y, Z. Residual variables are δ, ϵ, v, their estimates are d, e, u. Alternate forms of the variables are denoted by diacritic symbols ($y^*, \tilde{y}, \dot{y}, \ddot{y}, \hat{y}, \bar{y}, \breve{y}$). The version with tilde ($\tilde{y}$) usually denotes the conditional expectation. Depending on the context, the symbol y can denote:

(i)	a single random variable:	y or y_k or y_{kn}
(ii)	a vector of random variables:	$\boldsymbol{y}_{[K\times 1]} = [y_k]$
(iii)	a matrix of random variables:	$\boldsymbol{y}_{[K\times N]} = [y_{kn}]$
(j)	a single data point (case value):	y_{kn} or y_n
(jj)	a data vector:	$\boldsymbol{y}_{[K\times 1]} = [y_k]$ or $\boldsymbol{y}'_{[N\times 1]} = [y_n]$
(jjj)	a data matrix:	$\boldsymbol{y}_{[K\times N]} = [y_{kn}]$

Average. The notation of the average operator is constructed so as to avoid spelling out the range of the summation index and so as to resemble the expectation operator for random variable $\mathrm{E}(y_k)$:

$$\bar{y}_k = \frac{1}{N}\sum_{n=1}^{N} y_{kn} = \bar{\sum_{n}} y_{kn} = y_{k\bar{n}} \tag{1.14}$$

All notations for the average in Eq.1.14 are considered as correct and unambiguous, the second expression as the full form, the third expression as the abbreviated form, and the fourth expression as the shortest form. As concerns the shortest form: (i) The summation index is unambiguous because n does not appear on the left-hand side of Eq.1.14. (ii) The range of n is usually explained in the context. (iii) The cardinality N used as divisor is also clear from the context.

The operators mean(·) and cov(·;·). For typographical convenience, no strict distinction is made between the expectation $\bar{y}_k = \mathrm{E}(y_k)$ of a random variable, the mean $\bar{y}_k = y_{k\bar{n}}$ of a data vector, and to cover both cases, sometimes $\bar{y}_k = \mathrm{mean}(y_k)$ is written.

The operators var(·) and cov(·; ·) are applied to random variables and data in the same way. If the argument list of the covariance operator

$$\boldsymbol{S} = \mathrm{cov}(\boldsymbol{y}; \boldsymbol{x}) = \mathrm{cov}(y_1, y_2; x_4, x_5) = \begin{bmatrix} s_{14} & s_{15} \\ s_{24} & s_{25} \end{bmatrix} \tag{1.15}$$

contains a semicolon, the resulting covariance matrix is rectangular and in general non-symmetric, of dimension 1×1 at least. If it contains no semicolon, $\boldsymbol{S}$ is symmetric, of dimension 2×2 at least:

$$\boldsymbol{S} = \mathrm{cov}(y, x) = \begin{bmatrix} s_{xx} & s_{xy} \\ s_{yx} & s_{yy} \end{bmatrix} \tag{1.16}$$

Apologia. Words and symbols obtain different meanings through tradition and usage. The American word "I apologize" means "I express my regrets, because I'm wrong", and the Greek word "ἀπολογέομαι" means "I defend my behaviour, because I'm right." The traditional usage of Greek symbols in statistics parallels υ (= upsilon) with u instead of y, and η with y instead of h or $\bar{e}$, and I introduce $Q \mathrel{\hat{=}} \Psi$, $C \mathrel{\hat{=}} \Theta$, $W \mathrel{\hat{=}} \Omega$. To my Greek teacher: "I apologize." To my statistics teachers: "ἀπολογέομαι."

Chapter 2

The Basic and the Extended PLS Method

This chapter is devoted to the description of the PLS method and will be followed by chapter 3 which describes the theoretical foundations of the method. The first subchapter, Sc.2.1, gives a very short outline of what Herman Wold calls his Basic Method of PLS Soft Modeling (Wold 1979c, 1980c, 1982d). Because PLS is a rigorously founded, simply structured, and comprehensibly described methodology, it has provoked extensions, generalizations, re-interpretation and reformulations, e.g. by Bookstein (1980, 1982a, 1982c), Noonan and Wold (1982, 1983, 1986), Hui (1978), Apel (1978a, 1978b, 1979), Apel and Wold (1982) and Knepel (1980, 1981). For overviews cf. Falk (1988) Fornell (1982), and Valette-Florence (1986). The extensions proposed by Lohmöller (1979b, 1981a) will be outlined here in detail, starting in Sc.2.2 with the description of the extended model, followed in Sc.2.3 by the extended estimation facilities and in Sc.2.4 with methods of assessing the estimation results, and closing in Sc.2.5 by an application.

The Extended Method of LV path analysis (Lohmöller 1979b) allows for more differentiated and complicated models than the basic method. It facilitates determining several LVs for the same block of MVs simultaneously, assigning a different number of LVs for different blocks, selecting MVs as indicators of LVs for the same block, and thus having more flexible path models than the basic method, where the path model must be the same for all dimensions.

The flexibility of the extended method is not without disadvantages. The basic method is the most secure way of constructing identified models. The non-identity and orthogonality of the dimensions of LVs are assured by the deflation technique. In contrast, in the extended method special measures must be taken to achieve the non-identity of the LVs. Optimally, orthogonality restrictions are imposed on LVs within the same block. This is performed by the rotation algorithm outlined in Sc.2.3.4.

2.1 Wold's Basic Method of Soft Modeling

Herman Wold has introduced a type of modeling with latent variables which he calls "Soft Modeling". The name indicates that this sort of model building applies when the theoretical knowledge is scarce and stringent distributional assumptions are not applicable.

2.1.1 Model Specification

Variables. A soft model involves manifest variables (MVs) and latent variables (LVs), related by linear equations. The MVs (directly observed, observables, indicators) are partitioned into non-overlapping subsets or blocks y_{k_j}, $k_j = 1 \ldots K_j$, each block being indicative of one LV $\eta_j, j = 1 \ldots J$.

Inner model. The LVs are assumed to be interconnected by one or more linear relations. The basic method requires the LVs to form a recursive path model (a causal chain),

$$\eta_j = \beta_{j0} + \sum_i \beta_{ji}\eta_i + v_j, \tag{2.1}$$

where β_{ji} is a path coefficient, v_j is the inner residual variable, and the index i ranges over all predictors of η_j. Predictor specification is adopted,

$$\mathrm{E}(\eta_j|\forall\eta_i) = \beta_{j0} + \sum_i \beta_{ji}\eta_i, \tag{2.2}$$

i.e. the LV is assumed to be a linear function of its predictors η_i there remaining no linear relationship between predictors and the residual,

$$\mathrm{cov}(\eta_i; v_j) = 0. \tag{2.3}$$

Outer model. Each MV y_{k_j} is assumed to be generated as a linear function of its LV η_j and its outer residual variable ϵ_{k_j},

$$y_{k_j} = \pi_{k_j0} + \pi_{k_j}\eta_j + \epsilon_{k_j}, \tag{2.4}$$

where π_{k_j} is the loading coefficient (pattern coefficient). The outer relations are subject to predictor specification also,

$$\mathrm{E}(y_{k_j}|\eta_j) = \pi_{k_j0} + \pi_{k_j}\eta_j, \tag{2.5}$$

which implies that the outer residual ϵ_{k_j} is uncorrelated with the LV of the same block,

$$\mathrm{cov}(\epsilon_{k_j}; \eta_j) = 0. \tag{2.6}$$

Estimated model. In the estimated form the theoretical constants β_{ji}, π_{k_j} and the theoretical variables $\eta_j, v_j, \epsilon_{k_j}$ are replaced by their estimates which are called $b_{ji}, p_{k_j}, Y_j, u_j, e_{k_j}$.

Weight relations. As a vehicle for the estimation of the model parameters, the LVs are estimated as weighted aggregates of their indicators,

$$Y_j = \sum_{k_j} w_{k_j} y_{k_j}, \tag{2.7}$$

where w_{k_j} is a weight coefficient. The weights are estimated by least squares methods; there are, however, two different versions. In the first version (called ModeA, outwards directed, or factor mode) the MVs are regressed on an instrumental LV $\tilde{Y}_j$ (the so-called inside approximation, to be explained later) and the variances of the outer residuals are minimized:

$$y_{k_j} = \tilde{w}_{k_j 0} + \tilde{w}_{k_j} \tilde{Y}_j + \tilde{e}_{k_j} \tag{2.8}$$

In the second version (called ModeB, inwards directed, or regression mode) the instrument $\tilde{Y}_j$ is regressed on the MVs:

$$\tilde{Y}_j = \sum_{k_j} \tilde{w}_{k_j} y_{k_j} + \tilde{d}_j \tag{2.9}$$

The weights w_{k_j} in Eq.2.7 are rescaled versions of the provisional weights $\tilde{w}_{k_j}$ and provide that the LV in Eq.2.7 has unit variance.

Table 2.1: The PLS algorithm for the basic method of LVP analysis

Stage 1:	Iterative estimation of weights and LV scores Starting at step #4, repeat steps #1 to #4 until convergence is obtained.	
#1	Inner weights $v_{ji} = \begin{cases} \operatorname{sign}\operatorname{cov}(Y_j; Y_i) & \text{if } Y_j \text{ and } Y_i \text{ are adjacent} \\ 0 & \text{otherwise} \end{cases}$	
#2	Inside approximation $\tilde{Y}_j := \sum_i v_{ji} Y_i$	
#3	Outer weights; solve for w_{k_j} in $\tilde{Y}_{jn} = \sum_{k_j} \tilde{w}_{k_j} y_{k_j n} + d_{jn}$ $y_{k_j n} = \tilde{w}_{k_j} \tilde{Y}_{jn} + e_{k_j n}$	 in a ModeA block in a ModeB block
#4	Outside approximation $Y_{jn} := f_j \sum_{k_j} \tilde{w}_{k_j} y_{k_j n}$	
Stage 2:	Estimation of path and loading coefficients	
Stage 3:	Estimation of location parameters	

2.1.2 The Basic PLS Algorithm

The algorithm for estimating the unknowns (LV scores, paths, loadings, covariances) of the model proceeds in three stages. In the first two stages the MVs and LVs are treated as deviations from their means; in the third stage the LV means and the location parameters are estimated.

First stage. The LVs are estimated, which means that the LV scores are estimated if the raw data matrix is used as input. If the MV covariance matrix is used as input, the MV-LV-covariance is estimated. Given a raw data matrix y_{k_jn} for $n = 1 \dots N$ cases and an initial estimate w_{k_j} for the weights (say an appropriate 0-1 matrix), the algorithm consists of the iterative repetition of the four steps shown in Table 2.1.

The algorithm starts with step #4, then proceeds to steps #1, #2, and so on. In step #4, the LV is approximated by a weighted aggregate of the MVs, the so-called outside approximation. The scalar f_j assures that $\text{var}(Y_j) = 1$. In step #1 the inner weights v_{ji} are equal to -1 or $+1$, if Y_j is directly connected to Y_i, and $v_{ji} = 0$, if Y_j and Y_i are not adjacent in the path diagram. The sign of v_{ji} is identical to the sign of the correlation $r_{ji} = \text{cov}(Y_j; Y_i)$. In step #2 the inside approximation $\tilde{Y}_j$ is computed as the sign-weighted sum of the adjacent variables. In step #3 new estimates for the weights are found either by multiple regression (ModeB) or single regression (ModeA), using $\tilde{Y}_j$ as an instrumental variable. These weights are used to find a new outside approximation of the LVs in step #4, after which a new iteration cycle starts. The iteration is stopped when the weights do not change anymore.

Second stage. Once the LVs are estimated, the loadings and the path coefficients are estimated by ordinary least squares (OLS) regression. The specifications Eqs.2.1,..., 2.3 lead to the computation of the path coefficients and the specifications Eqs.2.4,..., 2.6 to the loading coefficients.

Special aspects of the relations between the MVs and the LVs may be clarified by the elimination of an LV and replacing it by its predictor,

$$y_{k_j} = p_{k_j} \left[\sum_i b_{ji} Y_i + u_j \right] + e_{k_j}, \tag{2.10}$$

or by elimination of all LVs:

$$y_{k_j} = p_{k_j} \left[\sum_i b_{ji} \left(\sum_{\ell_i} w_{\ell_i} y_{\ell_i} \right) + u_j \right] + e_{k_j}. \tag{2.11}$$

Third stage. The means and the location parameters for MVs and LVs are estimated by:

$$m_{k_j} = \bar{\sum_n} y_{k_jn} \tag{2.12}$$

$$n_j \equiv \bar{\sum_n} Y_{jn} = \sum_{k_j} w_{k_j} m_{k_j} \tag{2.13}$$

$$p_{k_j0} = m_{k_j} - p_{k_j} n_j \tag{2.14}$$

$$b_{j0} = \ell_j - \sum_i b_{ji} \ell_i \tag{2.15}$$

2.1.3 Extensions and Properties

LVs in two or more dimensions can be estimated by the deflation technique, whereby the latent variables of the first dimension are extracted from the MVs and the residuals e_{k_jn} are used as new input to the algorithm for the estimation of the second dimension. Allowance for overlapping subsets of indicators extends the practical applicability of the modeling technique; in this respect Wold's method will be somewhat changed, cf. Sc.2.3.5.

Due to the linear predictor specification all relationships between the MVs are summarized in the first and the second moments of the data matrix, i.e. by the mean vector and the covariance matrix. Hence the estimation can be done by using the moments instead of the raw data matrix; only the LV scores are lost by this method but can be computed immediately when the raw data are available.

For model testing Wold adopts the Stone-Geisser test of predictive relevance which gives as a by-product Jackknife standard deviation for the parameters. Whether the information summarized by the parameters of the model is more than just noise can be evaluated by this test, which requires no assumption beyond that of the homogeneity of the data matrix with respect to the model.

Wold (1982c) discusses the convergence of the iterative first stage of the PLS algorithm, the coherence of the estimation criteria, the consistency of the estimates in a large-sample sense and the relative gain and loss in the explicit estimation of the LV scores. As a least squares method PLS requires no assumption regarding the independence of the observations and the distribution. Beyond the specification that the conditional expectation is linear, PLS is distribution free. All extensions presented in the following sections are based on the foundation laid by Wold.

2.2 The Extended Method: Specification and Properties

2.2.1 Formal Specification

Taking the notation as listed in Table 1.2, the Latent Variables Path Model (LVPM) is given by the inner relation

$$\boldsymbol{\eta} = \boldsymbol{\beta}_0 + \mathbf{B}\boldsymbol{\eta} + \boldsymbol{v} \tag{2.16}$$

and by the outer relation

$$\boldsymbol{y} = \boldsymbol{\pi}_0 + \boldsymbol{\Pi}\boldsymbol{\eta} + \boldsymbol{\epsilon}. \tag{2.17}$$

The model is completed by the generalized weight relation:

$$\boldsymbol{\eta} = \boldsymbol{\omega}_0 + \boldsymbol{\Omega}'\boldsymbol{y} + \boldsymbol{\delta}. \tag{2.18}$$

Without loss of generality, the location parameters $\boldsymbol{\beta}_0, \boldsymbol{\pi}_0, \boldsymbol{\omega}_0$ will be disregarded in the following and all variables treated as deviations from their means.

The inner relation (inner model, structural model, core model, epistemological model, path model in a narrower sense) shows the LVs as dependent on each other and on the inner residual variables; the coefficients $\mathbf{B}$ are the path coefficients. The outer relation (loading relation, outer model, factor model, measurement model) shows the MVs as depending on the LVs and on the outer residual variables; the coefficients $\boldsymbol{\Pi}$ are the loadings (loading pattern coefficients). The generalized weight relation shows the LVs as depending on the MVs and on the "validity residual" $\boldsymbol{\delta}$, i.e. the part of the LV that is not measured by the MVs.

The inner relation Eq.2.16 is given in a formulation that is in accordance with the structure of the substantive interpretation. This form is called the structural form (SF) of the model. By subtraction of $\mathbf{B}\boldsymbol{\eta}$ on both sides, $\boldsymbol{\eta} - \mathbf{B}\boldsymbol{\eta} = \boldsymbol{v}$, $(\mathbf{I} - \mathbf{B})\boldsymbol{\eta} = \boldsymbol{v}$, and by premultiplication by $\mathbf{B}^* \equiv (\mathbf{I} - \mathbf{B})^{-1}$, one obtains the reduced form (RF) of the inner relation, also called the total-effect form:

$$\boldsymbol{\eta} = (\mathbf{I} - \mathbf{B})^{-1}\boldsymbol{v} \equiv \mathbf{B}^*\boldsymbol{v} \tag{2.19}$$

The algebraic treatment of Eq.2.16 is often simpler if the variables $\boldsymbol{\eta}$ do not appear on both sides of the relation. The outer model relation Eq.2.17 is transformed into the total-effect form by replacing the LVs by their reduced-form expression from Eq.2.19:

$$\boldsymbol{y} = \boldsymbol{\Pi}\boldsymbol{\eta} + \boldsymbol{\epsilon} = \boldsymbol{\Pi}\mathbf{B}^*\boldsymbol{v} + \boldsymbol{\epsilon} \equiv \boldsymbol{\Pi}^*\boldsymbol{v} + \boldsymbol{\epsilon} \tag{2.20}$$

2.2.2 Deductive Properties of the LVP Model

Once it is specified that the relationships between the variables are all linear, it follows that the covariances of the variables on the left-hand side of Eqs.2.16, 2.17 can be deduced when the covariances of the variables on the right-hand sides are known. More exactly, the covariances of the residual variables $\boldsymbol{v}$ and $\boldsymbol{\epsilon}$ must be known, together with the coefficients matrices $\mathbf{B}$ and $\boldsymbol{\Pi}$ (see Lohmöller 1979d).

Let the residual variables $\boldsymbol{v}$ and $\boldsymbol{\epsilon}$ be stacked together in a vector with zero mean and covariance matrix

$$\operatorname{cov}\begin{bmatrix} \boldsymbol{v} \\ \boldsymbol{\epsilon} \end{bmatrix} = \begin{bmatrix} \operatorname{cov}(\boldsymbol{v}) & \operatorname{cov}(\boldsymbol{v};\boldsymbol{\epsilon}) \\ \operatorname{cov}(\boldsymbol{\epsilon};\boldsymbol{v}) & \operatorname{cov}(\boldsymbol{\epsilon}) \end{bmatrix} \equiv \begin{bmatrix} \boldsymbol{\Psi} & \boldsymbol{\Gamma}' \\ \boldsymbol{\Gamma} & \boldsymbol{\Theta} \end{bmatrix}. \tag{2.21}$$

Let all elements of Eq.2.21 be known, as well as the loading pattern matrix $\boldsymbol{\Pi}$ and the path matrix $\mathbf{B}$. Then it follows:

$$\mathbf{P} \equiv \operatorname{cov}(\boldsymbol{\eta}) = \mathrm{E}\,[\mathbf{B}^*\boldsymbol{v}\boldsymbol{v}'\mathbf{B}^{*\prime}] = \mathbf{B}^*\boldsymbol{\Psi}\mathbf{B}^{*\prime} \tag{2.22}$$

$$\operatorname{cov}(\boldsymbol{\eta};\boldsymbol{v}) = \mathrm{E}\,[\mathbf{B}^*\boldsymbol{v}\boldsymbol{v}'] = \mathbf{B}^*\boldsymbol{\Psi} \tag{2.23}$$

$$\operatorname{cov}(\boldsymbol{y};\boldsymbol{v}) = \mathrm{E}\,[(\boldsymbol{\Pi}^*\boldsymbol{v} + \boldsymbol{\epsilon})\boldsymbol{v}'] = \boldsymbol{\Pi}^*\boldsymbol{\Psi} + \boldsymbol{\Gamma} \tag{2.24}$$

$$\begin{aligned}
\text{cov}(\boldsymbol{y};\boldsymbol{\epsilon}) &= \mathrm{E}\,[(\boldsymbol{\Pi}^*\boldsymbol{v}+\boldsymbol{\epsilon})\boldsymbol{\epsilon}'] = \boldsymbol{\Pi}^*\boldsymbol{\Gamma}'+\boldsymbol{\Theta} && (2.25)\\
\mathbf{T} \equiv \text{cov}(\boldsymbol{y};\boldsymbol{\eta}) &= \mathrm{E}\,[(\boldsymbol{\Pi}^*\boldsymbol{v}+\boldsymbol{\epsilon})(\boldsymbol{v}'\mathbf{B}^{*\prime})] = (\boldsymbol{\Pi}^*\boldsymbol{\Psi}+\boldsymbol{\Gamma})\mathbf{B}^{*\prime} && (2.26)\\
\boldsymbol{\Sigma} \equiv \text{cov}(\boldsymbol{y}) &= \mathrm{E}\,[(\boldsymbol{\Pi}^*\boldsymbol{v}+\boldsymbol{\epsilon})(\boldsymbol{v}'\boldsymbol{\Pi}^{*\prime}+\boldsymbol{\epsilon}')]\\
&= \mathrm{E}\,[\boldsymbol{\Pi}^*\boldsymbol{v}\boldsymbol{v}'\boldsymbol{\Pi}^{*\prime}+\boldsymbol{\Pi}^*\boldsymbol{v}\boldsymbol{\epsilon}'+\boldsymbol{\epsilon}\boldsymbol{v}'\boldsymbol{\Pi}^{*\prime}+\boldsymbol{\epsilon}\boldsymbol{\epsilon}']\\
&= \boldsymbol{\Pi}^*\boldsymbol{\Psi}\boldsymbol{\Pi}^{*\prime}+\boldsymbol{\Pi}^*\boldsymbol{\Gamma}'+\boldsymbol{\Gamma}\boldsymbol{\Pi}^{*\prime}+\boldsymbol{\Theta} && (2.27)
\end{aligned}$$

All variables being collected into one vector, the results of Eqs.2.22,..., 2.27 can be collected into one the covariance matrix:

$$\text{cov}\begin{bmatrix}\boldsymbol{v}\\ \boldsymbol{\epsilon}\\ \boldsymbol{\eta}\\ \boldsymbol{y}\end{bmatrix} = \begin{bmatrix}\boldsymbol{\Psi} & \cdot & \cdot & \text{sym.}\\ \boldsymbol{\Gamma} & \boldsymbol{\Theta} & \cdot & \cdot\\ \mathbf{B}^*\boldsymbol{\Psi} & \mathbf{B}^*\boldsymbol{\Gamma}' & \mathbf{P}=\mathbf{B}^*\boldsymbol{\Psi}\mathbf{B}^{*\prime} & \cdot\\ \boldsymbol{\Pi}^*\boldsymbol{\Psi}+\boldsymbol{\Gamma} & \boldsymbol{\Pi}^*\boldsymbol{\Gamma}+\boldsymbol{\Theta} & \mathbf{T}=(\boldsymbol{\Pi}^*\boldsymbol{\Psi}+\boldsymbol{\Gamma})\mathbf{B}^{*\prime} & \begin{bmatrix}\boldsymbol{\Sigma}=\\ \text{Eq.2.27}\end{bmatrix}\end{bmatrix} \tag{2.28}$$

Eq.2.28 demonstrates that the covariances of the model variables $\boldsymbol{y}$ and $\boldsymbol{\eta}$ can be derived only when the residual covariances $\boldsymbol{\Psi}, \boldsymbol{\Gamma}, \boldsymbol{\Theta}$ are specified. The formulae become much simpler if $\boldsymbol{\Gamma} = \mathbf{0}$, as is assumed usually.

The form of the LVP model as presented in this section is called the generative, or deductive, form, because it permits the deduction of the manifest variables,

$$\boldsymbol{y} := \boldsymbol{y}(\boldsymbol{v}, \boldsymbol{\epsilon}, \mathbf{B}, \boldsymbol{\Pi}), \tag{2.29}$$

or, given the covariances of the residual variables, the deduction of the covariances of the model variables:

$$(\boldsymbol{\Sigma}, \mathbf{P}, \mathbf{T}, \boldsymbol{\Sigma}_{\boldsymbol{y}}, \boldsymbol{\Sigma}_{\boldsymbol{y}\boldsymbol{\epsilon}}, \boldsymbol{\Sigma}_{\boldsymbol{y}\boldsymbol{v}}, \boldsymbol{\Sigma}_{\boldsymbol{\eta}\boldsymbol{v}}) := \boldsymbol{\Sigma}(\boldsymbol{\Psi}, \boldsymbol{\Gamma}, \boldsymbol{\Theta}, \mathbf{B}, \boldsymbol{\Pi}). \tag{2.30}$$

2.2.3 Inductive Properties of the LVP Model

In the weight relation Eq.2.18 the residual variables $\boldsymbol{\delta}$ are designed so as to account for those parts of the LVs which are not measured by the MVs. The $\boldsymbol{\delta}$-variables represent the partiality of operationalization of the constructs $\boldsymbol{\eta}$, or equivalently the lack of validity of the indicators $\boldsymbol{y}$. It is the characteristic feature of component models — as contrasted to Common Factor models — that these residuals are deliberately assumed to be zero,

$$\text{cov}(\boldsymbol{\delta}) = \mathbf{0}, \tag{2.31}$$

and the generalized weight relation Eq.2.18 is simplified to the ordinary weight relation:

$$\boldsymbol{\eta} = \boldsymbol{\Omega}'\boldsymbol{y}. \tag{2.32}$$

Let $\boldsymbol{y}, \boldsymbol{\Omega}, \boldsymbol{\Pi}, \mathbf{B}$ be given. Then the residual variables are expressed as:

$$\begin{aligned}
\boldsymbol{\epsilon} &= \boldsymbol{y} - \boldsymbol{\Pi}\boldsymbol{\eta} = (\mathbf{I} - \boldsymbol{\Pi}\boldsymbol{\Omega}')\boldsymbol{y} && (2.33)\\
\boldsymbol{v} &= \boldsymbol{\eta} - \mathbf{B}\boldsymbol{\eta} = (\mathbf{I} - \mathbf{B})\boldsymbol{\eta} = (\mathbf{I} - \mathbf{B})\boldsymbol{\Omega}'\boldsymbol{y}. && (2.34)
\end{aligned}$$

Furthermore, all submatrices of Eq.2.28 can be derived:

$$\begin{aligned}
\mathbf{T} \equiv \ \mathrm{cov}(\boldsymbol{y};\boldsymbol{\eta}) &= \mathrm{E}\,[\boldsymbol{y}\boldsymbol{y}'\boldsymbol{\Omega}] = \mathrm{E}(\boldsymbol{y}\boldsymbol{y}')\boldsymbol{\Omega} \\
&= \boldsymbol{\Sigma}\boldsymbol{\Omega} && (2.35)\\
\mathrm{cov}(\boldsymbol{y};\boldsymbol{\epsilon}) &= \mathrm{E}\,[\boldsymbol{y}\boldsymbol{y}'(\mathbf{I}-\boldsymbol{\Omega}\boldsymbol{\Pi}')] = \boldsymbol{\Sigma}-\boldsymbol{\Sigma}\boldsymbol{\Omega}\boldsymbol{\Pi}' \\
&= \boldsymbol{\Sigma}-\mathbf{T}\boldsymbol{\Pi}' && (2.36)\\
\mathrm{cov}(\boldsymbol{y};\boldsymbol{v}) &= \mathrm{E}\,[\boldsymbol{y}\boldsymbol{y}'\boldsymbol{\Omega}(\mathbf{I}-\mathbf{B})'] = \boldsymbol{\Sigma}\boldsymbol{\Omega}(\mathbf{I}-\mathbf{B})' \\
&= \mathbf{T}(\mathbf{I}-\mathbf{B})' && (2.37)\\
\mathbf{P} \equiv \ \mathrm{cov}(\boldsymbol{\eta}) &= \mathrm{E}\,[\boldsymbol{\Omega}'\boldsymbol{y}\boldsymbol{y}'\boldsymbol{\Omega}] \\
&= \boldsymbol{\Omega}'\boldsymbol{\Sigma}\boldsymbol{\Omega} && (2.38)\\
\mathrm{cov}(\boldsymbol{\eta};\boldsymbol{\epsilon}) &= \mathrm{E}\,[\boldsymbol{\Omega}'\boldsymbol{y}\boldsymbol{y}'(\mathbf{I}-\boldsymbol{\Omega}\boldsymbol{\Pi}')] \\
&= \mathbf{T}'-\mathbf{P}\boldsymbol{\Pi}' && (2.39)\\
\mathrm{cov}(\boldsymbol{\eta},\boldsymbol{v}) &= \mathrm{E}\,[\boldsymbol{\Omega}'\boldsymbol{y}\boldsymbol{y}'\boldsymbol{\Omega}(\mathbf{I}-\mathbf{B})'] \\
&= \mathbf{P}(\mathbf{I}-\mathbf{B})' && (2.40)\\
\boldsymbol{\Theta} \equiv \ \mathrm{cov}(\boldsymbol{\epsilon}) &= \mathrm{E}\,[(\mathbf{I}-\boldsymbol{\Pi}\boldsymbol{\Omega}')\boldsymbol{y}\boldsymbol{y}'(\mathbf{I}-\boldsymbol{\Omega}\boldsymbol{\Pi}')] \\
&= \boldsymbol{\Sigma}-\mathbf{T}\boldsymbol{\Pi}'-\boldsymbol{\Pi}\mathbf{T}'-\boldsymbol{\Pi}\mathbf{P}\boldsymbol{\Pi}' && (2.41)\\
\boldsymbol{\Gamma} \equiv \ \mathrm{cov}(\boldsymbol{\epsilon};\boldsymbol{v}) &= \mathrm{E}\,[(\mathbf{I}-\boldsymbol{\Pi}\boldsymbol{\Omega}')\boldsymbol{y}\boldsymbol{y}'\boldsymbol{\Omega}(\mathbf{I}-\mathbf{B})'] \\
&= (\mathbf{T}-\boldsymbol{\Pi}\mathbf{P})(\mathbf{I}-\mathbf{B})' && (2.42)\\
\boldsymbol{\Psi} \equiv \ \mathrm{cov}(\boldsymbol{v}) &= \mathrm{E}\,[(\mathbf{I}-\mathbf{B})\boldsymbol{\Omega}'\boldsymbol{y}\boldsymbol{y}'\boldsymbol{\Omega}(\mathbf{I}-\mathbf{B})'] \\
&= (\mathbf{I}-\mathbf{B})\mathbf{P}(\mathbf{I}-\mathbf{B})' && (2.43)
\end{aligned}$$

$$\mathrm{cov}\begin{bmatrix}\boldsymbol{v}\\ \boldsymbol{\epsilon}\\ \boldsymbol{\eta}\\ \boldsymbol{y}\end{bmatrix} = \begin{bmatrix}(\mathbf{I}-\mathbf{B})\mathbf{P}(\mathbf{I}-\mathbf{B})' & & & \text{sym.}\\ (\mathbf{T}-\boldsymbol{\Pi}\mathbf{P})(\mathbf{I}-\mathbf{B})' & \boldsymbol{\Sigma}-\mathbf{T}\boldsymbol{\Pi}'-\boldsymbol{\Pi}\mathbf{T}'-\boldsymbol{\Pi}\mathbf{P}\boldsymbol{\Pi}' & & \\ \mathbf{P}(\mathbf{I}-\mathbf{B})' & \mathbf{T}'-\mathbf{P}\boldsymbol{\Pi}' & \boldsymbol{\Omega}'\boldsymbol{\Sigma}\boldsymbol{\Omega} & \\ \mathbf{T}(\mathbf{I}-\mathbf{B})' & \boldsymbol{\Sigma}-\mathbf{T}\boldsymbol{\Pi}' & \boldsymbol{\Sigma}\boldsymbol{\Omega} & \boldsymbol{\Sigma}\end{bmatrix} \tag{2.44}$$

The form of the LV path model as presented in this section is called the inductive form, because it allows inductive conclusions to be drawn from the manifest on the latent variables:

$$(\boldsymbol{v},\boldsymbol{\epsilon},\boldsymbol{\eta}) := (\boldsymbol{y},\boldsymbol{\Omega},\boldsymbol{\Pi},\mathbf{B}) \tag{2.45}$$

Given the covariance matrix of the MVs and the parameter matrices, the covariances of all unobserved variables can be computed:

$$(\mathbf{T},\mathbf{P},\boldsymbol{\Theta},\boldsymbol{\Gamma},\boldsymbol{\Psi},\boldsymbol{\Sigma}_{y\epsilon},\boldsymbol{\Sigma}_{yv},\boldsymbol{\Sigma}_{\eta\epsilon},\boldsymbol{\Sigma}_{\eta v}) := (\boldsymbol{\Sigma},\boldsymbol{\Omega},\boldsymbol{\Pi},\mathbf{B}). \tag{2.46}$$

When comparing Eq.2.28 and Eq.2.44 it is found that, besides $\mathbf{B}$ and $\boldsymbol{\Pi}$ either the weight matrix $\boldsymbol{\Omega}$ or the residual covariances $\boldsymbol{\Psi},\boldsymbol{\Gamma},\boldsymbol{\Theta}$ must be known in order to derive the remaining unknowns of the model. As an early remark on LISREL (see Sc.5) it is noted that by including the weight matrix $\boldsymbol{\Omega}$ into the set of parameters, the residual covariance matrices are determined not by assumptions but by computations (or, if $\boldsymbol{\Omega}$ is estimated, by estimation).

2.2.4 Specification of an LVP Model

Partition of variables. The set of MVs $\boldsymbol{y}_{[K\times N]}$ is partitioned into non-overlapping subsets (called blocks) $\boldsymbol{y}_{j[K_j\times N]}, j = 1 \ldots J$. Each block is assumed to measure a set of (one or more) LVs $\boldsymbol{\eta}_{j[H_j\times N]}$. The single MVs are indexed either within the total set ($y_k, k = 1 \ldots K$) or in a non-overlapping way, according to the partition in blocks ($y_{k_j}, k_j = 1 \ldots K_j, j = 1 \ldots J$), or in a possibly overlapping way, according to their connection to an LV ($y_{kh}, k_h = 1 \ldots K_h, h = 1 \ldots H$). Similarly, the single LVs are indexed either within the total set ($\eta_h, h = 1 \ldots H$), in a non-overlapping way within subsets ($\eta_{h_j}, h_j = 1 \ldots H_j, j = 1 \ldots J$) or in a possibly overlapping way according to their connection to the MVs ($\eta_{h_k}, h_k = 1 \ldots H_k, k = 1 \ldots K$), just as is convenient.

Outer design matrix. The investigator has to specify which MV measures which LV. This is done by means of a Boolean matrix Design$(\boldsymbol{\Pi})_{[K\times H]}$, called the outer design matrix, with elements

$$\text{Design}(\pi_{kh}) = \begin{cases} 1 & \text{if } y_k \text{ is directly connected to } \eta_h \\ 0 & \text{otherwise} \end{cases} \tag{2.47}$$

The outer design matrix is used to specify that elements of the loading (and the weight) matrix are zero,

$$\pi_{kh} = \text{Design}(\pi_{kh})\pi_{kh} \qquad \text{-or-} \qquad \boldsymbol{\Pi} = \text{Design}(\boldsymbol{\Pi}) * \boldsymbol{\Pi}, \tag{2.48}$$

where the star denotes the Hadamard (elementwise) product. The specification for the weight matrix is:

$$\boldsymbol{\Omega} = \text{Design}(\boldsymbol{\Pi}) * \boldsymbol{\Omega}. \tag{2.49}$$

To preserve the block partitioning, Design$(\boldsymbol{\Pi})$ must have a block-diagonal form, i.e. all submatrices Design$(\boldsymbol{\Pi}_{ji})$ relating a block of MVs $\boldsymbol{y}_j$ to a block of LVs $\boldsymbol{\eta}_i$ for $j \neq i$ must be zero matrices. In the extended method further restrictions within Design$(\boldsymbol{\Pi}_{ji})$ are possible: If, for two LVs η_{1_j} and η_{2_j} belonging to the same block, some design coefficients Design$(\pi_{k_j1_j})$ or Design$(\pi_{k_j2_j})$, $(k_j = 1 \ldots K_j)$ are zero, then the indicators for these two LVs are partly overlapping and partly different. This provides a greater flexibility for model construction.

Path design matrix. The dependency structure between the LVs is specified by a Boolean matrix Design$(\mathbf{B})_{[H\times H]}$ called the path design matrix, with elements:

$$\text{Design}(\beta_{hg}) = \begin{cases} 1 & \text{if } \eta_g \text{ is predictor of } \eta_h \\ 0 & \text{otherwise} \end{cases} \tag{2.50}$$

The path design matrix is used to express the assumption that the specified elements of the path matrix $\mathbf{B}$ are zero:

$$\mathbf{B} = \text{Design}(\mathbf{B}) * \mathbf{B}. \tag{2.51}$$

The design matrix of a recursive path model is subdiagonal, i.e.:

$$(\text{Design}(\beta_{hg}) = 0 \quad \forall g > h) \qquad \Longrightarrow \qquad \text{recursive path model} \tag{2.52}$$

Further restrictions may be imposed on the path model.

LV correlation design matrix. In the extended method it is possible to require specified pairs of LVs to be uncorrelated. This is done by a Boolean matrix $\text{Design}(\boldsymbol{\rho})_{[H\times H,\ \text{symmetric}]}$ with elements:

$$\text{Design}(\rho_{hg}) = \begin{cases} 1 & \text{if } \eta_g \text{ and } \eta_h \text{ are free to be correlated} \\ 0 & \text{cov}(\eta_g;\eta_h) = 0 \end{cases} \tag{2.53}$$

The restriction is expressed as:

$$\boldsymbol{\rho} = \text{Design}(\boldsymbol{\rho}) * \boldsymbol{\rho}. \tag{2.54}$$

Predictor specification. For the inner model part, the conditional expectations of the LVs are specified to be linear function of the variables selected by the path design matrix:

$$\text{E}(\eta_h | \text{Design}(\beta_{hg})\eta_g) = \sum_g \beta_{hg}\eta_g. \tag{2.55}$$

This specification implies that predictors and residuals are uncorrelated:

$$\text{Design}(\mathbf{B}) * \text{cov}(\boldsymbol{\eta};\boldsymbol{v}) = \mathbf{0}. \tag{2.56}$$

The predictor specification for the outer relations is

$$\text{E}(y_k | \text{Design}(\pi_{kh})\eta_h) = \sum_h \pi_{kh}\eta_h\ , \tag{2.57}$$

which implies that predictors $\boldsymbol{\eta}$ and residuals $\boldsymbol{\epsilon}$ are uncorrelated:

$$\text{Design}(\boldsymbol{\Pi}) * \text{cov}(\boldsymbol{\eta};\boldsymbol{\epsilon}) = \mathbf{0}. \tag{2.58}$$

Weighting mode. The predictor specification for the weight relations can have two different forms, depending on the "direction" of the MV-LV-relationships which is specified by means of the WeightMode vector with elements:

$$\text{WeightMode}(j) = \begin{cases} 0 & \text{if block } \boldsymbol{y}_j \text{ is outwards directed (ModeA)} \\ 1 & \text{block } \boldsymbol{y}_j \text{ is inwards directed (ModeB)} \end{cases} \tag{2.59}$$

On the rationale of the choice see Wold (1980a), Lohmöller (1981a:2-02). In an outwards directed block (ModeA, WeightMode$(j) = 0$) the LV is specified to be a linear predictor of the MVs:

$$\text{E}(y_k | \text{Design}(\boldsymbol{\pi}_k) * \boldsymbol{\eta}) = \sum_h \omega_{kh}\eta_h \tag{2.60}$$

In an inwards directed block (ModeB, WeightMode$(j) = 1$) the MVs are specified to be linear predictors for the LVs:

$$\text{E}(\eta_h | \text{Design}(\boldsymbol{\pi}_h) * \boldsymbol{y}) = \sum_k \omega_{kh} y_k \tag{2.61}$$

It may be noted that, so far, no assumption of uncorrelated residual variables is made, i.e. no element of $\mathbf{\Psi}, \mathbf{\Gamma}, \mathbf{\Theta}$ is restricted. Predictor specification implies that several covariances between the predictor LVs and the (inner and outer) residuals are zero. There are as many zero correlations of this type implied as parameters which have to be estimated. The properties of conditional expectations will be discussed in more detail in Sc.3.1.

The extensions of the LVP models, as presented in this section, are:

(a) More than one LV per block

(b) Zero-weight restrictions: additional restrictions on the MV-LV relationships.

(c) Zero-$\boldsymbol{\rho}$ restrictions: LV-correlations restricted to zero.

The next section Sc.2.3 will introduce extensions of the PLS estimation algorithm for models which make use of the extended specifications.

2.3 Estimation in the Extended Method

Except in simulation studies, the latent variables $\boldsymbol{\eta}$, the residual variables $\boldsymbol{\epsilon}, \boldsymbol{\upsilon}$, the parameters $\mathbf{B}, \mathbf{\Pi}, \mathbf{\Omega}$ and the covariances $\mathbf{P}, \mathbf{T}, \mathbf{\Theta}, \mathbf{\Psi}$ are unknown. Known are the scores for the observed variables y_{kn} or the sample estimate $\boldsymbol{S}$ of the MV-covariance. When the parameters $\mathbf{B}, \mathbf{\Pi}, \mathbf{\Omega}$ are estimated, all other unknowns of the model can be computed, using the inductive formula Eqs.2.28, 2.44. The estimates are denoted by $\boldsymbol{B}, \boldsymbol{P}, \boldsymbol{W}$ etc., as outlined in Table 2.2.

In this chapter, Sc.2.3, two extensions and one modifications of the PLS algorithm will be presented and discussed. The two extensions are related to the two extensions of the LVP model noted in Sc.2.2, the zero-$\boldsymbol{\rho}$ restriction and the zero-weight restriction. The modification takes into account the fact that the inner model is specified as consisting of directed relationships between the LVs, which is neglected in the first step of the basic estimation method.

Tb.2.2 shows a rough sketch of the algorithm for the extended method, as it is implemented in the program PLSC (Lohmöller 1981a). Stage #2 of the algorithm is the iterative estimation of the weights and the LV scores. In contrast to the basic method (Tb.03), two more steps are introduced, steps #21 and #24. In these steps the orthonormality restrictions on the LV correlations are implemented (to be discussed in Sc.2.3.4). For the inner weights (Step #22) three different "inner weighting schemes" will be discussed in Sc.2.3.2 and Sc.2.3.3. The additional restrictions on the MV-LV relationships require a more differentiated handling of the weights (step #25) and loadings (step #31) and will be discussed in Sc.2.3.5. For overly restricted models conflicts between the estimation criteria (Sc.2.3.1) and the restrictions may arise; so the re-estimated solutions (Step #4), as discussed in Sc.2.3.6, may be required.

2.3.1 Partial Least Squares

When the model is specified, the next step is to establish an estimation procedure for the parameters. Often this is done by defining a scalar criterion and the parameters are estimated in such a manner that this criterion is optimized. In

Table 2.2: The PLS algorithm for the extended PLS method

#10	Model specification
#11	Read J, $[K_j]$, $[H_j]$, [WeightMode$_j$]
#12	Read Design($\mathbf{B}$), Design($\boldsymbol{\rho}$), Design($\boldsymbol{\Pi}$), starting values $\boldsymbol{W}$
#13	Read MV covariance matrix $\boldsymbol{S}$
#14	Check correctness and features of the model
#15	Print model specification
#20	Iterative estimation of the LV weights for solution no.1
	Starting at step #26, repeat steps #21 to #26 until convergence
#21	Patterned orthogonalization of $\boldsymbol{Y}, \boldsymbol{Y} := \boldsymbol{OY}$
	$\boldsymbol{O} :=$ Pattor($\boldsymbol{R}$, Design($\boldsymbol{\rho}$))
	$\boldsymbol{R} := \boldsymbol{ORO}'$
	$\boldsymbol{B} :=$ Path($\boldsymbol{R}$, Design($\mathbf{B}$))
#22	Inner weights $\boldsymbol{V}$
	$\boldsymbol{V} := \boldsymbol{V}(\boldsymbol{R}, \boldsymbol{B}$, Design($\mathbf{B}$), Inner Weighting Mode)
#23	Inside approximation, $\boldsymbol{Y}^* := \boldsymbol{VY}$
	$\boldsymbol{R} := \boldsymbol{VRV}'; \boldsymbol{T} := \boldsymbol{TV}'$
#24	Patterned orthogonalization of $\boldsymbol{Y}^*, \boldsymbol{Y}^* := \boldsymbol{OY}^*$
	$\boldsymbol{O} :=$ Pattor($\boldsymbol{R}$, Design($\boldsymbol{\rho}$))
	$\boldsymbol{R} := \boldsymbol{ORO}'$
	$\boldsymbol{B} :=$ Path Design($\mathbf{B}$))
#25	Outer weights $\boldsymbol{W}$
	$\boldsymbol{W} := \boldsymbol{W}(\boldsymbol{R}, \boldsymbol{S}, \boldsymbol{T}$, Design($\boldsymbol{\Pi}$), WeightMode)
#26	Outside approximation, $\boldsymbol{Y} := \boldsymbol{W}'\boldsymbol{y}$
	$\boldsymbol{R} := \boldsymbol{W}'\boldsymbol{SW}$
	$\boldsymbol{T} := \boldsymbol{SW}$
#30	Final computations and output
#31	Compute loadings, path coefficients, etc.
#32	Print $\boldsymbol{B}, \boldsymbol{R}, \boldsymbol{W}, \boldsymbol{P}, \boldsymbol{T}$
#33	Compute and print residual covariance matrices
#34	Compute and print summary table
#35	Punch LISREL model specification
#40	Re-estimation of weights for solution No.2 and No.3
#41	Solution No.2, if zero-$\boldsymbol{R}$ restrictions are imposed on pairs of LVs in the same outwards block:
	$\boldsymbol{O} :=$ Pattor($\boldsymbol{R}$, Design($\boldsymbol{\rho}_2$)); $\boldsymbol{T} = \boldsymbol{TO}'; \boldsymbol{W}_j := \boldsymbol{S}_j^{-1}\boldsymbol{T}_j$
	Repeat steps #26 to #34
#42	Solution No.3, if zero-$\boldsymbol{R}$ restrictions are imposed on pairs of LVs with different indicators
	$\boldsymbol{O} :=$ Pattor($\boldsymbol{R}$, Design($\boldsymbol{\rho}$)); $\boldsymbol{T} := \boldsymbol{TO}'; \boldsymbol{W} := \boldsymbol{S}^{-1}\boldsymbol{T}$
	Repeat steps #26 to #34

many well-known models, e.g. the multiple regression, the principal components model, the canonical correlation model, a criterion based on the variance of the residual variables is adopted (cf. Glang 1988). However, in the LVP model there are different kinds of residual variables $\boldsymbol{v}, \boldsymbol{\epsilon}, \boldsymbol{\delta}$. The combining of their variances into a single criterion has some arbitrariness. Summing up all residual variances implies that equal importance has been assigned to each variable.

Wold (1966a, 1966b) has proposed treating each residual variable separately, by defining a set of partial least squares (PLS) criteria (or local LS criteria):

$$q_h^2 \equiv \text{var}(u_h) \stackrel{!}{=} \min \ \forall \text{ endogenous LVs } h \tag{2.62}$$

$$c_k^2 \equiv \text{var}(e_k) \stackrel{!}{=} \min \ \forall \text{ outward directed blocks } k \tag{2.63}$$

$$\text{var}(d_h) \stackrel{!}{=} \min \ \forall \text{ inward directed blocks } h. \tag{2.64}$$

Minimizing the first criterion is identical to maximizing the squared multiple correlation $R_h^2 = 1 - q_h^2$ of the endogenous variables. The second and third criteria are related to the estimation of LV weights and loadings.

As no single criterion has been established, no algebraic solution of the parameter estimation is available, except for models with one or two LVs (Wold 1975b:332). Thus, an iterative algorithm to apply to the local criteria must be found.

The first basic idea of the PLS technique is that one part of the parameters is considered as known and held fixed, while the other part is estimated. There are only so many parameters included in the part to be estimated simultaneously, that a direct least squares solution is available. The term "Partial Least Squares" refers to the partitioning of the parameters in estimable subsets. Well-known procedures solving for a subset of parameters under LS criteria can be used as steps (LS modules) within a PLS iteration cycle. Some LS modules will be discussed in the next section, others are listed in Lohmöller and Wold (1980).

The second basic idea of the PLS technique is related to the fact that some LS modules are iterative themselves and require sub-iterations within the PLS main iteration cycle. Therefore the computing steps can be reordered, so that within a PLS iteration cycle (the outer loop) each sub-iteration (the inner loop) is pushed ahead only one step forward on its way to convergence. Thereby, the steps of the outer iteration become a linear sequence of the sub-iteration steps.

The ordering of the modules within the PLS iteration cycle entails no difficulties in Wold's basic method (Sc.2.1). But in the extended method some ambiguities arise, which are discussed by Lohmöller (1981a).

2.3.2 LS Modules for PLS Algorithms

In this section four models with one LV are presented. Algebraic solutions exist for these models (except for the centroid model), though the numerical solution may be iterative. The estimation procedures for these models can be used as LS modules within PLS algorithms. The choice of LS modules in this section is guided by the ensuing discussion of the inner weight relations.

The centroid. Z is called the centroid factor of a set of variables y_k, $k = 1 \ldots K$,

$$Z = \sum_k w_k y_k, \tag{2.65}$$

if the weights w_k are equal to $+1$ or -1 and chosen such that $\mathrm{var}(Z) =$ maximum. Usually, the weights are found by $w_k = \mathrm{sign}\,\mathrm{cov}(Z; y_k)$ (Woodward & Bentler 1979).

The multiple regression. If Z is the target variable (predictand) and $x_\ell, (\ell = 1 \ldots L)$ is a set of source variables (predictors), then the LS approximation to Z is known to be

$$Z^* = \sum_\ell w_\ell x_\ell, \tag{2.66}$$

where the weights w_ℓ are found by multiple regression.

The principal component. If Z is the principal component of a set of target variables y_k and ${}_cZ$ is an approximation to Z, then this approximation is improved by the linear combination

$${}_{c+1}Z = {}_cf \sum_k {}_cp_k y_k, \tag{2.67}$$

where ${}_cp_k = \mathrm{cov}({}_cZ; y_k)$ and ${}_cf$ is a scale factor that makes $\mathrm{var}({}_{c+1}Z) = 1$. To obtain the principal component, the computation Eq.2.67 must be repeated for iteration cycles $c = 1, 2, \ldots$ until the difference ${}_{c+1}p_k - {}_cp_k, \forall k$, is judged to be negligible (cf. Bock 1975:93).

The MIMIC variable. Consider a model with multiple indicators (effect variables, target variables) $y_k (k = 1 \ldots K)$ and multiple causes (source variables) x_ℓ $(\ell = 1 \ldots L)$ of a single latent variable Z (the MIMIC model, see Sc.3.4.4). If ${}_cZ$ is an approximation to Z, then it is improved by the linear combination

$${}_{c+1}Z = {}_cf(\sum_\ell {}_cw_\ell x_\ell + \sum_k {}_cp_k y_k), \tag{2.68}$$

where ${}_cp_k = \mathrm{cov}({}_cZ; y_k)$ is a covariance, ${}_cw_k = w({}_cZ; x_\ell)$ is a multiple regression coefficient and ${}_cf$ is a standardization factor. To obtain the LS estimator of Z, the computation Eq.2.68 must be repeated for iteration cycles $c = 1, 2, \ldots$ until convergence is obtained (Wold 1975c:327).

In short, the linear composite Z will tend to become

- the centroid factor, if sign r_{hg} is used as weight (Eq.2.65)
- the most predictable criterion, if the regression coefficients are used as weights (Eq.2.66)
- the best predictor, i.e. the principal component, if the covariance coefficients are used as weights (Eq.2.67)
- the best mediating (explanatory) variable, if the regression coefficients are used as weights for the source variables and the covariance coefficients for the target variables (Eq.2.68).

Estimating problems like these occur as partial problems in LVP models which are estimated by PLS. The multiple regression problem applies to the estimation of LVs in inward directed blocks and the principal components problem in outward directed blocks; all four types of problems occur in the inner part of the model.

2.3.3 The Inner Weighting Modes

After the model specification comes the iterative estimation of the LVs, the second stage of the PLS algorithm, as sketched in Tb.2.2. Each LV is involved in two equation systems, the inner system Eq.2.16 and the outer system Eq.2.17. In each iteration cycle $c = 1, 2, \ldots$ both equation systems are used to find approximations to the "true" LVs, denoted by $\boldsymbol{\eta} = [\eta_h]$. The one approximation is the outside approximation, which is a weighted aggregate of the MVs and denoted by $\boldsymbol{Y} = [Y_h]$, and the other is the inside approximation, which is a weighted aggregate of the adjacent LVs and denoted by $\tilde{\boldsymbol{Y}} = [\tilde{Y}_h]$.

Inside approximation. The inner equation system is used to find an inside approximation

$$ {}_c\tilde{Y} = {}_c\boldsymbol{V}\ {}_cY, \tag{2.69} $$

where the inner weight matrix $\boldsymbol{V}$ is a function of the LV correlations, path coefficients, the path design matrix and of an option to be explained below:

$$ {}_c\boldsymbol{V} = \boldsymbol{V}({}_c\boldsymbol{R}, {}_c\boldsymbol{B}, \text{Design}(\boldsymbol{B}), \text{Inner Weighting Scheme}). \tag{2.70} $$

The inner weight matrix $\boldsymbol{V}$ has a pattern of zeros so that the inside approximation $\tilde{Y}_h$ is a linear combination of the directly adjacent LVs. $\tilde{Y}_h$ is estimated to be the "most suitable neighbor" in its own vicinity.

Outside approximation. The outer equation system is used to find an outside approximation

$$ {}_cY = {}_{c-1}\boldsymbol{W}'\boldsymbol{y}, \tag{2.71} $$

where ${}_{c-1}\boldsymbol{W}$ is the weight matrix as determined in the preceding iteration cycle $c - 1$. The outside approximation Y_h is a linear combination of the MVs and is estimated to be the best approximation to the inside approximation $\tilde{Y}_h$.

A regression between inside approximation $\tilde{Y}$ and MVs gives new weights ${}_c\boldsymbol{W}$ and Eq.2.71 is applied anew in the next iteration cycle. The new outside approximation ${}_{c+1}Y$ is, then, at the same time a linear function of the MVs and an LS approximation to the most suitable neighbor $\tilde{Y}$. In this way, the "true" LV η_h is encircled from two sides. Finally, the outside approximation Y_h is taken as the estimate of η_h. This estimate will be the variable fitting optimally into both the inner and the outer equation system.

It remains to explain how to choose the inner weights v_{hg} in such a way that $\tilde{Y}_h$ is the best neighbor, in sense to be defined more precisely. This topic is discussed by Apel (1978b), Hui (1978), Noonan and Wold (1982). Three inner weighting schemes for $\boldsymbol{V}$ are listed in Table 2.3.

Table 2.3: Weighting Schemes for inside approximation

Inner weighting scheme		$v_{hg} = 0$, but if Y_g is source of Y_g, then $v_{hg} =$	target of Y_g, then $v_{hg} =$
#1	Path weighting scheme	b_{hg}	r_{hg}
#2	Centroid weighting scheme	$\operatorname{sign} r_{hg}$	$\operatorname{sign} r_{hg}$
#3	Factor weighting scheme	r_{hg}	r_{hg}

Centroid weighting scheme. In his basic method Wold (1982d) favours the computationally simplest of the weighting schemes, here called the centroid weighting scheme (see Tb.2.3, line #2). This scheme utilizes neither the directedness nor the strength of the relationships between the LVs. As we know from Eq.2.65 the LV of each block will become similar to the centroid factor.

Factor weighting scheme. The so-called factor weighting scheme gives an inside approximation which is one step on the way of becoming the principal component of its neighbors in the path diagram (see Tb.2.3, line #3, and Eq.2.67). Also, the outside approximation is determined in a way such that it contributes to the principal component of the neighboring LVs. The procedure can be interpreted in terms of a hierarchic factor model where the MVs are the first-level variables, the LVs the second level variables and the implicitly assumed principal component of the LVs the third-level variable. The factor weighting scheme maximizes the variance of the principal component of the LVs when the number of the LVs goes to infinity — clearly an unrealistic design, but a useful consideration within the framework of multiple set factor analysis which will be pursued in more detail in Sc.3.4.

Path weighting scheme. With this scheme (see Tb.2.3, line #1) the source variables of Y_h are weighted by the path coefficients ${}_cb_{hg}$ and the target variables by the correlations ${}_cr_{hg}$ in order to form the inside approximation $\tilde{Y}_h$. Thus, from the argument in Sc.2.3.2, it follows that $\tilde{Y}_h$ tends to become the MIMIC variable, the best mediating variable between the source and target variables and, in this sense, the best neighbor in the vicinity. If Y_h is exogenous, $\tilde{Y}_h$ is an approximation to the best predictor. If Y_h is the last member of a causal chain, $\tilde{Y}_h$ is an approximation to the best predictable criterion.

Hence, the path weighting scheme is adequate for an LV path model; it is the only scheme to take account of the fact that the relationships between the LVs are specified to be directed. When the theory contains no proposition about the direction of the LV relationships, then the factor weighting scheme may be used; the parameters are those of a multiple set factor model, which may or may not be restricted.

It has been found that the choice of inner weighting scheme has only little influence on the results and on the computing time when the model is a realistic one (Noonan & Wold 1982).

The centroid weighting scheme may have more advantages than the path weighting scheme when the LV correlation matrix is singular due, e.g., to insufficient sample size. On the other hand, if an LV correlation coefficient is close to zero and oscillates during the iteration from small positive to small negative values, then these values are magnified to -1 and $+1$ and have more impact on the results than e.g. the factor weighting scheme.

2.3.4 Patterned Orthogonalization of LVs

Typically, factor rotation algorithms start with orthogonal factors and search for a transformation giving a simple-structured loading matrix, maybe by allowing for correlated factors. The algorithm presented in this section starts with correlated factors $\boldsymbol{Y}$ with correlation matrix $\boldsymbol{R} = \text{cor}(\boldsymbol{Y})$ and determines a transformation matrix $\boldsymbol{O}$,

$$\boldsymbol{Y}^* = \boldsymbol{O}\boldsymbol{Y}, \tag{2.72}$$

which gives a simple-structured factor correlation matrix $\boldsymbol{R}^* = \text{cor}(\boldsymbol{Y}^*)$. More exactly, $\boldsymbol{R}^*$ is restricted in order to have zeros in positions specified by the design matrix

$$\boldsymbol{R}^* = \text{Design}(\boldsymbol{\rho}) * \boldsymbol{R}^*. \tag{2.73}$$

The method will be called "Patterned Orthogonalization" (Pattor) rotation (Lohmöller 1979b).

Jacobi transformation. The Pattor rotation resembles the Jacobi algorithm for diagonalizing a symmetric matrix (cf. Bock 1975:96-98), which consists of the iterative application of (2×2) transformation matrices to all pairs of variables. Let $\boldsymbol{R}_{[H \times H]}$ be the correlation matrix of a set of variables $\boldsymbol{Y} = [Y_1..Y_g, Y_h, Y_i..Y_H]'$. Then an orthonormal transformation $\boldsymbol{O}_{gh[2\times 2]}$ is applied to Y_g and Y_h to obtain uncorrelated substitutes Y_g^* and Y_h^*:

$$\begin{bmatrix} Y_g^* \\ Y_h^* \end{bmatrix} = \boldsymbol{O}_{gh} \begin{bmatrix} Y_g \\ Y_h \end{bmatrix} \tag{2.74}$$

But performing the orthogonalization on a pair Y_g, Y_h can destroy the zero covariance between the pairs Y_g, Y_i or Y_h, Y_i:

$$\begin{bmatrix} r_{gi}^* \\ r_{hi}^* \end{bmatrix} = \boldsymbol{O}_{gh} \begin{bmatrix} r_{gi} \\ r_{hi} \end{bmatrix} \neq 0. \tag{2.75}$$

Thus, if the transformation of all different pairs is completed, a new cycle of transformations must be initiated, starting afresh with the first pair of variables. After several iteration cycles the process converges, and the covariance between each pair of transformed variables differs from zero by not more than a specified small value δ. All (2×2) transformations in pairs are collected into one transformation matrix $\boldsymbol{O}_{[H\times H]}$ for all variables. Let ${}_c\boldsymbol{O}_{gh[H\times H]}$ denote an identity matrix, where the elements with indices $(g, g), (g, h), (h, g), (h, h)$ are replaced by the elements of the transformation matrix $O_{gh[2\times 2]}$, as applied to Y_g, Y_h in iteration cycle c. The total transformation matrix then is given by the product of all the transformations:

$$O = \prod_c \prod_g \prod_h {}_cO_{gh}. \tag{2.76}$$

The total transformation can be expressed now as:

$$Y^* = OY, \qquad R^* = ORO'. \tag{2.77}$$

O contains the eigenvectors of R; R^* is diagonal and contains the eigenvalues.

The Pattor rotation differs from the Jacobi method in two respects:

(i) A transformation is applied only to those pairs of variables whose correlation is restricted by Design(ρ).

(ii) The method is applied to a correlation matrix and nonsingular (instead of orthonormal) transformations O_{gh} are used which preserve the unity diagonal of the correlation matrix.

From the infinite possibilities of transformation O_{gh} fulfilling this requirement three special cases will be considered, two semi-partial and one symmetric transformation.

The first semipartial transformation partials out the first variable Y_g and leaves the second variable Y_h unaltered:

$${}_1O_{gh} = \begin{bmatrix} \frac{1}{\sqrt{1-r_{gh}^2}} & \frac{-r_{gh}}{\sqrt{1-r_{gh}^2}} \\ 0 & 1 \end{bmatrix} \tag{2.78}$$

The second semipartial transformation partials out the second variable Y_h and leaves the first one unaltered:

$${}_2O_{gh} = \begin{bmatrix} 1 & 0 \\ \frac{-r_{gh}}{\sqrt{1-r_{gh}^2}} & \frac{1}{\sqrt{1-r_{gh}^2}} \end{bmatrix} \tag{2.79}$$

The symmetric transformation transforms both variables so that the transformed variables Y_g^*, Y_h^* are uncorrelated least-squares approximations to their originals Y_g, Y_h (cf. Kaiser 1967),

$${}_sO_{gh} = \begin{bmatrix} p+q & p-q \\ p-q & p+q \end{bmatrix}, \tag{2.80}$$

where $p = 0.5/\sqrt{1+r_{gh}^2}$ and $q = 0.5/\sqrt{1-r_{gh}^2}$.

The Pattor algorithm converges whenever the Jacobi algorithm would converge, because it has to fulfil fewer restrictions. That is, if it is possible to diagonalize the LV correlation matrix R, then it is also possible to permit some non-diagonal elements of R to be not necessarily equal to zero.

Let $\boldsymbol{y}$ denote the vector of MVs, related to the LVs by $\boldsymbol{y} = \boldsymbol{P}\boldsymbol{Y} + \boldsymbol{e}$, $\boldsymbol{Y} = \boldsymbol{W}'\boldsymbol{y}$, $\boldsymbol{T} = \text{cov}(\boldsymbol{y}; \boldsymbol{Y})$, then the MVs are related to the rotated LVs by $\boldsymbol{T}^* = \boldsymbol{T}\boldsymbol{O}'$, $\boldsymbol{W}^* = \boldsymbol{W}\boldsymbol{O}'$, $\boldsymbol{P}^* = \boldsymbol{P}\boldsymbol{O}^{-1}$. In Tb.2.2 a shortcut notation for the Pattor invocation is used,

$$\boldsymbol{O} := \text{Pattor}(\boldsymbol{R}, \text{Design}(\boldsymbol{\rho}), \text{choice of rotation}), \tag{2.81}$$

which shows that the rotation matrix is a function of $\boldsymbol{R}$, of the design matrix and an option for the choice of first semipartial, second semipartial or symmetric rotation.

Pattor and PLS. In connection with the PLS algorithm the three rotation methods produce different results. With the first semipartial transformation the second LV often resembles a general factor (high loadings with equal signs) and the first LV resembles a contrast factor (loadings with different signs). The second semipartial transformation gives the first (untransformed) LV the appearance of a general factor, the second (transformed) LV is the contrast factor. The symmetric transformation often results in LVs with simple structured loadings, i.e. the loadings in each row are either very high or very low (Lohmöller 1981a:2-10).

It should be noted that the consequences of the choice of the rotation have not been thoroughly explored. Depending on the choice of the rotation option, the loadings may vary markedly (and also the speed of convergence), whereas the sum of squared multiple correlations of the LVs remains constant. In other models not even the constancy property holds. In the example in Sc.4.2.3 the rotation improved the R^2. In other models the rotation interferes with the main PLS procedure by destroying in each iteration cycle the improvement made by the main procedure, thus blocking the convergence.

Fortunately, often the roation gives hint of how to split a big multidimensional block of MVs into unidimensional blocks — and we are back in the comfortable realm of Wold's Basic Design.

Pattor and factor analysis. The Patterned Orthogonalization rotation, here shown as a module within a PLS algorithm, can be regarded as a rotation method in its own right and as complementary to well-established rotation methods. A sequence of three rotations may be performed:

(i) Varimax rotation is applied to find an orthogonal simple structure.

(ii) Promax rotation is applied to improve the simple structure of the loadings at the expense of the simple structure, i.e. orthogonality, of the factor correlations.

(iii) Finally, Pattor rotation is applied to small factor correlation coefficients which would be interpreted as zeros anyway. An appropriate matrix Design($\boldsymbol{\rho}$) may be empirically defined:

$$\text{Design}(\rho_{gh}) = \begin{cases} 0 & \text{if } |r_{gh}| < r_0 = 0.10, \text{ say} \\ 1 & \text{otherwise.} \end{cases} \tag{2.82}$$

Table 2.4: Constrained loadings and weights

	Constrained loadings	
$Y_h \in \boldsymbol{Y}_k$	$\Longleftrightarrow \text{Design}(\pi_{kh}) = 1$	$\boldsymbol{Y}_k$ is the set of LVs connected to y_k
H_k	$:= \sum_k \text{Design}(\pi_{kh})$	number of LVs directly connected to y_k
$\boldsymbol{R}_k$	$:= \text{cov}(\boldsymbol{Y}_k)$	$(H_k \times H_k)$ LV correlation matrix of $\boldsymbol{Y}_k$
$\boldsymbol{T}_k$	$:= \text{cov}(y_k; \boldsymbol{Y}_k)$	$(1 \times H_k)$ LV structure matrix
$\boldsymbol{P}_k$	$:= \boldsymbol{T}_k \boldsymbol{R}_k^{-1}$	$(1 \times H_k)$ LV pattern matrix
	Constrained weights	
$y_k \in \boldsymbol{y}_h$	$\Longleftrightarrow \text{Design}(\pi_{kh}) = 1$	$\boldsymbol{y}_h$ is the set of MVs connected to Y_h
K_h	$:= \sum_h \text{Design}(\pi_{kh})$	number of MVs directly connected to Y_h
$\boldsymbol{S}_h$	$:= \text{cov}(\boldsymbol{y}_h)$	$(K_h \times K_h)$ MV covariance matrix of $\boldsymbol{y}_h$
$\boldsymbol{T}_h$	$:= \text{cov}(\boldsymbol{y}_h; Y_h)$	$(K_h \times 1)$ LV structure matrix
$\boldsymbol{W}_h$	$:= \boldsymbol{S}_h^{-1} \boldsymbol{T}_h$	$(K_h \times 1)$ LV weight matrix

2.3.5 Weights and Loadings

Estimation of LV weights and loadings in ModeA and ModeB blocks differs in the basic method (Sc.2.1) and becomes more differentiated in the extended method. The additional constraints on the MV-LV relationships as well as the multidimensional structure of a single block require refinements and redefinitions of the estimation procedure.

The relationship between loading structure $\boldsymbol{T} = \text{cov}(\boldsymbol{y}; \boldsymbol{Y})$ and loadings pattern $\boldsymbol{y} = \boldsymbol{PY} + \boldsymbol{e}$ is very simple, as long as the LVs are uncorrelated, $\boldsymbol{R} = \mathbf{I}$, or no overlap in the indicators is permitted:

$$\text{Design}(\boldsymbol{\Pi})' \, \text{Design}(\boldsymbol{\Pi}) = \text{diagonal}. \tag{2.83}$$

In these cases the pattern $\boldsymbol{P}$ is found by replacing the specified elements of $\boldsymbol{T}$ by zeros:

$$\boldsymbol{P} = \text{Design}(\boldsymbol{\Pi}) * \boldsymbol{T} \tag{2.84}$$

If, however, correlated LVs are based on overlapping subsets of MVs, the difference between $\boldsymbol{T}$ and $\boldsymbol{P}$ must be observed and the estimation of the LV pattern coefficients must be done separately for each row of the matrix $\boldsymbol{P}$ (see Table 2.4, upper part). The loadings are found by multiple regression of the MV $\boldsymbol{y}_k$ on all LVs directly connected to it and the resulting elements of the row vector $\boldsymbol{P}_k$ must be inserted in the appropriate position of the total loading pattern matrix $\boldsymbol{P}$.

In a ModeA block the column $\boldsymbol{P}_j$ of $\boldsymbol{P}$ is used as weight, after the rescaling that gives unit variance to the LV. (Had the loading structure $\boldsymbol{T}_j$ been used for the weights ($\boldsymbol{W}_j = \text{Design}(\boldsymbol{P}_j) * \boldsymbol{T}_j$), then all LVs in this block would tend to become the first principal component, subject to the restrictions imposed on the design matrix, and the result would be a set of highly correlated LVs.) By using the factor patterns as LV weights, the LVs are determined in a way such that they contribute independently to the prediction of the MVs.

In a ModeB block the weights are found by multiple regression of the inside approximation $\tilde{Y}_h$ on a selected set of MVs, as shown in Table 2.4, lower part. The resulting elements of the column vector $\boldsymbol{W}_h$ must be inserted into the appropriate positions of the total weight matrix $\boldsymbol{W}$. The "advanced orthogonalization" of the MVs in inward directed block (see Eq.3.532; Lohmöller 1979b) is possible only if there are no restrictions on the weights within this block.

2.3.6 Three Solutions for Conflicting Constraints

It is a basic feature of the PLS parameter estimation technique for LVP models that the LVs are estimated as weighted aggregates of subsets of MVs. The extended class of LVP models estimable under PLS allows for a multidimensional LV for any block of MVs. When it comes to the estimation, it must be secured that the two presumptively different LVs, Y_h and Y_g, related to the same set of MVs, do not collapse and become identical, i.e. $r_{gh} = 1$. The non-identity of LVs can be assured by different types of assumptions:

(i) The LVs are specified as uncorrelated, Design(ρ_{gh}) = 0.

(ii) The LVs are differentiated by different relations to the MVs, Design($\boldsymbol{\pi}_h$) $\neq$ Design($\boldsymbol{\pi}_g$).

(iii) The LVs are differentiated by different relations to the other LVs, Design($\boldsymbol{\beta}_h$) $\neq$ Design($\boldsymbol{\beta}_g$).

It has been found that type (i) restrictions work best. Restrictions of type (ii) and (iii) often result in highly correlated LVs and are, therefore, not very satisfactory.

There are some objections to type (i) restrictions for endogenous LVs. If they have some predictors in common, they are expected to correlate positively. In fact, in this case the restriction $r_{hg} = 0$ must be compensated by negatively correlated inner residuals, $q_{hg} = \text{cov}(u_h; u_g) < 0$. A restriction $q_{hg} = 0$, instead of $r_{hg} = 0$, would be more in line with the conventional procedure. On the other hand, this is a restriction on the residual variables $\boldsymbol{u}$ which are of second-order latency, while the LVs are of first-order latency. PLS deals with situations where theoretical knowledge (i.e. about the distribution of the latent variables) is scarce and the estimates must, therefore, be closely tied to the data. If the assumption $q_{hg} = 0$ is, in fact, applicable, a final parameter estimation could be done using LISREL.

Conflicts can arise between constraints on the outer relations and the LV correlations. Orthogonality is possible between LVs when they are

(i) based on the same set of LVs, or

(ii) based on two different sets only in the canonical correlation model.

In general, however, it is not possible to determine uncorrelated weighted aggregates if they are not based on the same set of aggregands. This is to say that zero-$\boldsymbol{\rho}$ restrictions cannot be exactly fulfilled if they are imposed on LVs of different blocks, or if additional constraints are imposed on LVs of the same block. In these situations the correlations of the LVs may be forced to be low but not to be completely zero. If they are forced to zero by the Pattor step #25 (Tb.2.2), in the next step #26 the zero correlation will be destroyed by an amount dictated by the data.

When the model constraints conflict in this way, several solutions to the parameter estimation are possible, depending on which constraints are observed and which are abandoned. Three solutions will be considered:

Solution #1 adheres to the principle that the LVs are estimated as linear aggregates of the MVs, where the weight matrix has to fulfil the block-diagonal form and to observe additional constraints on the MV-LV relationships. It may turn out that the LV correlations do not fulfil the specified orthogonality pattern Design($\boldsymbol{\rho}$).

Solution #2 (see Tb.2.2, step #41) furnishes the required zero LV correlations within blocks, though at the expense of the minimization of the outer residual variances. This solution requires a Pattor rotation using a redefined design matrix Design($\boldsymbol{\rho}_2$) with elements:

$$\text{Design}(\rho_{hg}^{(2)}) = \begin{cases} 0 & \text{if } \text{Design}(\rho_{hg}) = 0 \text{ and} \\ & \quad Y_h, Y_g \text{ are directly related} \\ & \quad \text{to the same set of MVs} \\ 1 & \text{otherwise} \end{cases} \tag{2.85}$$

Then the weights are re-estimated using the multiple regression of all LVs on all MVs of that block, whether they are specified as inward or outward. (An outward-directed estimation of the weights would induce new correlations between the LVs.) The solution requires that $\boldsymbol{S}_j$ be nonsingular.

Solution #3 (see Tb.2.2, step #42) gives all the required LV correlations at the expense, however, of the block-diagonal form of the weight matrix. After performing the Pattor rotation (as in step #21 in Tb.2.2), the weight matrix is found by an unrestricted LS approximation to all MVs:

$$\boldsymbol{W} = \boldsymbol{S}^{-1}\boldsymbol{P} \tag{2.86}$$

Here the total MV covariance matrix is required to be invertible. If the model is consistent with the data, the weights of LVs in the non-connected MVs will be small.

The first of the three solutions is most in line with the pragmatic basic PLS assumption that the solution should be tied more closely to the data than to assumptions about the data. It is up to the model builder to decide which restrictions in his model are the most important ones and consequently which solution provides the best answer to his problem.

2.4 Assessment of Results

Estimates for a complex model have many aspects regarding their quality, sufficiency to explain the data, congruence with substantive expectations, precision and confidentiality. The assessment of the quality of the results must take all these different aspects into account.

Completely omitted from this section is the Blindfolding procedure (Lohmöller 1981a:5-09, Lohmöller & Wold 1984), which gives, as first result, the Stone-Geisser test criterion for predictive relevance of the model (Stone 1974, Geisser 1974, Abrahamsen 1987, Allen 1971, Apel 1977a, 1977b, Wold 1982d, S.Wold 1978) and, as a second result, Jackknife standard deviations for the parameter estimates (Miller 1974, Bergström & Wold 1982). The Blindfolding procedure and the related statistics work without any distributional assumptions. Hence they fit PLS methods in the same way as likelihood ratio tests fit ML methods. The Blindfolding procedure requires the raw data.

This subchapter, Sc.2.4 (strongly relying on Lohmöller 1981a:1-05 to 1-10), proposes an array of scalar-, vector- and matrix-valued measures to assess the results. Sc.2.4.1 focuses on the effect decomposition, residual variances and covariances, also variance ratios like squared multiple correlation, communality and redundancy coefficients. In Sc.2.4.3 several overall measures of fit are defined. No single index can tell everything about the quality and some of these indices (like the χ^2- and reliability indices) are developed more to facilitate a comparison to LISREL results (following in Sc.5) than as an appropriate PLS index (Sc.2.4.4).

2.4.1 Information from the Model

For the sake of easy reference, the first five equations of Sc.2.2, which give the structural and the reduced form of the LVP model, are here repeated (location parameters omitted):

$$\boldsymbol{Y} = \boldsymbol{B}\boldsymbol{Y} + \boldsymbol{u} \tag{2.87}$$

$$\boldsymbol{y} = \boldsymbol{P}\boldsymbol{Y} + \boldsymbol{e} \tag{2.88}$$

$$\boldsymbol{Y} = \boldsymbol{W}'\boldsymbol{y} + \boldsymbol{d} \tag{2.89}$$

$$\boldsymbol{Y} = (\mathbf{I} - \boldsymbol{B})^{-1}\boldsymbol{u} \equiv \boldsymbol{B}^{*}\boldsymbol{u} \tag{2.90}$$

$$\boldsymbol{y} = \boldsymbol{P}(\mathbf{I} - \boldsymbol{B})^{-1}\boldsymbol{u} + \boldsymbol{e} \equiv \boldsymbol{P}^{*}\boldsymbol{u} + \boldsymbol{e}. \tag{2.91}$$

Also Eq.2.16 is recalled showing how the covariances between all unobserved variables $(\boldsymbol{Y}, \boldsymbol{u}, \boldsymbol{e})$ are computed when $\boldsymbol{W}, \boldsymbol{P}, \boldsymbol{B}$ are estimated.

Direct and indirect paths. The path matrix $\boldsymbol{B}$ shows the direct effects of the LVs on each other, i.e. the influences by direct connections, by pathways consisting of only one arrow. The total-effect path matrix $\boldsymbol{B}^{*}$ shows the total influence of the LVs on each other, i.e. the sum of the effects of pathways consisting of one arrow, of two, three and upwards. The difference

$$\boldsymbol{B}^{+} = \boldsymbol{B}^{*} - \boldsymbol{B} \tag{2.92}$$

then shows the indirect effects, i.e. the influence through pathways which consist of more than one arrow (cf. Knepel 1981, Graff & Schmidt 1982). The name "reduced form coefficients" for $\boldsymbol{B}^*$ is used in a slightly different way than traditionally in econometrics, where the distinction between exogenous variables $\boldsymbol{X}$ and endogenous variables $\boldsymbol{Y}$ is made explicit. To be precise: the structural form is written as $\boldsymbol{Y} = \ddot{\boldsymbol{B}}\boldsymbol{Y} + \boldsymbol{G}\boldsymbol{X} + \boldsymbol{u}$, the total-effect form as $\boldsymbol{Y} = \ddot{\boldsymbol{B}}^*\boldsymbol{G}\boldsymbol{X} + \ddot{\boldsymbol{B}}^*\boldsymbol{u}$ and the reduced form as $\ddot{\boldsymbol{Y}} = \ddot{\boldsymbol{B}}^*\boldsymbol{G}\boldsymbol{X}$.

Direct and indirect loadings. The loading pattern matrix $\boldsymbol{P}$ shows the direct influence of the LVs on the directly attached MVs. The total-effect form $\boldsymbol{P}^*$ traces the influence of the LVs on the MVs back to the very first causes, to the inner residuals $\boldsymbol{u}$. The difference between total effect $\boldsymbol{P}^*$ and direct effect $\boldsymbol{P}$ is given by the indirect loading effect matrix

$$\boldsymbol{P}^+ = \boldsymbol{P}^* - \boldsymbol{P} \tag{2.93}$$

which is zero for exogenous variables (Knepel 1981).

2.4.2 Five Predictions, Five Residuals

There are different sorts of predictions, and hence different sorts of residuals, which are assembled in this diagram:

$$\begin{array}{ccccccc} & & \mathrm{d} & & \mathrm{u} & & \mathrm{e} \\ & & \downarrow & & \downarrow & & \downarrow \\ \mathrm{y} & \xrightarrow{\mathrm{W'}} & \mathrm{Y} & \xrightarrow{\mathrm{B}} & \mathrm{Y} & \xrightarrow{\mathrm{P}} & \mathrm{y} \end{array} \tag{2.94}$$

valid (y → Y); structural (Y → Y); communal (Y → y); redundant (Y → Y → y); operative (y → Y → Y → y)

Communality prediction. The manifest variables $\boldsymbol{y}$ are predicted by the latent variable Y that belongs to the same block.

$$\begin{aligned} \boldsymbol{y} &= \boldsymbol{P}\boldsymbol{Y} + \boldsymbol{e} \\ &= \text{common part} + \text{measurement residual} \end{aligned} \tag{2.95}$$

Eq.2.95 shows the MVs as consisting of the systematic part $\boldsymbol{y}^* = \boldsymbol{P}\boldsymbol{Y}$ (in traditional factor analysis called the common part) and residual part $\boldsymbol{e}$ (sometimes interpreted as measurement error), giving the covariance matrix of $\boldsymbol{y}^*$:

$$\boldsymbol{H} \equiv \mathrm{cov}(\boldsymbol{y}^*) = \mathrm{cov}(\boldsymbol{P}\boldsymbol{Y}) = \boldsymbol{P}\boldsymbol{R}\boldsymbol{P}' \tag{2.96}$$

The diagonal of $\boldsymbol{H}$, divided by the MV variances, is a diagonal matrix and contains the *communalities*:

$$\boldsymbol{H}^2 = (\mathbf{I} * \boldsymbol{P}\boldsymbol{R}\boldsymbol{P}')(\mathbf{I} * \boldsymbol{S})^{-1}, \tag{2.97}$$

H_k^2 is the proportion of the variance of y_k which is reproduced by the LVs directly connected to them. The average of the communalities within block j is denoted by H_j^2, the average over all variables by H^2.

Structural prediction. The latent variables $\boldsymbol{Y}$ are predicted by one or more other latent variables:

$$\begin{aligned}\boldsymbol{Y} &= \boldsymbol{BY} + \boldsymbol{u} \\ &= \text{structural part} + \text{specification residual} \end{aligned} \tag{2.98}$$

Eq.2.98 shows the LVs as consisting of the residual part $\boldsymbol{u}$ and the systematic part $\boldsymbol{Y}^* = \boldsymbol{BY}$, with covariance matrix:

$$\boldsymbol{R}^* \equiv \text{cov}(\boldsymbol{Y}^*) = \text{cov}(\boldsymbol{BY}) = \boldsymbol{BRB}'. \tag{2.99}$$

The diagonal of $\boldsymbol{R}^*$

$$\boldsymbol{R}^2 = \mathbf{I} * (\boldsymbol{BRB}') \tag{2.100}$$

contains the *squared multiple correlation* coefficients of the endogenous LVs with their predictors. R_h^2 is the proportion of the variance of Y_h that can be reproduced by these predictors.

Validity prediction. The latent variable $\boldsymbol{Y}$ are predicted by their manifest variables, as shown in the generalized weight relation:

$$\begin{aligned}\boldsymbol{Y} &= \boldsymbol{W}'\boldsymbol{y} + \boldsymbol{d} \\ &= \text{valid part} + \text{validity residual} \end{aligned} \tag{2.101}$$

The parameters of the linear relation are denoted by $\boldsymbol{W}$ and called the LV weight coefficients. The residual variable $\boldsymbol{d}$ represents the unmeasured part of the LV, and the squared multiple correlation of Y_h with its indicators is called the validity. In component aproaches like PLS the validity error of Equation 2.101 is set to zero, the validity is set to unity.

Redundancy prediction. If we substitute the $\boldsymbol{Y}$ in Eq.2.95 from Eq.2.98, we eliminate the directly predicting LVs from our prediction approach.

$$\begin{aligned}\boldsymbol{y} &= \boldsymbol{PBY} + \boldsymbol{Pu} + \boldsymbol{e} \equiv \boldsymbol{y}^{**} + (\boldsymbol{Pu} + \boldsymbol{e}) \\ &= \text{redundant part} + \text{nonredundant}. \end{aligned} \tag{2.102}$$

The common part $\boldsymbol{PY}$ is split up into the unique part $\boldsymbol{Pu}$ and the redundant part $\boldsymbol{PBY}$. The unique part can be reproduced only by the LVs (or the respective residuals $\boldsymbol{u}$) which are directly connected to y_k. The redundant part y_k^{**} is represented twice in the LVs; it can be predicted either by the LVs directly connected to y_k, i.e. its own LVs, or by the predictor LVs of its own LVs. If the directly connected LVs $\boldsymbol{Y}$ are eliminated from $\boldsymbol{PY}$ and substituted by their predictable (systematic) part $\boldsymbol{BY}$, then that which remains reproducible is $\boldsymbol{y}^{**} = \boldsymbol{PBY}$, i.e. the redundant part (or predictable common, or substitutable common part). The covariance of the substituted LVs with the redundant part of the MVs is:

$$\boldsymbol{T}^* \equiv \text{cov}(\boldsymbol{y}^{**}; \boldsymbol{Y}^*) = \text{cov}(\boldsymbol{PBY}; \boldsymbol{BY}) = \boldsymbol{PBRB}'. \tag{2.103}$$

The covariance matrix of the redundant parts of the MVs is

$$\boldsymbol{F} \equiv \text{cov}(\boldsymbol{y}^{**}) = \text{cov}(\boldsymbol{PBY}) = \boldsymbol{PBRB'P'}, \tag{2.104}$$

where the diagonal, divided by the MV variances, contains the *redundancy* coefficients:

$$\boldsymbol{F}^2 = (\mathbf{I} * \boldsymbol{PBRB'P'})(\mathbf{I} * \boldsymbol{S})^{-1} . \tag{2.105}$$

F_k^2 is the proportion of the variance s_k^2 of the MV y_k which is reproduced by the predictor LVs of its own LVs. The redundancies of the exogenous MVs are zero.

Operative prediction. If we substitute the $\boldsymbol{Y}$ in Equation 2.102 from Equation 2.101,

$$\begin{aligned} \boldsymbol{y} &= \boldsymbol{PBW'y} + \boldsymbol{PBd} + \boldsymbol{Pu} + \boldsymbol{e} \equiv \boldsymbol{y}^{***} + (\boldsymbol{PBd} + \boldsymbol{Pu} + \boldsymbol{e}) \\ &= \text{operative part} + \text{non-operative}, \end{aligned} \tag{2.106}$$

then all LVs are eliminated and the effects are traced back to the predictor MVs. The variance of $\boldsymbol{y}^{***}$ has been called the operative variance by Fornell and Bookstein (1981):

$$\text{cov}(\boldsymbol{y}^{***}) = \boldsymbol{PBW'SWB'P'}. \tag{2.107}$$

For PLS models the predictions $\boldsymbol{y}^{***}$ and $\boldsymbol{y}^{**}$ are identical and the operative variance Eq.2.107 is equal to the redundancy Eq.2.105, because the validity residual $\boldsymbol{d}$ is deliberately set equal to zero and also because the LV estimates have perfect multiple correlation with their MVs. This is different in LISREL-type models, where Eq.2.101 for the estimated LVs contains unmeasured parts $\boldsymbol{d}$.

2.4.3 Fit Indices

No single fit index for the estimation is defined; however, several different indices may be used to evaluate different aspects of the estimates. The indices are averaged variances, averaged variance proportions or root mean squared (rms) covariances. The operator rms$(\cdot)$, as in

$$\text{rms}\,\boldsymbol{A} = \sqrt{\bar{\sum_i}\bar{\sum_j} a_{ij}^2}, \tag{2.108}$$

is used as a shortcut notation for the root mean squared function of a matrix $\boldsymbol{A}$, where the diagonal is excluded if $\boldsymbol{A}$ is symmetric.

The indices defined in the following are all scalar quantities, even if they are denoted by the same symbol as the matrix they summarize:

$$H^2 := \text{trace}\,\boldsymbol{H}^2/K \qquad \text{communality} \tag{2.109}$$

$$F^2 := \text{trace}\,\boldsymbol{F}^2/K \qquad \text{redundancy} \tag{2.110}$$

$$R^2 := \text{trace}\,\boldsymbol{R}^2/H \qquad \text{SMC} \tag{2.111}$$

$$s^2 := \text{trace}\,\boldsymbol{S}/K \qquad = 1 \text{ if } \boldsymbol{S} = \text{cov}(\boldsymbol{y}) = \text{cor}(\boldsymbol{y}) \tag{2.112}$$

$$r^2 := \operatorname{trace} \boldsymbol{R}/H \qquad = 1 \qquad (2.113)$$
$$c^2 := \operatorname{trace} \boldsymbol{C}/K \qquad \text{mean outer residual variance} \qquad (2.114)$$
$$q^2 := \operatorname{trace} \boldsymbol{Q}/H \qquad \text{mean inner residual variance} \qquad (2.115)$$
$$s := \operatorname{rms} \boldsymbol{S} \qquad \text{rms outer covariance} \qquad (2.116)$$
$$r := \operatorname{rms} \boldsymbol{R} \qquad \text{rms inner correlation} \qquad (2.117)$$
$$c := \operatorname{rms} \boldsymbol{C} \qquad \text{rms outer residual covariance} \qquad (2.118)$$
$$q := \operatorname{rms} \boldsymbol{Q} \qquad \text{rms inner residual covariance} \qquad (2.119)$$
$$s_{\boldsymbol{eY}} := \operatorname{rms} \operatorname{cov}(\boldsymbol{e}; \boldsymbol{Y}) \qquad (2.120)$$
$$s_{\boldsymbol{eu}} := \operatorname{rms} \operatorname{cov}(\boldsymbol{e}; \boldsymbol{u}) \qquad (2.121)$$
$$t := \operatorname{rms} \boldsymbol{T} \qquad \text{rms LV structure} \qquad (2.122)$$
$$r_{abs} := \bar{\sum_h} \bar{\sum_g} |r_{hg}| \,. \qquad (2.123)$$

Note that H^2, F^2, R^2 and $q^2 = r^2 - R^2 = 1 - R^2$ are variance proportions. When the MVs are standardized, then $c^2 = s^2 - H^2 = 1 - H^2$ also is a variance proportion. All variance proportions vary between 0 and 1.

2.4.4 Reliability Indices

ML function. Three more fit indices are derived from the ML function:

$$L = \log|\boldsymbol{\Sigma}| + \operatorname{trace}(\boldsymbol{S}\boldsymbol{\Sigma}^{-1}) - \log|\boldsymbol{S}| - K, \qquad (2.124)$$

where K is the number of MVs, $\boldsymbol{S}$ is the empirical covariance matrix and $\boldsymbol{\Sigma}$ the theoretical covariance matrix. $\boldsymbol{\Sigma}$ can be replaced by an estimate $\hat{\boldsymbol{\Sigma}}$, which is deduced from estimated parameters (cf. Eq.2.27). Assuming a multinormal distribution, the logarithm of the likelihood function of the observations y_{kn} is $-(N-1)L/2$. If the model holds, $(N-1)L$ is a χ^2-variable with f_t degrees of freedom:

$$(N-1)L \sim \chi^2_{f_t}, \qquad f_t = \frac{(K^2 + K)}{2} - t, \qquad (2.125)$$

where t is the number of independet parameters used to construct $\hat{\boldsymbol{\Sigma}}$. Three different models for $\hat{\boldsymbol{\Sigma}}$ will be considered:

Baseline model M_0. In the most restricted model the MVs are assumed to be uncorrelated with the variances as observed:

$$M_0 : \boldsymbol{\Sigma}_0 = \mathbf{I} * \boldsymbol{S}. \qquad (2.126)$$

The ML function Eq.2.124 simplifies, because $\boldsymbol{S}\hat{\boldsymbol{\Sigma}}_0^{-1}$ is an identity matrix with trace equal to K:

$$L_0 = \log|\mathbf{I} * \boldsymbol{S}| - \log|\boldsymbol{S}| \,. \qquad (2.127)$$

Degrees of freedoms are $f_0 = (K^2 - K)/2$. Model Eq.2.126 is the worst fitting model with highest $\chi^2_0 = (N-1)L_0$ and it can be used as a baseline model for all better fitting models.

Common factor model M_1. The next model is a strongly restricted common factor model with diagonal residual covariance $\mathbf{\Theta}_1$:

$$M_1 : \mathbf{\Theta}_1 = \mathbf{I} * \mathbf{\Theta}_1, \qquad \mathbf{\Sigma}_1 = \boldsymbol{PRP}' + \mathbf{\Theta}_1, \tag{2.128}$$

with loss function L_1, degrees of freedom f_1 and $\chi_1^2 = (N-1)L_1$.

Factor model M_2. The third model is a less restricted factor model with block-diagonal residual covariance $\mathbf{\Theta}_2$:

$$M_2 : \mathbf{\Theta}_2 = [\text{Design}(\mathbf{\Pi})\,\text{Design}(\mathbf{\Pi})'] * \mathbf{\Theta}_2, \tag{2.129}$$

$$\mathbf{\Sigma}_2 = \boldsymbol{PRP}' + \mathbf{\Theta}_2,$$

with loss function M_2, degrees of freedom f_2 and $\chi_2^2 = (N-1)L_2$.

Indices. Let M_s and M_r denote any pair of the three models considered above. The difference of the fit of two models is measured by the reliability index of Tucker and Lewis (1973):

$$r_{\text{TL}} = \frac{L_s/f_s - L_r/f_r}{L_0/f_0 - 1/N} \tag{2.130}$$

or by the reliability index by Bentler and Bonett (1980)

$$r_{\text{BB}} = \frac{L_s - L_r}{L_0} = \frac{\chi_s^2 - \chi_r^2}{\chi_0^2}. \tag{2.131}$$

If M_s is the baseline model M_0, it serves as a yardstick for comparison and the reliability coefficients indicate the total fit of M_s. If M_s and M_r are two different models, like M_1 and M_2, the reliability coefficients indicate the difference of the fit. For the logic of incremental fit indices see Bentler and Bonett (1980). The program PLSC computes both reliablity indices, comparing model Eq.2.129 with the baseline model Eq.2.126 (the sample-size dependent term 1/N in Eq.2.130 is replaced by zero).

Notice that the models Eqs.2.128, 2.129 are by no means PLS models, because they imply assumptions on the residual covariances which have not been made for the estimation. Hence the indices Eqs.2.125, 2.128,2.129 are not appropriate for the assessment of fit for PLS models. They are described here and implemented in the program PLSC only to facilitate the comparison — and to show how different the results for PLS and LISREL can be. This will be further pursued in Sc.5.5.

The Tucker-Lewis index relies on the property of χ^2-variable that $\text{E}(\chi^2) = f$, hence $\text{E}(L_r/f_r) = 1/N$, hence $\text{E}(r_{\text{TL}}) = 1$ for a fitting model. The Bentler-Bonett index can be used for any fit measures L_r, L_s, L_0. With the interpretation given above it is the ratio of two variances and hence the reciprocal of Fisher's F. Other reliability indices which have the structure of the Bentler-Bonett index may be defined on the basis of the fit indices defined in Sc.2.4.3:

$$r_{s^2} = \frac{s^2 - c^2}{s^2} = 1 - \frac{c^2}{s^2} \qquad \text{explained MV variance} \tag{2.132}$$

$$r_s = \frac{s-c}{s} = 1 - \frac{c}{s} \qquad \text{explained MV covariance} \tag{2.133}$$

$$r_r = \frac{r-q}{r} = 1 - \frac{q}{r} \qquad \text{explained LV correlation} \tag{2.134}$$

If $\boldsymbol{S}$ is a correlation, $r_{s^2} = H^2$.

ULS function. Yet another fit index is based on the Unweighted Least Squares (ULS) function

$$\text{ULS} = \text{rms}(\boldsymbol{S} - \boldsymbol{\Sigma}), \tag{2.135}$$

where $\boldsymbol{\Sigma}$ is to be replaced by an estimate for any of the models M_0, M_1, M_2. The index takes the form of a Bentler-Bonett index:

$$g = 1 - \frac{\text{rms}(\boldsymbol{S} - \hat{\boldsymbol{\Sigma}}_1)}{\text{rms}(\boldsymbol{S})} = 1 - \frac{\text{rms}(\boldsymbol{S} - \hat{\boldsymbol{\Sigma}}_1)}{s}. \tag{2.136}$$

This index will be used in Sc.5.5 as an off-diagonal fit measure.

2.4.5 Some Advice

The fit of the outer model part can be judged to be satisfactory,
- if the unexplained part of the MV variance $1 - H^2 \approx c^2/s^2$ is low enough
- if the unexplained part of the MV covariance c/s is low enough
- if in outwards directed blocks the communalities H^2 are high enough and the residual covariances c_{kl} are low enough
- if the residual covariances between blocks are near zero
- if the covariances $\text{cov}(\boldsymbol{e}; \boldsymbol{Y})$ between outer residuals and the LVs are near zero.

Due to the fact that the LVs are estimated as weighted aggregates of the MVs, some peculiarities must be taken into account:
- If there is just one MV in a block, then the residual variance is equal to zero
- If there are just two MVs in a block, then the residual variables will be perfectly correlated.
- If there are only few LVs in a block, then the LV usually "overextracts" the MVs, leaving the residuals negatively correlated. This holds especially in outwards directed blocks. More on this in Sc.5.2.

The fit of the inner model part can be judged as satisfactory,
- if the unexplained part of the LV variance $1 - R^2$ is low enough
- if the unexplained part of the LV covariance q/r is low enough
- if no single residual covariance q_{hg} between endogenous variables is higher than a path coefficient.

The fit of the total model can be judged as satisfactory,
- if the redundancy coefficient F^2 is high enough
- if the covariance $\text{cov}(\boldsymbol{e}, \boldsymbol{u})$ of inner and outer residuals is low enough.

Table 2.5: Environment and personality (Marjoribanks 1974:100)

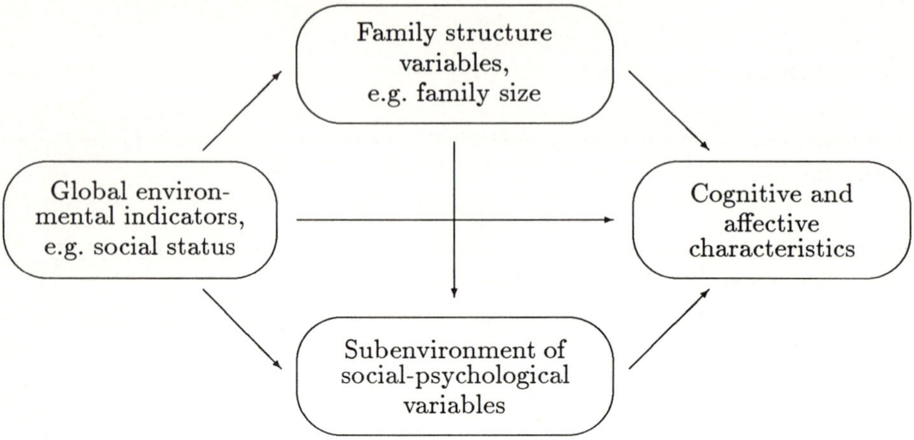

2.5 Application: Home Environment and Intelligence

The hypothesis that intellectual achievement, educational opportunities and success of a student depend on his or her social background has in the last decades been the subject of causal speculations and empirical research in the educational field. The simple hypothesis that occupational and educational status of the parents (Social-Economic Status, SES) determine the intellectual achievement,

$$\text{SES} \quad \rightarrow \quad \text{Intelligence}, \tag{2.137}$$

was enlarged by mediatory variables introduced to explain the influence of social status of the parents on the achievement of the child. Explicative constructs of this type are e.g. class-specific language (restricted code) and class-specific socialization. Marjoribanks (1972) studies characteristics of the learning environment at home as explanatory variables:

$$\text{SES} \quad \rightarrow \quad \text{Home Environment} \quad \rightarrow \quad \text{Intelligence}. \tag{2.138}$$

The explanatory value of the mediating variable, Home Environment, is satisfactory if, besides the direct influences depicted by arrows in Eq.2.138, no direct influence of Social Status on Intelligence must be postulated; that is, if the total influence of social class is channelled and mediated by the explanatory variable, Home Environment.

Though Marjoribanks (1972:103) started with the intention of developing a refined measure of home learning environment, he very soon embarked on complex causal and explanatory post-facto modeling. Other investigators subjected his 18-variables correlation matrix to various statistical techniques. However, none these techniques was capable of coping adequately with all aspects of the model Eq.2.138.

Table 2.6: Learning environment and ability

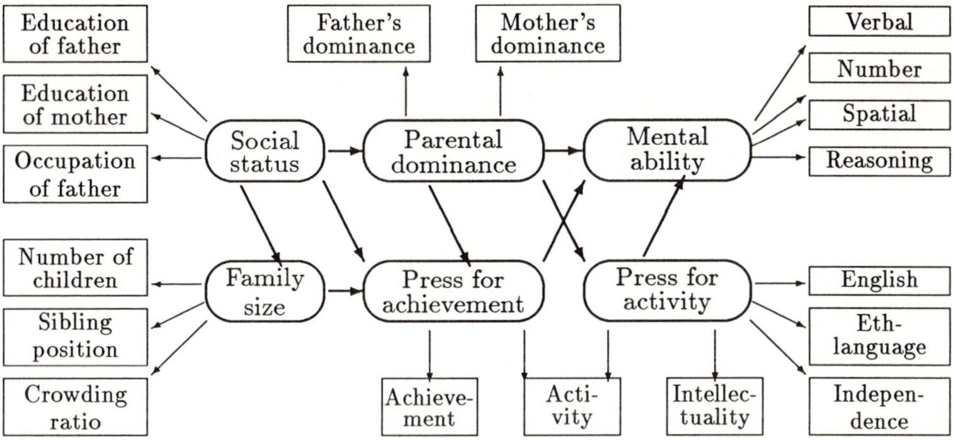

Multi-block models. Marjoribanks (1974:99-100) presents a detailed version of the model Eq.2.138 which is repeated here in Table 2.5. This model includes as a second background variable the family structure (e.g. family size); also the target variable is extended by the affective characteristics of the child. Marjoribanks proves (in his 1974 reader) the validity of the model Tb.2.5 by checking bivariate parts of the model. A closed and systematic analysis of the total model (affective characteristics are excluded, because they are not measured) is possible and has been undertaken several times with, however, inadequate statistical means.

Under the point of view of LVP modeling an analysis of the model should possess three properties:

(i) Model Tb.2.5 is defined in terms of "constructs" or latent variables; the measurements are related to these theoretical variables in the same way as MVs to LVs. Thus only a model with latent variables is adequate.

(ii) Model Eq.2.138 has three, model Tb.2.5 has four distinct blocks of variables. Any analysis that reduces the model to 2 blocks and sacrifices the mediating variables is inadequate.

(iii) Model Tb.2.5 assumes directed relationships between LVs. Hence any correlational model would be inadequate.

Table 2.6 (from Lohmöller 1979b, 1981a) shows the LV path model to be analysed here. As distinct from the model Tb.2.5, the environment block is here split into two blocks, Parental Dominance and Educational Pressure; the fourth block contains two LVs, giving a model with 5 blocks and 6 LVs in all. Before the different analyses of the data are reviewed, the data must be described.

The data were collected by Marjoribanks (1972) and the correlation matrix (Table 2.7) was published by Brandstätter (1976). The sample consists of 185 eleven-year old boys and their parents. The boys were tested by the Primary

Table 2.7: Marjoribanks (1972) data: Correlation matrix

Social Status																		
Education of father	100																	
Education of mother	63	100																
Occupation of father	79	60	100															
Family Size																		
Number of children	-9	-25	-13	100														
Sibling position	-11	-26	-12	87	100													
Crowding ratio	-17	-34	-25	83	73	100												
Parental Domn.																		
Father's dominance	26	30	20	-13	-22	-22	100											
Mother's dominance	23	21	20	-1	4	1	-48	100										
Press for ...																		
achievement	50	52	56	-46	-40	-51	32	26	100									
activities	48	49	45	-30	-26	-33	43	30	72	100								
intellectuality	50	54	51	-35	-31	-40	38	37	80	80	100							
independence	37	32	36	-6	-8	-7	11	35	40	46	48	100						
English	32	43	34	-16	-15	-15	23	32	52	63	61	49	100					
ethlanguage	44	46	36	-22	-22	-31	34	21	51	64	62	41	44	100				
PMA																		
PMA, verbal	29	38	43	-32	-27	-34	16	21	66	52	61	42	50	35	100			
PMA, number	26	32	30	-33	-26	-34	10	15	66	41	53	34	27	24	66	100		
PMA, spatial	26	20	31	-4	-4	-7	9	4	28	22	26	10	18	9	34	38	100	
PMA, reasoning	22	15	29	-3	-4	-9	11	4	39	26	30	22	28	18	39	44	43	100

Mental Ability Test (PMA). Their parents were interviewed in a semi-structured interview concerning social status (block 1), family structure (block 2) and the home environment (blocks 3 and 4). The environment variables consisted originally of 188 items and are condensed in a two-stage procedure, first on 23, then on 8 variables. The variable Father's Dominance contains the two middle-level variables "father's involvement in child's activities" and "father's role in family decision making", and similarly the variable Mother's Dominance.

Previous analyses. In his first analysis of the data, Marjoribanks 1972 applies multiple regression and uses incremental R^2 to demonstrate, curiously enough, the additional explanatory power of the home environment over the SES:

$$R^2(\text{PMA}|\text{SES,Home}) = R^2(\text{PMA}|\text{SES}) + R^2(\text{PMA}|\text{Home–SES}), \qquad (2.139)$$

instead of showing that SES has no explanatory power over the home environment:

$$R^2(\text{PMA}|\text{SES,Home}) = R^2(\text{PMA}|\text{Home}) + R^2(\text{PMA}|\text{SES–Home}), \qquad (2.140)$$

which would be in line with the model Eq.2.138.

Brandstätter (1976) uses path analysis with ten manifest variables for a refined version of the model Tb.2.5: with the first eight variables of Tb.2.7, one of the six Press variables and one of the four PMA variables. This gives a total of $6 \times 4 = 24$ analyses. With 28 parameters for the eight background variables and 24×17

Table 2.8: Marjoribanks (1972) data: Outer model parameters

Blocks of Manifest variables	Weight matrix W Y_1	Y_2	Y_3	Y_4	Y_5	Y_6	Pattern matrix P Y_1	Y_2	Y_3	Y_4	Y_5	Y_6	H^2	F^2
Social Status														
Education of father	11	.	.	.	.	.	81	.	.	.	.	.	65	.
Education of mother	64	.	.	.	.	.	93	.	.	.	.	.	87	.
Occupation of father	38	.	.	.	.	.	84	.	.	.	.	.	71	.
Family Size														
Number of children	.	−10	.	.	.	.	.	−82	.	.	.	.	68	8
Sibling position	.	−10	.	.	.	.	.	−75	.	.	.	.	56	6
Crowding ratio	.	−101	.	.	.	.	.	−100	.	.	.	.	100	11
Parental Dominance														
Father's dominance	.	.	99	.	.	.	.	.	53	.	.	.	28	7
Mother's dominance	.	.	97	.	.	.	.	.	49	.	.	.	24	7
Press for ...														
achievement	.	.	.	130	−99	.	.	.	.	95	28	.	99	56
activities	.	.	.	−41	75	.	.	.	.	48	78	.	84	56
intellectuality	.	.	.	−9	79	.	.	.	.	65	72	.	93	65
independence	.	.	.	1	16	.	.	.	.	31	52	.	37	25
English	.	.	.	−3	12	.	.	.	.	36	61	.	51	34
ethlanguage	.	.	.	6	−6	.	.	.	.	40	53	.	44	30
PMA														
PMA, verbal	.	.	.	.	.	42	.	.	.	.	.	86	73	39
PMA, number	.	.	.	.	.	42	.	.	.	.	.	87	76	41
PMA, spatial	.	.	.	.	.	18	.	.	.	.	.	59	35	19
PMA, reasoning	.	.	.	.	.	25	.	.	.	.	.	67	45	24

All table entries multiplied by 100. Dots indicate zeros.

parameters for the two dependent variables (a total of 436 parameters), this is not truly a reduction of the 153 correlation coefficients under analysis and not in keeping with the property (i) stated above.

Walberg and Marjoribanks (1973, 1974, 1976) and Marjoribanks (1974) present a canonical correlation analysis, with the four PMA variables on one side, and the 14 other variables on the other side. The model underlying this analysis conforms to none of the properties discussed above; as opposed to Eq.2.139, SES and Environment are joined into one block, which is in contradiction to the property (ii) requested above.

Harris and McArthur (1974) approach the question of how many LVs are required to explain the relationships between environmental and mental variables by an Interbattery Factor model (see Sc.3.3.4). One may be surprised how to name and interpret the variable which at the same time produces the socials status of the parents and the mental abilities of the child. The causal direction introduced by Harris and McArthur goes from the higher-order Interbattery factor down to the lower-order block-specific factors Environment and Mental Ability, which is clearly different from all causal models introduced so far and also contradicts property (iii).

PLS model, outer part. Table 2.8 shows the outer model parameters, i.e. the weights and loadings, together with the communalities H^2 and the redundancies F^2. The weights of the variables Education of Father, Number of Children, Sibling Position and Press for Ethlanguage (referring to the correct usage of any language spoken at home other than English) are all very low; hence these variables are not necessary to measure the latent variable and could have been omitted. But the loadings and communalities are of substantial size, indicating that the attached LV is sufficient to explain a good deal of the variance of the MV.

The variables Father's Dominance and Mother's Dominance are negatively correlated ($s_{7,8} = -.48$, Tb.2.7), but both contribute with positive weights and loadings to the LV Parental Dominance. The common factor of both MVs should have loadings with different signs such that $r_{7,8} = p_7 \times p_8$ is negative. The explanation of this fact can be found in the MV correlation matrix, Tb.2.7: Both variables correlate, not with different, but with equal signs with most of the other variables. Hence an LV Parental Dominance is a better mediating and explanatory variable in the path model than the common factor Mother's Dominance vs. Father's Dominance. It is not a question of father vs. mother but of parents vs. children, of dominance vs. toleration.

The two dimensions in the fourth block come out as uncorrelated and are labelled by the highest-loading MVs as Y_4 = Press for Achievement and Y_5 = Press for Activities. All loadings are positive. The weights are quite different from the loadings, e.g. $w_{9,5} = -.99$ and $p_{9,5} = 0.28$, $w_{10,4} = -.41$ and $p_{10,4} = .48$, which goes to show how difficult it was to make the two LVs uncorrelated.

PLS model, inner part. Table 2.9 shows the inner model parameters. The path model was "empirically" restricted. The restrictions were chosen so that no path coefficient is less than 0.10 and no residual covariance greater than 0.10. The only exception is $q_{4,5} = -.25$, the covariance of the two residual variables from the same block (cf. Sc.2.3.4).

The correlation between SES and Intelligence, usually found to be around 0.30, comes out relatively high here: $r_{1,6} = 0.43$. As the residual covariance matrix shows, this correlation is completely ($q_{1,6} = 0.01$) accounted for by the parameters, though the model provides no direct connection between the two background factors, SES and Family Size, and the target variable, PMA. All the influences of the background are channelled and mediated by the three home environment variables. Marjoribanks has, then, realized the prime aim of his study, the measurement of the explanatory variables.

The direct influence of Parental Dominance on Intelligence is negative, $b_{6,3} = -.29$, but the correlation is positive, $r_{6,3} = 0.31$. As the covariance decomposition at the bottom of Tb.2.9 shows, the positive relationship is partly due to the joint background influence of SES and Family Size contributing 0.18, and partly to the mediating influence of the Press variables contributing 0.41 to the total correlation. Dominant behaviour of the parents, if it is just plain and direct, has a negative influence on the intellectual development of the child. But if determined and decisive behaviour on the part of the parents is instrumented and expressed

Table 2.9: Marjoribanks (1972) data: Inner model parameters

Latent variables	Y_1	Y_2	Y_3	Y_4	Y_5	Y_6	Y_1	Y_2	Y_3	Y_4	Y_5	Y_6
	Path matrix $\boldsymbol{B}$						Reduced path matrix $\boldsymbol{B}^*$					
Y_1 Social status	.	.	.	.	.	.	100	.	.	.	.	.
Y_2 Familiy size	33	.	.	.	.	.	33	100	.	.	.	.
Y_3 Parental dominance	52	.	.	.	.	.	52	.	100	.	.	.
Y_4 Press f.achievemnt	33	36	15	.	.	.	53	36	15	100	.	.
Y_5 Press f.activity	.	.	66	.	.	.	34	.	66	.	100	.
Y_6 PMA	.	.	−29	78	44	.	42	28	12	78	44	100
	LV correlation $\boldsymbol{R}$						Inner residual cov. $\boldsymbol{Q}$					
Y_1 Social status	100						100					
Y_2 Familiy size	33	100					0	89				
Y_3 Parental dominance	52	21	100				0	4	73			
Y_4 Press f.achievemnt	53	51	40	100			0	0	0	58		
Y_5 Press f.activity	36	7	66	.	100		1	−7	−1	−25	56	
Y_6 PMA	43	32	31	66	25	100	1	−5	−1	1	0	46

Effect decomposition for cor(PMA,Dominance)			
Direct effect (direct path) b			
PMA←-.29←Dom		-.29	
Indirect effects			
PMA←.78←Ach←.15←Dom	.12		
PMA←.44←Act←.66←Dom	.29		
Sum of indirect effects	.41	.41	
Reduced path b^*		.12	.12
Background effects			
PMA←.78←Ach←.33←SES→.52→Dom	.13		
PMA←.78←Ach←.36←Fam←.33←SES→.52→Dom	.05		
Sum of background effects	.18		.18
Sum of all effects = cor(PMA;Dominance) =			.31

in concrete demands, appeals and proposals and linked to substantial education goals, it has a beneficial influence.

The two Press factors show quite different relationships to their neighboring LVs. The Press for Achievement is dependent on both background factors and highly predictive for mental abilities, whereas the Press for Activity is dependant on Parental Dominance and not so predictive for the IQ.

Chapter 3

Foundations of Partial Least Squares

The description of the Latent Variables Path (LVP) model and the Partial Least Squares (PLS) estimation procedure is now in chapter 3 followed by the reflection on the statistical foundations of the method. Thereby, simple cases of LVP model serve as examples for general features of the PLS method.

The first section, Sc.3.1, reviews the notion of predictor specification which links the considerations about cause-effect relations and prediction (Sc.1.2) to the least squares methods. The next three sections explore properties and special cases of models with one (Sc.3.2, Sc.3.3) and two blocks of variables (Sc.3.4). In these sections it is shown that the predictor specification for separate parts of the model leads to a general least squares expression which can be solved algebraically for the parameters. In Sc.3.5 the Split Principal Components theorem is introduced which gives the basis for the derivation of exact optimality properties for special multi-block models and asymptotic optimality properties for more general multi-block models. Sc.3.6 fails to establish a similarly fruitful theorem for Split Multiple Regression but, nevertheless, gives some insight in what it means to assume two separate blocks of predictors instead of one.

3.1 Conditional Expectation and Predictor Specification

In Sc.2.1.2 the distribution of the residual variables was specified only by the first and second order moments, the means and covariances. This was sufficient for the deduction of the covariances of the explicit variables. When it comes to the estimation, the distribution must be specified in more detail.

One possible assumption is the multinormality of the residual variables. From multinormality of the residuals it follows that the explicit variables as linear forms thereof are multinormal too, and that the moments of order higher than two (skewness, kurtosis, etc.) are zero. However, multinormality is a very hard assumption which is seldom tested and seldom justified.

The PLS technique of model building implies only very general, soft distributional assumptions which led to the name "Soft Modeling" as a short term for LV path modeling with PLS estimation (H.Wold 1975 onwards). It is not the concepts nor the models nor the estimation techniques which are "soft", only the distributional assumptions.

Prediction properties postulated. Linking up with Sc.1.2, the purpose of the models treated by PLS is the prediction of y_n, which is the effect / result / outcome score for case o_n, in terms of the cause / input score x_n, in symbols:

$$\text{pred } y_n = f(x_n).$$

The predictions provided by the PLS method possess three qualities, and one may start the discourse by postulating that the predictions should posses these properties:

- The function $f(x)$ used for prediction must have the formal property of being open to an interpretation as a cause-effect relationship.
- The prediction cannot be 100% correct for the every single case, but it must be correct in the long run, i.e. over the total collection of cases (unbiasedness).
- The prediction error $(y_n - \text{pred } y_n)$ cannot be zero for the single case, but the prediction must be chosen so that the error is as small as possible over the total collection of cases. "As small as possible" will be defined as $\sum_n (y_n - \text{pred } y_n)^2 = \min$ (least squares property).

The statistical concept of conditional expectation possesses these properties, as will be shown in the following subsections.

3.1.1 The Notion of Conditional Expectation

The joint distributions of left-hand and right-hand side variables of Eq.2.1, Eq.2.4, Eq.2.7 are described by:

$$\begin{aligned} \mathrm{E}(\eta_j|\eta_i) &= \beta_{jo} + \sum_i \beta_{ji}\eta_i \\ \mathrm{E}(y_{k_j}|\eta_j) &= \pi_{k_j o} + \pi_{k_j}\eta_j \\ \mathrm{E}(\eta_j|y_{k_j}) &= \omega_{k_j o} + \sum_{k_j} \omega_{k_j} y_{k_j}. \end{aligned}$$

These expressions are called conditional expectations, which will be explained in more detail now.

Expected value. We consider a variable y with discrete values $y_g, g = 1, \ldots, G$ and probabilities $\Pr(y = y_g) = \pi_g$. The (overall, unconditional) expected value of y is defined as

$$\mu \equiv \mathrm{E}(y) = \sum_g \pi_g y_g, \tag{3.101}$$

where $\mathrm{E}(\cdot)$ is the expectation operator and μ the resulting constant. For a continuous variable y with density function $\varphi(y)$ the expected value is defined as

$\mu = \int_{-\infty}^{+\infty} y\varphi(y)\, dy$. What remains of y after subtraction of the expected value,

$$\upsilon = y - \mu, \tag{3.102}$$

is the residual variable υ. Expected value and residual are assembled in an equation that models the variable y as consisting of two parts:

$$y = \mu + \upsilon. \tag{3.103}$$

Properties of the expected value. The properties of μ are described by two properties of υ: Firstly, the residual variable has expectation zero,

$$\mathrm{E}(\upsilon) = 0, \tag{3.104}$$

which for discrete y is proved by

$$\begin{aligned} \mathrm{E}(\upsilon) &= \mathrm{E}(y - \mu) \\ &= \sum_g \pi_g (y_g - \mu) \\ &= \sum_g \pi_g y_g - \mu \sum_g \pi_g \\ &= \mathrm{E}(y) - \mu 1 \\ &= \mu - \mu = 0 \qquad \textit{q.e.d.} \end{aligned} \tag{3.105}$$

Secondly, the variation around the mean μ is minimal, in relation to all other points μ_0 of the distribution of y (LS property):

$$\mathrm{E}(y - \mu_0)^2 = \min \quad \Longleftrightarrow \quad \mu_0 = \mu. \tag{3.106}$$

Proof: Consider μ_0 as the sum of μ and a real number c,

$$\mathrm{E}(y - \mu_0)^2 = \mathrm{E}(y - (\mu + c))^2 = \mathrm{E}(y - \mu)^2 + c^2. \tag{3.107}$$

Whether c is positive or negative, in any case c^2 will add to the value of $\mathrm{E}(y - \mu)^2 + c^2$. The minimum will be reached only if $c = 0$, i.e. if $\mu_0 = \mu$, *q.e.d.*

Conditional expected values. We consider two discrete variables x and y with joint probabilities $\Pr(x = x_k, y = y_g) = \pi_{kg}$ and define the expected value y for a given and fixed value of x, i.e. under the condition $x = x_k$:

$$\mu_{y|x=x_k} \equiv \mathrm{E}(y|x = x_k) = \frac{1}{\pi_k} \sum_g \pi_{kg} y_g, \qquad \forall k. \tag{3.108}$$

The residual variable is denoted by $\upsilon_{|x=x_k}$, and distribution of y under the condition $x = x_k$ is modeled as

$$y_{|x=x_k} = \mu_{y|x=x_k} + \upsilon_{|x=x_k}, \qquad \forall k. \tag{3.109}$$

The conditional expected values, again, have the two properties that the residual has minimum variance and zero expectation:

$$\mathrm{E}(\upsilon_{|x=x_k}) = 0, \qquad \forall k. \tag{3.110}$$

In Table 3.1a,b the five conditional expected values $\mu_{y|x=-2}, \mu_{y|x=-1}, \ldots, \mu_{y|x=+2}$ are marked by five short horizontal lines.

Table 3.1: Probability distribution and conditionals expectation

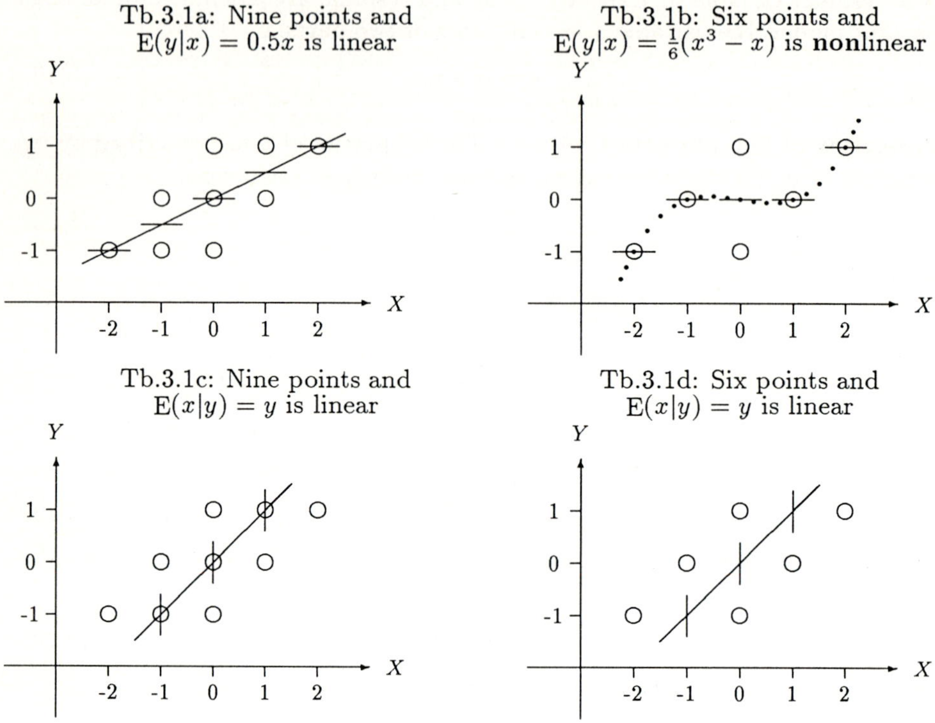

Conditional expectation. The conditional expectation $E(y|x)$ is a random variable. It is a function of x:

$$\tilde{y} \equiv E(y|x) = f(x). \tag{3.111}$$

What remains of y after subtraction of the conditional expectation is the residual variable v:

$$v = y - E(y|x). \tag{3.112}$$

Conditional expectation and residual are assembled in an equation that models the variable y as consisting of two parts:

$$y = E(y|x) + v = \tilde{y} + v. \tag{3.113}$$

If x is discrete, one may think of $E(y|x)$ as the union of all conditional expected values,

$$E(y|x) = \{E(y|x = x_k), \forall k\} = \{\mu_{y|x=x_k}, \forall k\},$$

and the residual variable is the collection of all residuals for given x:

$$v = \{v_{|x=x_k}, \forall k\}.$$

3.1.2 Properties of the Conditional Expectation

Theorem: Expected residual. The expected value of the residual is zero:

$$\mathrm{E}(\upsilon) = 0. \tag{3.114}$$

The proof starts from the fact that $\mathrm{E}(\upsilon|x = x_k) = 0, \forall k$. This holds for discrete x (finite range of k) as well as for continuous x (infinite range of k). For the latter we write $\mathrm{E}(\upsilon|x) = 0$. The proof for discrete x:

$$\begin{aligned}\mathrm{E}(\upsilon) &= \mathrm{E}(\{\upsilon_{|x=x_k}, \forall k\})\\ &= \sum_k \pi_k \, \mathrm{E}(\upsilon_{|x=x_k})\\ &= \sum_k \pi_k 0 = 0 \qquad q.e.d.\end{aligned} \tag{3.115}$$

For continuous x with distribution function $\Phi(x)$ one uses $\mathrm{E}(\upsilon|x) = 0$ to find that $\mathrm{E}(\upsilon) = \int \mathrm{E}(\upsilon|x) \; d\Phi(x) = \int 0 \; d\Phi(x) = 0$.

Means. The means of the four variables of Eq.3.113, $x, y, \tilde{y}, \upsilon$, are:

$$\mathrm{E}\begin{bmatrix}\upsilon\\ x\\ \tilde{y}\\ y\end{bmatrix} = \begin{bmatrix}0\\ \mu_x\\ \mu_y\\ \mu_y\end{bmatrix}. \tag{3.116}$$

The proof for $\mathrm{E}(\tilde{y})$:

$$\begin{aligned}\mathrm{E}(\tilde{y}) &= \mathrm{E}(y - \upsilon)\\ &= \mathrm{E}(y) - \mathrm{E}(\upsilon)\\ &= \mathrm{E}(y) - 0 = \mathrm{E}(y) \qquad q.e.d.\end{aligned} \tag{3.117}$$

Theorem: Uncorrelatedness. The residual υ and any function $f(x)$ of the condition x are uncorrelated:

$$\mathrm{E}[\upsilon f(x)] = 0. \tag{3.118}$$

The proof starts from the facts that $f(x|x = x_k) = f(x_k)$ is a constant and that $\mathrm{E}(\upsilon|x = x_k) = 0, \forall k$. The proof for discrete x:

$$\begin{aligned}\mathrm{E}[\upsilon f(x)] &= \{\mathrm{E}(\upsilon f(x)|x_k), \forall k\}\\ &= \sum_k \pi_k \, \mathrm{E}(\upsilon f(x)|x = x_k)\\ &= \sum_k \pi_k f(x_k) \, \mathrm{E}(\upsilon|x = x_k)\\ &= \sum_k \pi_k f(x_k) 0 = 0 \qquad q.e.d.\end{aligned} \tag{3.119}$$

For continuous x: $\mathrm{E}[\upsilon f(x)] = \int \mathrm{E}[\upsilon f(x)|x] \; d\Phi(x) = \int f(x) \, \mathrm{E}(\upsilon|x) \; d\Phi(x) = \int f(x) 0 \; d\Phi(x) = 0$.

Covariances. We denote the variances by $\psi^2 = \mathrm{var}(v) = \mathrm{E}(v^2)$ and $\tilde{\sigma}_y^2 = \mathrm{var}(\tilde{y})$. The covariance matrix of the four variables is:

$$\mathrm{cov}\begin{bmatrix} v \\ x \\ \tilde{y} \\ y \end{bmatrix} = \begin{bmatrix} \psi^2 & . & . & \text{sym.} \\ 0 & \sigma_x^2 & . & . \\ 0 & \sigma_{xy} & \tilde{\sigma}_y^2 & . \\ \psi^2 & \sigma_{xy} & \tilde{\sigma}_y^2 & \tilde{\sigma}_y^2 + \psi^2 \end{bmatrix}. \tag{3.120}$$

Proofs: Both x and $\tilde{y}$ can be understood as functions of x, hence $\mathrm{E}(vx) = 0$ and $\mathrm{E}(v\tilde{y}) = 0$, due to theorem Eq.3.118.

$$\begin{aligned}
\mathrm{cov}(v; x) &= \mathrm{E}(v; x) - \mathrm{E}(v)\,\mathrm{E}(x) = 0 - 0\,\mathrm{E}(x) = 0 \\
\mathrm{cov}(\tilde{y}; x) &= \mathrm{E}(\tilde{y}; x) - \mathrm{E}(\tilde{y})\,\mathrm{E}(x) = 0 - 0\,\mathrm{E}(x) = 0 \\
\mathrm{E}(yv) &= \mathrm{E}[(\tilde{y} + v)v] = \mathrm{E}(\tilde{y}v + vv) = \mathrm{E}(\tilde{y}v) + \mathrm{E}(vv) = 0 + \mathrm{E}(vv) = \psi^2 \\
\mathrm{cov}(\tilde{y}; x) &= \mathrm{cov}(y - v; x) = \mathrm{cov}(y; x) - \mathrm{cov}(v; x) = \mathrm{cov}(y; x) + 0 = \sigma_{xy} \\
\mathrm{cov}(y; \tilde{y}) &= \mathrm{cov}(\tilde{y} + v; \tilde{y}) = \mathrm{cov}(\tilde{y}; \tilde{y}) + \mathrm{cov}(v; \tilde{y}) = \mathrm{var}(\tilde{y}) + 0 = \tilde{\sigma}_y^2 \\
\mathrm{var}(y) &= \mathrm{var}(\tilde{y} + v) = \mathrm{var}(\tilde{y}) + \mathrm{var}(v) + 2\,\mathrm{cov}(\tilde{y}; v) = \mathrm{var}(\tilde{y}) + \mathrm{var}(v) + 0 \\
&= \tilde{\sigma}_y^2 + \psi^2
\end{aligned}$$

Least squares property. The variation around a function $f_0(x)$ is minimal if and only if $f_0(x)$ is the conditional expectation:

$$\mathrm{E}[y - f_0(x)]^2 = \min \quad \Longleftrightarrow \quad f_0(x) = \mathrm{E}(y|x). \tag{3.121}$$

Any function $f_0(x)$ can be decomposed into $f_0(x) = \mathrm{E}(y|x) + f_+(x)$, and hence the variance is decomposed:

$$\begin{aligned}
\mathrm{E}[y - f_0(x)]^2 &= \mathrm{E}[y - \mathrm{E}(y|x) - f_+(x)]^2 \\
&= \mathrm{E}[v - f_+(x)]^2 \\
&= \mathrm{E}[v]^2 - 2\,\mathrm{E}[vf_+(x)] + \mathrm{E}[f_+(x)]^2.
\end{aligned}$$

As demonstrated in theorem 3.118, the second term $\mathrm{E}[vf_+(x)]$ is zero for any function of x. Being a square, the third term is always positive, and the total expression attains its minimum if the third expression is zero, i.e. if $f_+(x) = 0$, i.e. if $f_0 = \mathrm{E}(y|x)$, *q.e.d.*

Nonreversibility. An important property of the conditional expectation is that it cannot be simply reversed. The reverse conditional expectation

$$\tilde{x} = \mathrm{E}(x|y) = g(y) \tag{3.122}$$

is different from $\mathrm{E}(y|x)$, except in case of a deterministic relation. The difference is shown by two discrete bivariate distributions in Tb.3.1, and one may simply look at the linearity of the conditional expectations. For the nine-point distribution both conditional expectation, $\mathrm{E}(y|x)$ (Tb.3.1a) and $\mathrm{E}(x|y)$ (Tb.3.1c) are linear, whereas for the six-point distribution $\mathrm{E}(y|x)$ (Tb.3.1b) is nonlinear and $\mathrm{E}(x|y)$ (Tb.3.1d) is linear.

3.1.3 The Linear Expectation

Linear approximation. The conditional expectation $f(x)$ can be approximated by a polynomial of any degree:

$$\mathrm{E}(y|x) = f_{\text{true}}(x) = \breve{\beta}_0 + \breve{\beta}_1 x + \breve{\beta}_2 x^2 + \breve{\beta}_3 x^3 + \breve{\beta}_4 x^4 + \cdots . \tag{3.123}$$

As a special case, we consider the polynomial of degree one, the linear function,

$$f_{\text{linear}}(x) = \beta_0 + \beta_1 x \equiv \alpha + \beta x. \tag{3.124}$$

Tb.3.1a shows a probability distribution of nine points with probabilities $\pi_{kg} = \frac{1}{9}$, where the five conditional expected values are perfectly fitted by a straight line, giving a linear conditional expectation $\mathrm{E}(y|x) = 0.5x$. In Tb.3.1b three of the points are omitted, giving each point a probability $\pi_{kg} = \frac{1}{6}$, and the five conditional expected values are fitted by a third degree polynomial:

$$\tilde{y}_{\text{true}} \equiv \mathrm{E}(y|x) = \breve{\beta}_0 + \breve{\beta}_1 y + \breve{\beta}_3 x^3 = 0 - \frac{1}{6}x + \frac{1}{6}x^3.$$

The linear approximation for Tb.3.1b is $\tilde{y}_{\text{linear}} = \beta x = 0.5x$.

Specification of linear expectation. The conditional expectation, even if its algebraic and statistical properties are easily determined, is in general difficult to compute numerically. The researcher can postulate that the conditional expectation be linear:

$$\mathrm{E}(y|x) = \text{linear} = \alpha + \beta x. \tag{3.125}$$

However, this is not the narrow concept of the conditional expectation any more. The extension of the concept may be considered justified if essential properties of the narrow concept carry over to the broader concept. We will postulate that the properties stated in Sc.3.1.2 be retained.

As we will use only the linear expectation later on, we change the notation at this time so that the symbols $\tilde{y}_{\text{true}}$, υ_{true}, ψ^2_{true} stand for the true, full conditional expectation, $\tilde{y} \equiv \tilde{y}_{\text{linear}}$, $\upsilon \equiv \upsilon_{\text{linear}}$, $\psi^2 \equiv \psi^2_{\text{linear}}$ stand for the linear model, and $\upsilon_{\text{nonlinear}}$, $\psi^2_{\text{nonlinear}}$ for the nonlinearity residual. The relations between the variables are:

$$\begin{aligned} y &= (\alpha + \beta x) + \upsilon \\ &\equiv \tilde{y}_{\text{linear}} + \upsilon_{\text{linear}} \\ &= \tilde{y}_{\text{linear}} + \upsilon_{\text{nonlinear}} + \upsilon_{\text{true}} \\ &= \tilde{y}_{\text{true}} + \upsilon_{\text{true}}. \end{aligned} \tag{3.126}$$

In Tb.3.1b the residual υ_{true} is nonzero for the point $(x, y) = (0, -1)$ and $(0, +1)$ which are not hit by the cubic curve; the residual $\upsilon_{\text{nonlinear}}$ is nonzero for $\mathrm{E}(y|x = -1)$ and $\mathrm{E}(y|x = +1)$ which are the points of the cubic curve not being hit by a straight regression line $y = \beta x$; the total residual υ_{linear} is nonzero for all of these four points.

Properties of linear expectation. We will see that all essential properties of the conditional expectation are retained, if the three terms of Eq.3.126 are uncorrelated. Firstly, both $\tilde{y}_{\text{linear}}$ and $\upsilon_{\text{nonlinear}}$ are functions of x, hence they are uncorrelated with υ_{true}, due to Eq.3.118. Secondly, to make $\mathrm{E}(\upsilon_{\text{nonlinear}}) = 0$ and $\mathrm{cov}(\upsilon_{\text{nonlinear}}, \tilde{y}_{\text{linear}}) = 0$, we choose:

$$\beta = \frac{\sigma_{xy}}{\sigma_x^2} \tag{3.127}$$

$$\alpha = \mu_y - \beta\mu_x. \tag{3.128}$$

Proofs:

$$\begin{aligned}
\mathrm{E}(\upsilon_{\text{nonlinear}}) &= \mathrm{E}[\mathrm{E}(y|x) - (\alpha + \beta x)] = \mathrm{E}[\mathrm{E}(y|x)] - \mathrm{E}[\alpha] + \mathrm{E}[\beta x] \\
&= \mu_y - \alpha - \beta\mu_x = \mu_y - (\mu_y - \beta\mu_x) - \beta\mu_x = 0 \\
\mathrm{cov}(\upsilon_{\text{nonlinear}}, \tilde{y}_{\text{linear}}) &= \mathrm{cov}[\mathrm{E}(y|x) - (\alpha + \beta x); (\alpha + \beta x)] \\
&= \beta\,\mathrm{cov}[\mathrm{E}(y|x); x] - \beta^2\,\mathrm{cov}(x; x) = \beta\sigma_{xy} - \beta^2\sigma_x^2 \\
&= \frac{\sigma_{xy}}{\sigma_x^2}\sigma_{xy} - \left(\frac{\sigma_{xy}}{\sigma_x^2}\right)^2 \sigma_x^2 = 0,
\end{aligned}$$

using $\mathrm{cov}[\mathrm{E}(y|x); x] = \sigma_{xy}$ from Eq.3.120.

Now it is established that the three terms of Eq.3.126 are uncorrelated, and the variance of y can be decomposed as:

$$\begin{aligned}
\mathrm{var}(y) &= \mathrm{var}(\alpha + \beta x) + \mathrm{var}(\upsilon_{\text{nonlinear}}) + \mathrm{var}(\upsilon_{\text{true}}) \\
&= \beta^2\sigma_x^2 + \underbrace{\psi^2_{\text{nonlinear}} + \psi^2_{\text{true}}} \\
&= \beta^2\sigma_x^2 + \psi^2_{\text{linear}}
\end{aligned} \tag{3.129}$$

To sum up, the nonlinearity residual, being uncorrelated with υ_{true}, x, and $\tilde{y}_{\text{linear}}$, only adds to the residual variance. If a truly nonlinear conditional expectation is specified as linear, the "nonlinearity residual" (simplification residual, specification error) is joined with the "true residual" υ_{true} (to be augmented by sampling error and measurement error in case of an empirical distribution).

Means and covariance of linear expectation.

$$\mathrm{E}\begin{bmatrix} \upsilon \\ x \\ \tilde{y} \\ y \end{bmatrix} = \begin{bmatrix} 0 \\ \mu_x \\ \alpha + \beta\mu_x \\ \alpha + \beta\mu_x \end{bmatrix} \tag{3.130}$$

$$\mathrm{cov}\begin{bmatrix} \upsilon \\ x \\ \tilde{y} \\ y \end{bmatrix} = \begin{bmatrix} \psi^2 & . & . & \text{sym.} \\ 0 & \sigma_x^2 & . & . \\ 0 & \beta\sigma_x^2 & \beta^2\sigma_x^2 & . \\ \psi^2 & \beta\sigma_x^2 & \beta^2\sigma_x^2 & \beta^2\sigma_x^2 + \psi^2 \end{bmatrix} \tag{3.131}$$

Proofs:

$$\mathrm{E}(\upsilon) = \mathrm{E}(\upsilon_{\text{true}} + \upsilon_{\text{nonlinear}}) = \mathrm{E}(\upsilon_{\text{true}}) + \mathrm{E}(\upsilon_{\text{nonlinear}}) = 0 + 0$$

$$
\begin{aligned}
\mathrm{E}(\tilde{y}) &= \mathrm{E}(\alpha + \beta x) = \alpha + \beta\,\mathrm{E}(x) = \alpha + \beta\mu_x \\
\mathrm{E}(y) &= \mathrm{E}(\tilde{y} + \upsilon) = \mathrm{E}(\tilde{y}) + \mathrm{E}(\upsilon) = \alpha + \beta\mu_x + 0 \\
\mathrm{cov}(\upsilon; x) &= \mathrm{cov}(\upsilon_{\text{true}} + \upsilon_{\text{nonlinear}}; x) = \mathrm{cov}(\upsilon_{\text{true}}; x) + \mathrm{cov}(\upsilon_{\text{nonlinear}}; x) = 0 + 0 \\
\mathrm{cov}(\upsilon; \tilde{y}) &= \mathrm{cov}(\upsilon; \alpha + \beta x) = \beta\,\mathrm{cov}(\upsilon; x) = \beta 0 \\
\mathrm{cov}(\upsilon; y) &= \mathrm{cov}(\upsilon; \alpha + \beta x + \upsilon) = \beta\,\mathrm{cov}(ups; x) + \mathrm{cov}(\upsilon, \upsilon) = 0 + \psi^2 \\
\mathrm{cov}(x; y) &= \mathrm{cov}(x; \alpha + \beta x + \upsilon) = \beta\,\mathrm{cov}(x; x) + \mathrm{cov}(x; \upsilon) = \beta\sigma_x^2 + 0 \\
\mathrm{var}(y) &= \mathrm{cov}(\alpha + \beta x + \upsilon; \alpha + \beta x + \upsilon) \\
&= \beta^2\,\mathrm{cov}(x; x) + 2\beta\,\mathrm{cov}(x; \upsilon) + \mathrm{cov}(\upsilon; \upsilon) = \beta^2\sigma_x^2 + 0 + \psi^2
\end{aligned}
$$

Summary. The main implications of the linear conditional expectation are summarized in four lines:

$$
\begin{aligned}
y = \alpha + \beta x + \upsilon, \qquad & \tilde{y} \equiv \mathrm{E}(y|x) = \alpha + \beta x \\
& \Longrightarrow \mathrm{E}(\upsilon) = 0 \\
& \Longrightarrow \mathrm{cov}(\upsilon; x) = \mathrm{cov}(\upsilon; \tilde{y}) = 0 \\
& \Longrightarrow \mathrm{cov}(y; x) = \mathrm{cov}(\tilde{y}; x) = \beta\,\mathrm{var}(x).
\end{aligned}
\tag{3.132}
$$

This sort of summary will be used in the following sections a standard expression. We can use the same format for vector variables: $\boldsymbol{y}_{[K\times 1]}$, $\boldsymbol{x}_{[L\times 1]}$, $\boldsymbol{\upsilon}_{[K\times 1]}$, and $\mathbf{B}_{[K\times L]}$:

$$
\begin{aligned}
\boldsymbol{y} = \boldsymbol{\alpha} + \mathbf{B}\boldsymbol{x} + \boldsymbol{\upsilon}, \qquad & \tilde{\boldsymbol{y}} \equiv \mathrm{E}(\boldsymbol{y}|\boldsymbol{x}) = \boldsymbol{\alpha} + \mathbf{B}\boldsymbol{x} \\
& \Longrightarrow \mathrm{E}(\boldsymbol{\upsilon}) = \mathbf{0} \\
& \Longrightarrow \mathrm{cov}(\boldsymbol{\upsilon}; \boldsymbol{x}) = \mathrm{cov}(\boldsymbol{\upsilon}; \tilde{\boldsymbol{y}}) = \mathbf{0} \\
& \Longrightarrow \mathrm{cov}(\boldsymbol{y}; \boldsymbol{x}) = \mathrm{cov}(\tilde{\boldsymbol{y}}; \boldsymbol{x}) = \mathbf{B}\,\mathrm{var}(\boldsymbol{x}).
\end{aligned}
\tag{3.133}
$$

Observations. Up to now in this section we have omitted the index n for observations. We introduce it now for a collection of observational unit $o_n, n = 1, \ldots, N$:

$$
\begin{aligned}
y_n = \alpha + \beta x_n + \upsilon_n, \qquad & \tilde{y}_n \equiv \mathrm{E}(y_n|x_n) = \alpha + \beta x_n \\
& \Longrightarrow \mathrm{E}(\upsilon_n) = 0 \\
& \Longrightarrow \mathrm{cov}(\upsilon_n; x_n) = \mathrm{cov}(\upsilon_n; \tilde{y}_n) = 0 \\
& \Longrightarrow \mathrm{cov}(y_n; x_n) = \mathrm{cov}(\tilde{y}_n; x_n) = \beta\,\mathrm{var}(x_n).
\end{aligned}
\tag{3.134}
$$

Some remarks:

(i) Homogeneity with respect to the model: The parameters α, β are the same for all observations, whereas the variables x_n, υ_n are different. A data set (a collection of cases, a sample) is called homogeneous with respect to model Eq.3.134 if it does not contain cases with grossly different constant α and slope parameter β

(ii) In order to ensure that $\mathrm{cov}(\upsilon_n; x_n) = 0$ often it is specified that x_n is a constant, not a variable. This leads to the classical regression model with nonstochastic regressors.

(iii) No identical distributions required: For any two cases o_n and o_{n+1}, it is not assumed that the residuals υ_n and υ_{n+1} have the same distribution (besides $E(\upsilon_n) = E(\upsilon_{n+1}) = 0$), or that the conditions x_n and x_{n+1} have the same distribution. That is to say, no homoskedasticity of the residuals is required.

(iv) No independence of cases required: Nothing is said in Eq.3.134 about the correlation between two different cases, $\text{cor}(\upsilon_n; \upsilon_{n+1})$.

OLS consistency. The theoretical constants α, β are unknown in most cases, and they have to be estimated from observed data. The model in estimated form is:

$$y_n = a + bx_n + u_n. \quad (3.135)$$

The constants

$$b = \frac{\sum_{n=1}^{N} y_n(x_n - \bar{x})}{\sum_{n=1}^{N}(x_n - \bar{x})^2} \quad (3.136)$$

$$a = \bar{y} - b\bar{x} \quad (3.137)$$

are the Ordinary Least Squares (OLS) estimates, because they satisfy the condition

$$\sum_n u_n^2 = \sum_n (y_n - a - bx_n)^2 \stackrel{!}{=} \min \quad (3.138)$$

over all possible choices of $a, b \in \mathbb{R}$.

Under mild supplementary conditions the Ordinary Least Squares (OLS) estimates a, b are consistent estimators of the population parameters α and β, i.e.:

$$\text{prob} \lim_{N\to\infty} a = \alpha$$
$$\text{prob} \lim_{N\to\infty} b = \beta \quad (3.139)$$

The following condition is sufficient for consistency,

$$\sum_i |\,\text{cor}(\upsilon_n; \upsilon_{n+i})| < \infty, \quad (3.140)$$

i.e., the observations need not to be independent, but as the number of observations goes to infinity, the sum of the correlations between cases (the dependence of cases) must stay below infinity (Wold 1963, 1988b).

3.1.4 Predictor Specification

The term Predictor Specification is introduced to cover both the practical purpose and the statistical treatment of a scientific model. *"Presp [predictor specification] is imposed on relations that the investigator wants to use for prediction, be it in theoretical or estimated form, and Presp provides the ensuing predictions"* (Wold 1988b:589). Predictor specification is a shortcut term for the type of model building where the investigator

- starts with the purpose of prediction (not primarily a structural explanation, e.g.)

- sets up a system of relations, preferably linear, where the structure of the relations must be founded in the substance of the matter, and the predictive purpose should not jeopardize a structural-causal interpretation of the relation (causal-predictive relation),
- adopts the set of statistical assumptions on the relations which is summarized in the conditional expectation Eq.3.133 and implies the assumptions
  - that x is the predictor (cause, stimulus) of y, and not the other way round
  - that $\tilde{y}$ is the systematic part of y, with respect to x
  - that the systematic part $\tilde{y}$ is a linear function of x,

  which has the consequences that
  - that the theoretical prediction is given by $\alpha + \beta x$, with prediction error v
  - that the empirical prediction is given by $a + bx_n$, with prediction error u_n
  - that among linear models the variance of v is minimal (LS property, Eq.3.138),
  - that predictor x and residual v are uncorrelated
  - that systematic part $\tilde{y}$ and residual v are uncorrelated
  - that the variance of y is decomposed into two parts, $\text{var}(\tilde{y}) + \text{var}(v)$
  - that OLS regression is consistent, under mild supplementary conditions, Eq.3.139.

The contrast between predictive vs. structural/causal modeling is no absolute. Predictive optimality means minimum residual variance. Structural simplicity and parsimony is brought to the fore by zero correlations — which implies additivity and minimality of variances. For simple model both aspects come at the same time; for complex models there is a parting of the ways.

3.1.5 Estimation

The uncorrelatedness of predictor and residual can be used to identify (in the population) or to estimate (in the sample) the parameters, when the variances of x and y and their covariance are given. It requires, as a nondramatic additional assumption, that the variance of x be nonzero and finite, and in the case of multiple predictors, that the covariance of the predictors be non-singular (invertible). Until further notice, the location parameters will be ignored in the following, without loss of generality.

Using the multiple regression with two predictors x and z as an example, the estimation is developed in five steps:

Step 1: Model relation.

$$y = \beta x + \gamma z + v \tag{3.141}$$

Step 2: Predictor specification.

$$\text{E}(y|x, z) \equiv \tilde{y} = \beta x + \gamma z, \tag{3.142}$$

which implies that predictors and residual are uncorrelated:

$$\text{E}(xv) = \text{E}(zv) = 0 \tag{3.143}$$

Step 3: Covariance deduction. The model equation Eq.3.141 is multiplied by the predictor variables, and the expectations are taken. On the basis of Eq.3.143 the expressions simplify:

$$\begin{aligned}\sigma_{xy} \equiv \mathrm{E}(xy) &= \mathrm{E}[x(\beta x + \gamma z + v)]\\ &= \mathrm{E}(\beta xx) + \mathrm{E}(\gamma xz) + \mathrm{E}(xv)\\ &= \beta\,\mathrm{E}(xx) + \gamma\,\mathrm{E}(xz) + 0\\ &= \beta\sigma_{xx} + \gamma\sigma_{xz} \qquad (3.144)\\ \sigma_{zy} \equiv \mathrm{E}(zy) &= \mathrm{E}[z(\beta x + \gamma z + v)]\\ &= \mathrm{E}(\beta zx) + \mathrm{E}(\gamma zz) + \mathrm{E}(zv)\\ &= \beta\,\mathrm{E}(zx) + \gamma\,\mathrm{E}(zz) + 0\\ &= \beta\sigma_{zx} + \gamma\sigma_{zz} \qquad (3.145)\end{aligned}$$

The equations Eqs.3.144, 3.145 are assembled in a matrix expression, the so-called normal equation system:

$$\begin{bmatrix}\sigma_{xy}\\ \sigma_{zy}\end{bmatrix} = \begin{bmatrix}\sigma_{xx} & \sigma_{xz}\\ \sigma_{zx} & \sigma_{zz}\end{bmatrix}\begin{bmatrix}\beta\\ \gamma\end{bmatrix}. \qquad (3.146)$$

Step 4: Stationary relation. The deduced covariances between predictors and predictand are written in form of a stationary condition:

$$\begin{bmatrix}\sigma_{xy}\\ \sigma_{zy}\end{bmatrix} - \begin{bmatrix}\sigma_{xx} & \sigma_{xz}\\ \sigma_{zx} & \sigma_{zz}\end{bmatrix}\begin{bmatrix}\beta\\ \gamma\end{bmatrix} = \mathbf{0}. \qquad (3.147)$$

The equation system Eq.3.147 turns out as to be solvable by premultiplication of the inverse of the predictor covariance. In more complicated models the stationary relation may turn out to be an eigenvalue problem, or to have no closed-form solution at all.

Step 5: Solution. The direct solution includes matrix inversion:

$$\begin{bmatrix}\beta\\ \gamma\end{bmatrix} = \begin{bmatrix}\sigma_{xx} & \sigma_{xz}\\ \sigma_{zx} & \sigma_{zz}\end{bmatrix}^{-1}\begin{bmatrix}\sigma_{yx}\\ \sigma_{yz}\end{bmatrix} \qquad (3.148)$$

An iterative solution uses sequentially and repeatedly the following two equations which are reformulated from Eq.3.144 and Eq.3.145:

$$\beta := \frac{(\sigma_{yx} - \gamma\sigma_{zx})}{\sigma_{xx}} \qquad (3.149)$$

$$\gamma := \frac{(\sigma_{yz} - \beta\sigma_{xz})}{\sigma_{zz}}. \qquad (3.150)$$

This is the Gauß relaxation algorithm.

Remark 3⋆1 Notice that the number of free parameters in step 1 is identical to the number of zero correlations yielded in step 2; further, that each zero correlation is used once to deduce the predictor-predictand covariance. Thus there is the same number of zero residual correlations and free parameters.

Remark 3⋆2 Notice further that the derivation of the estimators includes no definition of a function to be maximized and that no first derivatives of the function with respect to the parameters are set to zero. The maximum property of LS and the zero predictor-residual correlation follow from the predictor specification. Thus the derivation of estimators can be simplified drastically when predictor specification can be used.

3.1.6 Interlocking Conditional Expectation

Consider a chain model with two equations where y, the predictand of the first equation, is the predictor of the second equation:

$$\begin{array}{ccccc} & & v & & \epsilon \\ & & \downarrow & & \downarrow \\ x & \xrightarrow{\beta} & y & \xrightarrow{\gamma} & z \end{array}$$

$$y = \beta x + v \tag{3.151}$$

$$z = \gamma y + \epsilon \tag{3.152}$$

Ideal conditions covariance. We assume ideal conditions and specify the three predetermined variables x, v, ϵ to be uncorrelated. Hence the covariance of the five variables is:

$$\operatorname{cov}\begin{bmatrix} x \\ v \\ \epsilon \\ y \\ z \end{bmatrix} = \begin{bmatrix} \sigma_x^2 & & & & \\ 0 & \sigma_v^2 & & & \\ 0 & 0 & \Theta & & \\ \beta\sigma_x^2 & \sigma_v^2 & 0 & \beta^2\sigma_x^2 + \sigma_v^2 & \\ \gamma\beta\sigma_x^2 & \gamma\sigma_v^2 & \Theta & \gamma(\beta^2\sigma_x^2 + \sigma_v^2) & \gamma(\beta^2\sigma_x^2 + \sigma_v^2)\gamma + \Theta \end{bmatrix} \tag{3.153}$$

$$= \Sigma(\sigma_x^2, \sigma_v^2, \Theta, \beta, \gamma) \tag{3.154}$$

The covariance matrix of the three "observed" variables x, y, and z which has six elements is deduced from five parameters. Hence it is a restricted model. A direct effect $z \leftarrow x$ is missing. Empirical data may contradict this model.

Structural form specification.

$$\begin{aligned} y = \beta x + v, \qquad & \tilde{y} \equiv \mathrm{E}(y|x) = \beta x \\ & \Longrightarrow \mathrm{E}(v) = 0 \\ & \Longrightarrow \operatorname{cov}(v; x) = \operatorname{cov}(v; \tilde{y}) = 0 \\ & \Longrightarrow \operatorname{cov}(y; x) = \operatorname{cov}(\tilde{y}; x) = \beta \operatorname{var}(x), \end{aligned} \tag{3.155}$$

$$\begin{aligned} z = \gamma y + \epsilon, \qquad & \tilde{z} \equiv \mathrm{E}(z|y) = \gamma y \\ & \Longrightarrow \mathrm{E}(\epsilon) = 0 \\ & \Longrightarrow \operatorname{cov}(\epsilon; y) = \operatorname{cov}(\epsilon; \tilde{z}) = 0 \\ & \Longrightarrow \operatorname{cov}(z; y) = \operatorname{cov}(\tilde{z}; y) = \gamma \operatorname{var}(y). \end{aligned} \tag{3.156}$$

The implied covariances of all five variables are:

$$\operatorname{cov}\begin{bmatrix} x \\ \upsilon \\ \epsilon \\ y \\ \boldsymbol{z} \end{bmatrix} = \begin{bmatrix} \sigma_x^2 & & & & \\ 0 & \sigma_\upsilon^2 & & & \\ \boldsymbol{\sigma}_{\epsilon x} & -\beta\boldsymbol{\sigma}_{\epsilon x} & \boldsymbol{\Theta} & & \\ \beta\sigma_x^2 & \sigma_\upsilon^2 & \mathbf{0} & \sigma_y^2 = \beta^2\sigma_x^2 + \sigma_\upsilon^2 & \\ \gamma\beta\sigma_x^2 + \boldsymbol{\sigma}_{\epsilon x} & \gamma\sigma_\upsilon^2 - \beta\boldsymbol{\sigma}_{\epsilon x} & \boldsymbol{\Theta} & \gamma\sigma_y^2 & \gamma\sigma_y^2\gamma' + \boldsymbol{\Theta} \end{bmatrix} \tag{3.157}$$

which is proved by:

$$\operatorname{cov}(\epsilon; \upsilon) = \operatorname{cov}(\epsilon; y - \beta x) = \operatorname{cov}(\epsilon; y) - \beta \operatorname{cov}(x; \epsilon) = 0 - \beta\boldsymbol{\sigma}_{\epsilon x}$$

From the four zeros in Eq.3.153 two have vanished.

Reduced form specification. We substitute y in the second equation from the first one:

$$\boldsymbol{z} = \gamma y + \epsilon = \gamma(\beta x + \upsilon) + \epsilon = (\gamma\beta)x + (\gamma\upsilon + \epsilon) \tag{3.158}$$

and specify conditional expectations on the first equation and on the substituted (reduced) form of the second equation.

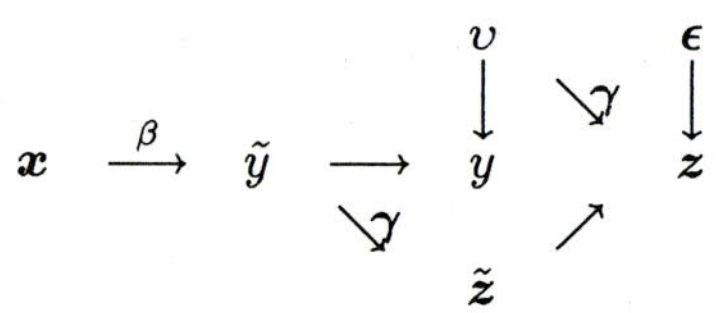

$$\begin{aligned} y = \beta x + \upsilon, \qquad & \tilde{y} \equiv \mathrm{E}(y|x) = \beta x \\ & \Longrightarrow \mathrm{E}(\upsilon) = 0 \\ & \Longrightarrow \operatorname{cov}(\upsilon; x) = \operatorname{cov}(\upsilon; \tilde{y}) = 0 \\ & \Longrightarrow \operatorname{cov}(y; x) = \operatorname{cov}(\tilde{y}; x) = \beta \operatorname{var}(x), \end{aligned} \tag{3.159}$$

$$\begin{aligned} \boldsymbol{z} = \gamma\beta x + \gamma\upsilon + \epsilon, \qquad & \tilde{\boldsymbol{z}} \equiv \mathrm{E}(\boldsymbol{z}|x) = \gamma\beta x \\ & \Longrightarrow \mathrm{E}(\gamma\upsilon + \epsilon) = \mathbf{0} \\ & \Longrightarrow \operatorname{cov}(\gamma\upsilon + \epsilon; x) = \operatorname{cov}(\gamma\upsilon + \epsilon; \tilde{\boldsymbol{z}}) = \mathbf{0} \\ & \Longrightarrow \operatorname{cov}(\boldsymbol{z}; x) = \operatorname{cov}(\tilde{\boldsymbol{z}}; x) = \gamma\beta \operatorname{var}(x). \end{aligned} \tag{3.160}$$

The covariances implied are:

$$\operatorname{cov}\begin{bmatrix} x \\ \upsilon \\ \epsilon \\ y \\ \boldsymbol{z} \end{bmatrix} = \begin{bmatrix} \sigma_x^2 & & & & \\ 0 & \sigma_\upsilon^2 & & & \\ \mathbf{0} & \boldsymbol{\sigma}_{\epsilon\upsilon} & \boldsymbol{\Theta} & & \\ \beta\sigma_x^2 & \sigma_\upsilon^2 & \boldsymbol{\sigma}_{\upsilon\epsilon} & \beta^2\sigma_x^2 + \sigma_\upsilon^2 & \\ \gamma\beta\sigma_x^2 & \gamma\sigma_\upsilon^2 + \boldsymbol{\sigma}_{\epsilon\upsilon} & \gamma\boldsymbol{\sigma}_{\epsilon\upsilon} + \boldsymbol{\Theta} & \gamma\sigma_y^2 + \boldsymbol{\sigma}_{\epsilon\upsilon} & \gamma(\sigma_y^2 + \boldsymbol{\sigma}_{\epsilon\upsilon})\gamma' + \boldsymbol{\Theta} \end{bmatrix} \tag{3.161}$$

which is proved by:

$$\begin{aligned}\operatorname{cov}(x;\epsilon) &= \operatorname{cov}(x;(\gamma v+\epsilon)-\gamma v)=\operatorname{cov}(x;\gamma v+\epsilon)-\gamma\operatorname{cov}(x;v)\\ &= \mathbf{0}-\gamma\mathbf{0}=\mathbf{0} \qquad (3.162)\\ \operatorname{cov}(y;\epsilon) &= \operatorname{cov}(\tilde{y}+v;\epsilon)=\operatorname{cov}(\tilde{y};\epsilon)+\operatorname{cov}(v;\epsilon)=\mathbf{0}+\operatorname{cov}(v;\epsilon). \qquad (3.163)\end{aligned}$$

Again, of the four zero covariance of the ideal covariance Eq.3.153 only two are implied by the reduced form specification. Besides the zero correlations that follow immediately from predictor specification, we get one more due to the interlocking character of the two predictor specifications. $\operatorname{cov}(x;\epsilon)$ is zero with the reduced form specification Eq.3.161 but not with the structural form specification Eq.3.157.

Incidentally, it has not been proved — and is in general not true — that $\operatorname{cov}(y;\epsilon) = \operatorname{cov}(v;\epsilon) = 0$. Eq.3.160 says that the criterion variable z is not regressed on its direct predecessor y, but on the pre-predecessor x.

Summary. Predictor specification on two equations provides some flexibility. There is a choice between maximum predictive power (structural form specification) and minimal indirect effects (reduced form specification), which stands for structural simplicity of the model.

3.1.7 Eigenvalue Problem and Power Method

One of the most simple algorithms for the numerical evaluation of an eigenvalue/eigenvector problem is the so-called power method (vector iteration method, power iteration, von Mises method, Hotelling method). This method is introduced here for later reference: If one can show that an algorithm has the structure of the power algorithm, it is proved that this algorithm converges, and that it arrives at a solution that is equivalent to an eigenvalue/eigenvector solution.

Symmetric eigenvalue problem. Given a symmetric matrix $\boldsymbol{S}$, the values λ^2 satisfying the equation

$$\left|\boldsymbol{S}-\lambda^2\mathbf{I}\right| = 0 \qquad (3.164)$$

are called the eigenvalues of $\boldsymbol{S}$; the vectors $\boldsymbol{w}$ satisfying the two equivalent relations

$$(\boldsymbol{S}-\lambda^2\mathbf{I})\boldsymbol{w}=\mathbf{0}, \qquad \boldsymbol{S}\boldsymbol{w}=\lambda^2\boldsymbol{w}, \qquad (3.165)$$

are called the corresponding eigenvectors. The eigenvector maximizes the Rayleigh quotient

$$\lambda^2 = \frac{\boldsymbol{w}'\boldsymbol{S}\boldsymbol{w}}{\boldsymbol{w}'\boldsymbol{w}} = \max_{\boldsymbol{w}} \qquad (3.166)$$

The ambiguity of the scale of $\boldsymbol{w}$ in Eq.3.165 is usually resolved by requesting $\boldsymbol{w}'\boldsymbol{w} = 1$ or $\boldsymbol{w}'\boldsymbol{S}\boldsymbol{w} = 1$. If the scaling $\boldsymbol{w}'\boldsymbol{w} = 1$ is applied, premultiplication of Eq.3.165 by $\boldsymbol{w}'$ yields:

$$\lambda^2 = \boldsymbol{w}'\boldsymbol{S}\boldsymbol{w}\ . \qquad (3.167)$$

With scaling $\boldsymbol{w}'\boldsymbol{S}\boldsymbol{w} = 1$ one gets $\boldsymbol{w}'\boldsymbol{w} = 1/\lambda^2$.

The power algorithm. The equation systems Eqs.3.164, 3.165 can be solved directly for λ^2 and $\boldsymbol{w}$, if $\boldsymbol{S}$ has order two or three, but for greater matrices other methods, and in general iterative methods, are to be applied. One of these iterative methods is known as the power iteration method. The iteration consists of the general cycle:

$$ {}_c\lambda^2 \; {}_c\boldsymbol{w} := \boldsymbol{S} \; {}_{c-1}\boldsymbol{w}, \qquad c = 1, 2, \ldots \text{cycles.} \tag{3.168}$$

or in more detail:

$$\begin{aligned} {}_c\breve{\boldsymbol{w}} &:= \boldsymbol{S} \; {}_{c-1}\boldsymbol{w} \\ {}_c\lambda^4 &:= {}_c\breve{\boldsymbol{w}}' \, {}_c\breve{\boldsymbol{w}} \\ {}_c\boldsymbol{w} &:= {}_c\breve{\boldsymbol{w}} / {}_c\lambda^2, \qquad c = 1, 2, \ldots \text{cycles.} \end{aligned}$$

(Pre-subscripts like ${}_c(\cdot)$, ${}_{c-1}(\cdot)$, ${}_0(\cdot)$ denote the iteration cycles.) Starting with (nearly) arbitrary values ${}_0\boldsymbol{w}$, the procedure converges to the principal eigenvalue and the corresponding eigenvector (cf. Sc.3.1.8),

$$\lim_{c\to\infty} {}_c\boldsymbol{w} = \lim_{c\to\infty} {}_{c-1}\boldsymbol{w} = \boldsymbol{w} \tag{3.169}$$

$$\lim_{c\to\infty} {}_c\lambda^2 = \lim_{c\to\infty} {}_{c-1}\lambda^2 = \lambda^2. \tag{3.170}$$

The procedure fails to converge if the two largest eigenvalues are identical, which nearly never occurs in empirical data, and arrives to the second largest eigenvalue if ${}_0\boldsymbol{w}'\boldsymbol{w} = 0$. Convergence is slow, if the first two eigenvalues λ_1^2 and λ_2^2 are nearly equal. Convergence speed is proportional to λ_1^2/λ_2^2.

If an iteration scheme can be shown to have the structure of Eq.3.168, its convergence is proved.

The first ones to describe the power method to full extend were von Mises and Pollaczek-Geiringer (1929). They proved the convergence and pointed out the self-correcting feature of the algorithm. Hotelling (1933) applied the algorithm to the correlation matrix of random variables.

Asymmetric eigenvalue problem. Given a rectangular matrix $\boldsymbol{A}$, the values λ satisfying the equation

$$|\boldsymbol{A} - \lambda \mathbf{I}| = 0 \tag{3.171}$$

are called the singular values of $\boldsymbol{A}$; the vectors $\boldsymbol{w}$ and $\boldsymbol{g}$ satisfying the equations

$$\boldsymbol{A}\boldsymbol{w} = \lambda \boldsymbol{g}, \qquad \boldsymbol{A}'\boldsymbol{g} = \lambda \boldsymbol{w}, \tag{3.172}$$

are called the corresponding left, and right, respectively, eigenvectors. The ambiguity of the scale of $\boldsymbol{w}, \boldsymbol{g}$ in Eq.3.172 is usually resolved by requesting $\boldsymbol{w}'\boldsymbol{w} = \boldsymbol{g}'\boldsymbol{g} = 1$. If the scaling $\boldsymbol{w}'\boldsymbol{w} = 1$ is applied, premultiplication of Eq.3.172a by $\boldsymbol{g}'$, and of Eq.3.172b by $\boldsymbol{w}'$, yields:

$$\lambda = \boldsymbol{g}'\boldsymbol{A}\boldsymbol{w} = \boldsymbol{w}'\boldsymbol{A}'\boldsymbol{g}. \tag{3.173}$$

The power algorithm for asymmetric eigenvalue problem. The iteration with general iteration cycle,

$$\begin{aligned} {}_c\breve{\lambda}\ {}_c\boldsymbol{g} &:= \boldsymbol{A}\ {}_{c-1}\boldsymbol{w} \\ {}_c\lambda\ {}_c\boldsymbol{w} &:= \boldsymbol{A}'\ {}_c\boldsymbol{g} \qquad c = 1, 2, \ldots \text{cycles}, \end{aligned} \tag{3.174}$$

converges to the singular value λ, the left eigenvector $\boldsymbol{g}$ and the right eigenvector $\boldsymbol{w}$ of $\boldsymbol{A}$. If the scaling $\boldsymbol{g}'\boldsymbol{g} = 1$ is imposed, the rectangular matrix eigenproblem can be translated into the symmetric, and the two iteration steps Eq.3.171 can be combined into one step which has the structure of Eq.3.168,

$$({}_c\breve{\lambda}\ {}_c\lambda)\,{}_c\boldsymbol{w} = \boldsymbol{A}'\boldsymbol{A}\ {}_{c-1}\boldsymbol{w}, \tag{3.175}$$

where $\boldsymbol{S} = \boldsymbol{A}'\boldsymbol{A}$ and, in the limit, ${}_c\breve{\lambda} = {}_c\lambda = \lambda$ and $\lambda^2_{\text{Eq.3.171}} = \lambda^2_{\text{Eq.3.164}}$.

The consideration of rectangular matrix eigenvalue problems is useful with respect to two-block models (Sc.3.4). If the solution is equivalent to an eigenvalue solution, and the one weight vector is found in an iterative way, the other weight vector is found by Eq.3.174.

Two-matrix eigenvalue problem. Given two symmetric real matrices $\boldsymbol{S}, \boldsymbol{D}$, the values λ satisfying

$$|\boldsymbol{S} - \lambda\boldsymbol{D}| = 0 \tag{3.176}$$

are called the eigenvalues of $\boldsymbol{S}$ with respect to (or in the metric of) $\boldsymbol{D}$. The corresponding eigenvectors satisfy the equation:

$$\boldsymbol{S}\boldsymbol{w} = \lambda^2\boldsymbol{D}\boldsymbol{w}, \qquad (\boldsymbol{S} - \lambda^2\boldsymbol{D})\boldsymbol{w} = 0. \tag{3.177}$$

The eigenvector maximizes the Rayleigh quotient

$$\lambda^2 = \frac{\boldsymbol{w}'\boldsymbol{S}\boldsymbol{w}}{\boldsymbol{w}'\boldsymbol{D}\boldsymbol{w}} = \max_{\boldsymbol{w}} \tag{3.178}$$

For a non-singular $\boldsymbol{D}$ the problem can be translated into a one-matrix asymmetric eigenproblem by

$$\begin{aligned} \boldsymbol{S}\boldsymbol{w} &= \lambda^2\boldsymbol{D}\boldsymbol{w} \\ \boldsymbol{S}(\boldsymbol{D}^{-1/2}\boldsymbol{D}^{1/2})\boldsymbol{w} &= \lambda^2\boldsymbol{D}^{1/2}\boldsymbol{D}^{1/2}\boldsymbol{w} \\ \boldsymbol{S}\boldsymbol{D}^{-1/2}\boldsymbol{w}^* &= \lambda^2\boldsymbol{D}^{1/2}\boldsymbol{w}^* \\ \boldsymbol{D}^{-1/2}\boldsymbol{S}\boldsymbol{D}^{-1/2}\boldsymbol{w}^* &= \boldsymbol{D}^{-1/2}\lambda^2\boldsymbol{D}^{1/2}\boldsymbol{w}^* \\ \boldsymbol{S}^*\boldsymbol{w}^* &= \lambda^2\boldsymbol{w}^*. \end{aligned} \tag{3.179}$$

Hence the two-matrix eigenvalue problem can be solved like the one-matrix symmetric eigenvalue problem. Considerations about numeric accuracy, however, have led to the development of specific algorithms for the different problems (cf. e.g. Golub & Van Loan 1983).

3.1.8 Convergence of Power Algorithm

Let $\boldsymbol{y}_{[K\times 1]}$ a set of variables with covariance $\boldsymbol{S} = \mathrm{E}(\boldsymbol{y}\boldsymbol{y}')$ and first principal component $\xi = \boldsymbol{w}'\boldsymbol{y}$, $\mathrm{E}(\boldsymbol{y}'\xi) = \boldsymbol{S}\boldsymbol{w} = \lambda^2\boldsymbol{w}$. Let ${}_0\boldsymbol{w}$ an approximation to the first eigenvector (weight vector) in iteration cycle 0 of the power algorithm, so that the approximation in the next cycle is: ${}_1\boldsymbol{w} := {}_0f\,\mathrm{E}(\boldsymbol{y}\,{}_0\xi)$, where ${}_0f$ is an appropriately chosen constant, e.g. ${}_0f^{-1} = \mathrm{var}({}_0\xi)$ or λ^2.

We consider the starting values ${}_0\boldsymbol{w}$ as consisting partly of the first eigenvector $\boldsymbol{w}$ and partly of the sum of all other eigenvectors. To treat the case of slowest convergence, we let the sum of all others consist of only the second eigenvector $\breve{\boldsymbol{w}}$. We denote the second largest eigenvalue by $\breve{\lambda}^2$, and the related second principal component by $\breve{\xi}$. At the beginning the two eigenvector are mixed in an unknown proportion $a : \breve{a}$.

$$
\begin{aligned}
{}_0\boldsymbol{w} &= a\boldsymbol{w} + \breve{a}\breve{\boldsymbol{w}} && (3.180)\\
{}_0\xi &= a\xi + \breve{a}\breve{\xi}, \qquad \mathrm{E}(\xi\breve{\xi}) = 0 && (3.181)\\
\mathrm{E}(\boldsymbol{y}\,{}_0\xi) &= a\lambda^2\boldsymbol{w} + \breve{a}\breve{\lambda}^2\boldsymbol{w}_k && (3.182)\\
{}_1\boldsymbol{w} &= {}_0f\,\mathrm{E}(\boldsymbol{y}\,{}_0\xi) = {}_0f(a\lambda^2\boldsymbol{w} + \breve{a}\breve{\lambda}^2\breve{\boldsymbol{w}}) && (3.183)
\end{aligned}
$$

If we choose the normalization constant ${}_0f$ equal to $1/\lambda^2$, the last equation becomes

$$
{}_1\boldsymbol{w} = a\boldsymbol{w} + \left(\frac{\breve{\lambda}^2}{\lambda^2}\right)\breve{a}\breve{\boldsymbol{w}}
$$

If the first eigenvalue (λ^2) is greater than the second, then $0 \leq (\breve{\lambda}^2/\lambda^2) < 1$, and the contribution of the second eigenvector $\breve{\boldsymbol{w}}$ to the approximation ${}_1\boldsymbol{w}$ has shrunken by this factor from iteration cycle 0 to cycle 1.

We repeat the computation Eqs.3.180,..., 3.183, but now with $\breve{\breve{a}} := (\breve{\lambda}^2/\lambda^2)\breve{a}$ in lieu of $\breve{a}$, and get the second approximation to the weights:

$$
{}_2\boldsymbol{w} = a\boldsymbol{w} + \left(\frac{\breve{\lambda}^2}{\lambda^2}\right)\breve{\breve{a}}\breve{\boldsymbol{w}} = a\boldsymbol{w} + \left(\frac{\breve{\lambda}^2}{\lambda^2}\right)^2\breve{a}\breve{\boldsymbol{w}}. \qquad (3.184)
$$

In general, in the cth iteration cycle the approximation becomes

$$
{}_c\boldsymbol{w} = a\boldsymbol{w} + \left(\frac{\breve{\lambda}^2}{\lambda^2}\right)^c\breve{a}\breve{\boldsymbol{w}}. \qquad (3.185)
$$

The ratio $\breve{\lambda}^2/\lambda^2$ is smaller than one, hence it vanishes with increasing iteration cycle c:

$$
\left(\frac{\breve{\lambda}^2}{\lambda^2}\right)^c \to 0 \qquad \text{as} \qquad c \to \infty
$$

and

$$
{}_c\boldsymbol{w} \to a\boldsymbol{w} \qquad \text{as} \qquad c \to \infty,
$$

which completes the proof of the convergence of the power algorithm for principal components. In Sc.3.5 we will present several proofs which follow exactly the line of the this proof.

Table 3.2: Aggregate, factor and component

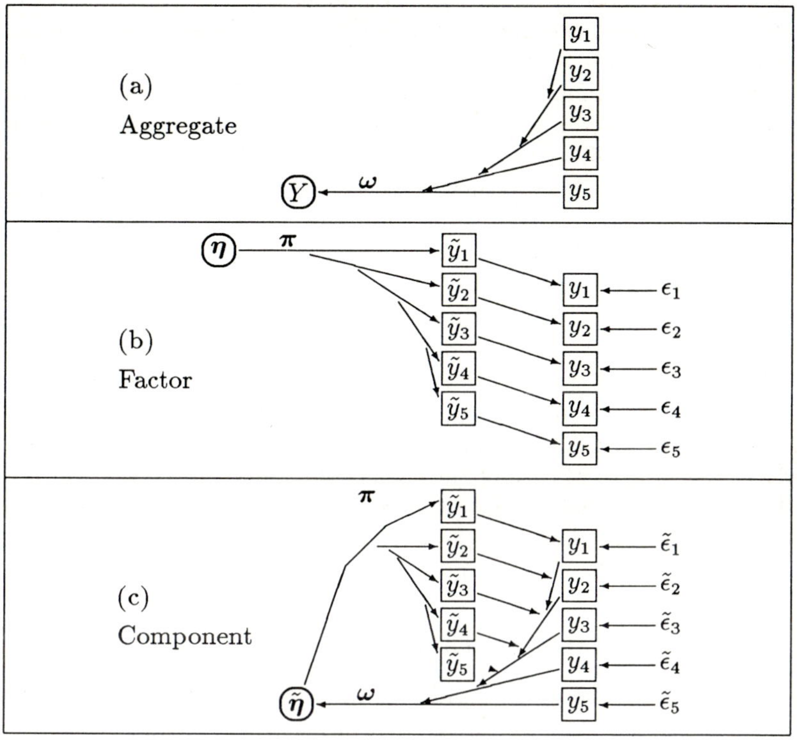

3.2 Principal and Other Components

Latent variables can be more or less latent.

(i) Firstly, they can be completely latent, unknown, hidden, invisible, undercover, unmanifested, and then their scientific purpose is as obscure as the LVs themselves.

(ii) Secondly, due to some assumptions on the nature of the LVs their deductive aspects may become patent, estimable, palpable, evident, and thus their scientific usefulness emerges. In a Common Factor model the factors are considered as "known" when the factor loadings, structures and correlations are known.

(iii) Thirdly, due to some more stringent assumptions the scores for the LVs become estimable, the total LV thus becoming known. In a Principal Component model the weights, scores and loadings are considered and estimated from the beginning, whereas the estimation of weights and scores in FA requires additional assumptions beyond the factor model (see the controversy between Hoppe 1981, 1982, and Falter & Lohmöller 1982).

The third meaning of LVs will be pursued in the following. It is in line with the usage of the term 'latent' in latent trait analysis which results in individual scores for the latent traits (Rasch), and in latent class analysis (Lazarsfeld), which results in assignments of observations in manifest classes to latent classes. Latent variables, in this sense, are latent only before the estimation takes place. Some aspects may remain latent (as in factor analysis), or it may be that no aspects remain latent after estimation (as in component analysis). The term LV is used to denote factors *sensu* factor analysis, where no assumptions about the weights and scores are made, as well as to denote components and linear forms where, after the estimation, the loadings, structures, weights and scores are approximatively known.

In this section Sc.3.2 some least-squares aspects of components and factors will be discussed. Only one block of MVs will be considered. As building blocks for arguments in later sections, the properties of linear aggregates and components are discussed in Sc.3.2.2 to Sc.3.2.4. Thereby, care is taken to introduce only one assumption at time, so that it can be demonstrated what properties are gained in the passage from aggregates to components and from there to principal components. In Sc.3.2.5 the principal component model will be introduced and explored, as a special case of the (general) component concept introduced in Sc.3.2.4. Sc.3.2.1 starts with conceptual and notational considerations on factors, components and composites. Table 3.2 gives a graphical representation of the different models considered in Sc.3.2.

3.2.1 Concepts and Notations

Different symbols will be used to indicate the different levels of latentness of the latent variables in the various stages of model building and estimating. For the first level, the obscure variable, no formal notation is used. For the two other levels:

$\boldsymbol{\eta}$	=	the true latent variable
$\tilde{\boldsymbol{\eta}} = \mathrm{E}(\boldsymbol{\eta}\vert\boldsymbol{y})$	=	the expectation of the latent variable, given the manifest variables
$\hat{\boldsymbol{\eta}} = \boldsymbol{\omega}'\boldsymbol{y}$	=	the theoretical estimate of the latent variable
$\boldsymbol{Y} = \boldsymbol{w}'\boldsymbol{y}$	=	the estimated latent variable, a linear composite, with estimated weights $\boldsymbol{w} \mathrel{\hat{=}} \boldsymbol{\omega}$.

Throughout Sc.3.2 the LVs can be understood as multidimensional, i.e. $\boldsymbol{\eta}$ is a set of variables with covariance matrix $\boldsymbol{\rho}$. The variables and the parameters of the LV model are:

$$\begin{aligned}\boldsymbol{y}_{[K\times 1]} &\sim \mathrm{M}(\mathbf{0}, \boldsymbol{\Sigma}, \ldots)\\ \boldsymbol{\eta}_{[J\times 1]} &\sim \mathrm{M}(\mathbf{0}, \boldsymbol{\rho}, \ldots)\\ \boldsymbol{\delta}_{[J\times 1]} &\sim \mathrm{M}(\mathbf{0}, \boldsymbol{\varphi}, \ldots)\\ \boldsymbol{\epsilon}_{[K\times 1]} &\sim \mathrm{M}(\mathbf{0}, \boldsymbol{\Theta}, \ldots),\end{aligned}$$

where the lower-order moments of the variables are denoted by $\boldsymbol{y} \sim$ M(mean, covariance,...). The parameters of the models are:

$$\boldsymbol{\pi}_{[K\times J]} \equiv [\pi_{kj}]$$

$$\omega_{[K\times J]} \equiv [\omega_{kj}],$$

and additionally we introduce the symbol s:

$$\gamma_{[K\times J]} \equiv \text{cov}(\epsilon;\boldsymbol{\delta})$$

$$\tau_{[K\times J]} \equiv \text{cov}(\boldsymbol{y};\boldsymbol{\eta}).$$

The MV relation. The relation of manifest variables to the latent variables is described by the equation:

$$\boldsymbol{y} = \tilde{\boldsymbol{y}} + \epsilon = \boldsymbol{\pi\eta} + \epsilon. \tag{3.201}$$

This equation describes the observation $\boldsymbol{y}$ as consisting of two parts, the systematic part $\tilde{\boldsymbol{y}} = \boldsymbol{\pi\eta}$ which is a function of the latent variables, and the residual part ϵ. Under the traditional factor-analytical interpretation, $\boldsymbol{\eta}$ has the true scores, and ϵ is the measurement error. The part of the variance of the observation that is due to the true scores is called the reliability (or communality), $\text{var}(\tilde{y}_k)/\,\text{var}(y_k)$, and the residual variance is due to unreliability of the measures $\boldsymbol{y}$ selected to measure the variables $\boldsymbol{\eta}$.

The LV relation. The relation of the LVs to the MVs is described by the equation:

$$\boldsymbol{\eta} = \tilde{\boldsymbol{\eta}} + \boldsymbol{\delta} = \boldsymbol{\omega}'\boldsymbol{y} + \boldsymbol{\delta}. \tag{3.202}$$

The LVs $\boldsymbol{\eta}$ are split up into two parts, the one part being accessible to observation and measurable, $\tilde{\boldsymbol{\eta}} = \boldsymbol{\omega}'\boldsymbol{y}$, and the other being the unobserved and unmeasured part $\boldsymbol{\delta}$. A residual $\boldsymbol{\delta}$ which is nonzero indicates that the "true" latent variable $\boldsymbol{\eta}$ is more than the observation, that the latent variable includes a theoretical surplus over the empirical content. If a given factor with a defined range of meaning and validity is not fully measured by its indicators, this is due to the insufficient validity of the MVs. Thus, $\boldsymbol{\delta}$ may be understood as the "lack of validity" of the MVs, or validity residual. From another point of view, the LVs can not be fully determined by the given set of MVs, and $\boldsymbol{\delta}$ is called the "indeterminateness" of the LVs.

Decomposition of $\boldsymbol{y}$. If Eq.3.202 is inserted into Eq.3.201, the manifest variables $\boldsymbol{y}$ are decomposed into three parts:

$$\boldsymbol{y} = \boldsymbol{\pi}(\boldsymbol{\omega}'\boldsymbol{y} + \boldsymbol{\delta}) + \epsilon = \boldsymbol{\pi\omega}'\boldsymbol{y} + \boldsymbol{\pi\delta} + \epsilon. \tag{3.203}$$

The first part is the "mirror image" of the MVs themselves, as it is reflected by the LVs. The second part $\boldsymbol{\pi\delta}$ works like a "one-way mirror": The influence of the "validity residuals" $\boldsymbol{\delta}$ can be traced down to the observation, but not the other way round. The third part is the outer residual ϵ which is uncorrelated with the LV.

Decomposition of $\boldsymbol{\eta}$. If Eq.3.201 is inserted into Eq.3.202, the latent variable $\boldsymbol{\eta}$ is decomposed into three parts:

$$\boldsymbol{\eta} = \boldsymbol{\omega}'(\boldsymbol{\pi\eta} + \boldsymbol{\epsilon}) + \boldsymbol{\delta} = \boldsymbol{\omega}'\boldsymbol{\pi\eta} + \boldsymbol{\omega}'\boldsymbol{\epsilon} + \boldsymbol{\delta}. \tag{3.204}$$

The first part $\boldsymbol{\omega}'\boldsymbol{\pi\eta}$ is the "mirror image" of the LV, as it is reflected by the MVs; it is the validated and reliably measured part of the theoretical concept; this part of the concept has left obscurity and has become measurable. The second part $\boldsymbol{\omega}'\boldsymbol{\epsilon}$ are "stains of the mirror" which are brought in by the outer residuals $\boldsymbol{\epsilon}$ and are to be understood as the unreliability of a linear aggregate for a given set of fallible measures $\boldsymbol{y}$. The third part are the validity residuals $\boldsymbol{\delta}$. The reduced form of Eq.3.204 shows how the LV is generated from the residuals:

$$\boldsymbol{\eta} = (\mathbf{I} - \boldsymbol{\omega}'\boldsymbol{\pi})^{-1}\boldsymbol{\delta} \ + (\mathbf{I} - \boldsymbol{\omega}'\boldsymbol{\pi})^{-1}\boldsymbol{\omega}'\boldsymbol{\epsilon}. \tag{3.205}$$

Decomposition of $\boldsymbol{\delta}$. If Eq.3.204 is solved for $\boldsymbol{\delta}$, then $\boldsymbol{\delta}$ is exhibited to consists of two parts:

$$\boldsymbol{\delta} = \boldsymbol{\eta} - \boldsymbol{\omega}'(\boldsymbol{\pi\eta} + \boldsymbol{\epsilon}) = (\mathbf{I} - \boldsymbol{\omega}'\boldsymbol{\pi})\boldsymbol{\eta} + \boldsymbol{\omega}'\boldsymbol{\epsilon}. \tag{3.206}$$

The first part of $\boldsymbol{\delta}$ is shared with the LVs; under the prerequisites of Eq.3.212 where $\hat{\boldsymbol{\eta}} \mathrel{\hat{=}} \boldsymbol{\eta}$ is a linear composite this first term would vanish due to $\boldsymbol{\omega}'\boldsymbol{\pi} = \mathbf{I}$. The second term shows the inner residuals $\boldsymbol{\delta}$ partly to be a linear function of the outer residuals, which is a quite disturbing expression. In order to keep the truly latent of the LV separate from the measurement error, one would like to apply further model specifications such that the theoretical surplus $\boldsymbol{\delta}$ is not a function of measurement error $\boldsymbol{\epsilon}$.

Forms of the model. The two model relations Eq.3.201 and Eq.3.202 allow for different forms, and, depending on what is treated as given and what can be derived, we will label four forms:

- Given is $\boldsymbol{\delta}, \boldsymbol{\epsilon}, \boldsymbol{\pi}, \boldsymbol{\omega}$: The generative or reduced form
- Given is $\boldsymbol{\delta}, \boldsymbol{\epsilon}, \boldsymbol{\eta}, \boldsymbol{\pi}$: The deductive or structural form
- Given is $\boldsymbol{y}, \text{cov}(\boldsymbol{y}; \boldsymbol{\eta})$: The pure inductive form
- Given is $\boldsymbol{y}, \text{cov}(\boldsymbol{y}; \boldsymbol{\eta}), \text{cov}(\boldsymbol{\eta})$: The full inductive form.

Rank of $\boldsymbol{y}$ and $\boldsymbol{\Sigma}$. Given a set of variables $\boldsymbol{y} = [y_k, k = 1 \ldots K + M]$, it may turn out that one of the variables y_k can be demonstrated to be a perfect linear combination of other variables of the same set, i.e. $R^2_{k|\text{others}} = 1$. The number of variables in $\boldsymbol{y}$ which cannot be represented by the others is called the dimensionality or the rank of $\boldsymbol{y}$, $K = \text{rank}(\boldsymbol{y})$. The concept of rank applies also to the product matrix $\boldsymbol{\Sigma} = \text{E}(\boldsymbol{yy}')$: if one of the columns/rows of $\boldsymbol{\Sigma}$ can be represented by a linear function of other columns/rows, $\boldsymbol{\Sigma}$ is said to be rank deficient. The rank of $\boldsymbol{y}$ and $\boldsymbol{\Sigma}$ are identical, as are the rank of $\boldsymbol{\eta}$ and $\boldsymbol{\rho}$ $(= \text{cov}(\boldsymbol{\eta}))$:

$$K = \text{rank}(\boldsymbol{y}) = \text{rank}(\boldsymbol{\Sigma}) \tag{3.207}$$

$$J = \text{rank}(\boldsymbol{\eta}) = \text{rank}(\boldsymbol{\rho}). \tag{3.208}$$

The rank is indicated by the number of eigenvalues of the product matrix ($\boldsymbol{\Sigma}$ or $\boldsymbol{\rho}$, resp.) which are greater than zero.

Joint rank of LVs and MVs. Bentler's (1982) definition of LV models requires that the rank of the joint space of MVs and LVs is greater than the rank of the MV space alone:

$$L = \text{rank}(\boldsymbol{y} \vee \boldsymbol{\eta}) > K = \text{rank}(\boldsymbol{y}) \qquad \text{Factor model.} \tag{3.209}$$

According to this definition, composites and components (which are linear aggregates of the MVs) are not latent variables, because the rank of component models is equal to the rank of the MVs:

$$L = \text{rank}(\boldsymbol{y} \vee \boldsymbol{\omega}'\boldsymbol{y}) = K = \text{rank}(\boldsymbol{y}) \qquad \text{Component model.} \tag{3.210}$$

Our predictive orientation (see Sc.5) leads us not to adopt Bentler's definition for LVs, which applies to our LVs of the second notion (see pg.81) but not to our estimated LVs of the third notion. However, it is useful to keep the rank of MVs, LVs and residuals in mind. If in the following a LV is labeled a factor, it is meant to be a dimension beyond the space of the observation. What makes the LVs reach beyond the space of observations is the theoretical surplus, introduced as $\boldsymbol{\delta}$ in Eq.3.202.

Rank of residuals. Two types of residuals can be distinguished, the factor residuals $\boldsymbol{\epsilon} = \boldsymbol{y} - \boldsymbol{\pi\eta}$ and the component residuals $\hat{\boldsymbol{\epsilon}} = \boldsymbol{y} - \boldsymbol{\pi}\hat{\boldsymbol{\eta}} = \boldsymbol{y} - \boldsymbol{\pi\omega}'\boldsymbol{y} = (\mathbf{I} - \boldsymbol{\pi\omega}')\boldsymbol{y}$. In both cases, the rank of the residuals is the joint rank of MVs and LVs, minus the rank of the LVs, i.e. $L - J$. In special, the rank of component residuals is necessarily smaller than the rank of the MVs, $L - J = K - J$. The factor residuals have the full rank of the MVs, if $L = K + J$, because $L - J = (K + J) - J = K$. This implies that the covariance matrix of factor residuals may be invertible, but the component residual covariance is definitely not invertible.

Unbiasedness of LV scores. Something goes always wrong when something is estimated which is beyond the space of the observation. What can go wrong with factor score estimation, is explored in more detail in Sc.3.3 and Sc.5.2. The problem can be posed with respect to the unobserved variables as well as to the covariances. With respect to the variables, the problem is whether the expectation of the unobserved variables, given the observation, coincides with the variables themselves; if yes, the estimates are called unbiased:

$$\begin{aligned} \text{E}(\boldsymbol{\eta}|\boldsymbol{y}) &= \boldsymbol{\eta} \\ \text{E}(\boldsymbol{\epsilon}|\boldsymbol{y}) &= \boldsymbol{\epsilon} \\ \text{E}(\boldsymbol{\delta}|\boldsymbol{y}) &= \boldsymbol{\delta}. \end{aligned} \tag{3.211}$$

With respect to the covariances, one may test the conditions

$$\begin{aligned} \text{E}(\boldsymbol{\eta\eta}') &= \text{E}(\tilde{\boldsymbol{\eta}}\boldsymbol{\eta}') = \text{E}(\tilde{\boldsymbol{\eta}}\tilde{\boldsymbol{\eta}}') \\ \text{E}(\boldsymbol{y\eta}') &= \text{E}(\boldsymbol{y}\tilde{\boldsymbol{\eta}}') = \text{E}(\tilde{\boldsymbol{y}}\boldsymbol{\eta}') = \text{E}(\tilde{\boldsymbol{y}}\tilde{\boldsymbol{\eta}}') \end{aligned}$$

3.2.2 Aggregates, Composites

The model. Let be given a set of variables $\boldsymbol{y}_{[K\times 1]} \sim \mathrm{M}(0, \boldsymbol{\Sigma}, \ldots)$ and an arbitrary but known vector $\boldsymbol{\omega}_{[K\times J]} \equiv [\omega_{kj}]$ of weight coefficients,which together produce a variable $\boldsymbol{Y}$, as shown in Tb.3.2a:

$$\boldsymbol{Y} = \boldsymbol{\omega}'\boldsymbol{y}. \tag{3.212}$$

A weighted sum of variables, as defined in Eq.3.212, is called a linear combination, or a linear aggregate, or a composite, and the variables $\boldsymbol{y}$ are called aggregands.

Implied covariance. The covariance matrix of the joint vector of variables $\boldsymbol{Y}$ and $\boldsymbol{y}$ is:

$$\mathrm{cov}\begin{bmatrix} \boldsymbol{Y} \\ \boldsymbol{y} \end{bmatrix} \equiv \begin{bmatrix} \boldsymbol{\rho} & \boldsymbol{\tau}' \\ \boldsymbol{\tau} & \boldsymbol{\Sigma} \end{bmatrix} = \begin{bmatrix} \boldsymbol{\omega}'\boldsymbol{\Sigma}\boldsymbol{\omega} & \boldsymbol{\omega}'\boldsymbol{\Sigma} \\ \boldsymbol{\Sigma}\boldsymbol{\omega} & \boldsymbol{\Sigma} \end{bmatrix}, \tag{3.213}$$

which is proved by:

$$\begin{aligned} \boldsymbol{\tau} \equiv \mathrm{cov}(\boldsymbol{y}; \boldsymbol{Y}) &= \mathrm{cov}(\boldsymbol{y}; \boldsymbol{\omega}'\boldsymbol{y}) = \mathrm{E}(\boldsymbol{y}\boldsymbol{y}'\boldsymbol{\omega}) = \mathrm{E}(\boldsymbol{y}\boldsymbol{y}')\boldsymbol{\omega} \\ &= \boldsymbol{\Sigma}\boldsymbol{\omega} \\ \boldsymbol{\rho} \equiv \mathrm{var}(\boldsymbol{Y}) &= \mathrm{cov}(\boldsymbol{\omega}'\boldsymbol{y}; \boldsymbol{\omega}'\boldsymbol{y}) = \boldsymbol{\omega}'\boldsymbol{\Sigma}\boldsymbol{\omega} = \boldsymbol{\tau}'\boldsymbol{\omega} \\ &= \boldsymbol{\omega}'\boldsymbol{\tau}. \end{aligned}$$

If, at variance to the prerequisites of Eq.3.212, the weights are unknown but the covariance $\boldsymbol{\tau}$ is known, and if furthermore $\boldsymbol{\Sigma}$ is invertible, the weights can be reconstructed uniquely by

$$\boldsymbol{\omega} = \boldsymbol{\Sigma}^{-1}\boldsymbol{\tau}. \tag{3.214}$$

Remark 3⋆3 If $\boldsymbol{\Sigma}$ is singular, the weights found by a general inverse $\boldsymbol{\Sigma}^{-}$ are not unique. If there is one $\breve{\boldsymbol{\omega}} = \boldsymbol{\Sigma}^{-}\boldsymbol{\tau}$, then there exists an infinite number of $\breve{\boldsymbol{\omega}}$s. All of them, however, produce the same aggregate, $\boldsymbol{Y} = \breve{\boldsymbol{\omega}}'\boldsymbol{y}$, the same $\boldsymbol{\tau}$ and the same $\boldsymbol{\rho}$. Even if the weights are not unique, the aggregate and the covariances are.

Remark 3⋆4 A implied covariance matrix like Eq.3.213 will be called an inductive expression. It starts with observed variables and works from bottom (MVs) to top (LVs) and gives the covariances of the LVs.

3.2.3 Factors

The factor model. Let be given the factor $\boldsymbol{\eta}_{[J\times 1]} \sim \mathrm{M}(0, \boldsymbol{\rho}, \ldots)$, the residuals $\boldsymbol{\epsilon}_{[K\times 1]} \sim \mathrm{M}(\mathbf{0}, \boldsymbol{\Theta}, \ldots)$ and a vector $\boldsymbol{\pi}_{[K\times J]} \equiv [\pi_{kj}]$ of loading coefficients. Let $\boldsymbol{\eta}$ be specified to be the predictor of $\boldsymbol{y}$ (see Tb.3.2b):

$$\begin{aligned} \boldsymbol{y} = \boldsymbol{\pi}\boldsymbol{\eta} + \boldsymbol{\epsilon}, \qquad & \tilde{\boldsymbol{y}} \equiv \mathrm{E}(\boldsymbol{y}|\boldsymbol{\eta}) = \boldsymbol{\pi}\boldsymbol{\eta} \\ & \Longrightarrow \mathrm{E}(\boldsymbol{\epsilon}) = \mathbf{0} \\ & \Longrightarrow \mathrm{cov}(\boldsymbol{\epsilon}; \boldsymbol{\eta}) = \mathrm{cov}(\boldsymbol{\epsilon}; \tilde{\boldsymbol{y}}) = \mathbf{0} \\ & \Longrightarrow \mathrm{cov}(\boldsymbol{y}; \boldsymbol{\eta}) = \mathrm{cov}(\tilde{\boldsymbol{y}}; \boldsymbol{\eta}) = \boldsymbol{\pi}\,\mathrm{var}(\boldsymbol{\eta}). \end{aligned} \tag{3.215}$$

Deductive form of covariances. This model implies that the covariance matrix of the joint vector of all variables can be expressed in terms of $\boldsymbol{\Theta}, \boldsymbol{\rho}$ and $\boldsymbol{\pi}$,

$$\operatorname{cov}\begin{bmatrix} \boldsymbol{\epsilon} \\ \boldsymbol{\eta} \\ \tilde{\boldsymbol{y}} \\ \boldsymbol{y} \end{bmatrix} = \begin{bmatrix} \boldsymbol{\Theta} & . & . & \text{sym.} \\ \mathbf{0} & \boldsymbol{\rho} & . & . \\ \mathbf{0} & \boldsymbol{\pi\rho} & \tilde{\boldsymbol{\Sigma}} \equiv \boldsymbol{\pi\rho\pi}' & . \\ \boldsymbol{\Theta} & \boldsymbol{\pi\rho} & \tilde{\boldsymbol{\Sigma}} & \boldsymbol{\Sigma} = \tilde{\boldsymbol{\Sigma}} + \boldsymbol{\Theta} \end{bmatrix} \tag{3.216}$$

which is proved by:

$$\begin{aligned}
\boldsymbol{\tau} \equiv \operatorname{cov}(\boldsymbol{y};\boldsymbol{\eta}) &= \operatorname{cov}(\boldsymbol{\pi\eta}+\boldsymbol{\epsilon};\boldsymbol{\eta}) = \boldsymbol{\pi}\operatorname{cov}(\boldsymbol{\eta};\boldsymbol{\eta}) + \operatorname{cov}(\boldsymbol{\epsilon};\boldsymbol{\eta}) \\
&= \boldsymbol{\pi\rho} + \mathbf{0} \\
\operatorname{cov}(\tilde{\boldsymbol{y}};\boldsymbol{\eta}) &= \operatorname{cov}(\boldsymbol{\pi\eta};\boldsymbol{\eta}) = \boldsymbol{\pi}\operatorname{cov}(\boldsymbol{\eta};\boldsymbol{\eta}) = \boldsymbol{\pi\rho} \\
\tilde{\boldsymbol{\Sigma}} \equiv \operatorname{cov}(\tilde{\boldsymbol{y}}) &= \operatorname{cov}(\boldsymbol{\pi\eta}) = \boldsymbol{\pi\rho\pi}' \\
\boldsymbol{\Sigma} \equiv \operatorname{cov}(\boldsymbol{y}) &= \operatorname{cov}(\boldsymbol{\pi\eta}+\boldsymbol{\epsilon};\boldsymbol{\pi\eta}+\boldsymbol{\epsilon}) \\
&= \boldsymbol{\pi}\operatorname{cov}(\boldsymbol{\eta};\boldsymbol{\eta}) + \boldsymbol{\pi}\operatorname{cov}(\boldsymbol{\eta};\boldsymbol{\epsilon}) + \operatorname{cov}(\boldsymbol{\epsilon};\boldsymbol{\eta})\boldsymbol{\pi}' + \operatorname{cov}(\boldsymbol{\epsilon};\boldsymbol{\epsilon}) \\
&= \boldsymbol{\pi\rho\pi}' + \mathbf{0} + \mathbf{0} + \boldsymbol{\Theta} = \tilde{\boldsymbol{\Sigma}} + \boldsymbol{\Theta} \\
\operatorname{cov}(\boldsymbol{y};\boldsymbol{\epsilon}) &= \operatorname{cov}(\boldsymbol{\pi\eta}+\boldsymbol{\epsilon};\boldsymbol{\epsilon}) \\
&= \boldsymbol{\pi}\operatorname{cov}(\boldsymbol{\eta};\boldsymbol{\epsilon}) + \operatorname{cov}(\boldsymbol{\epsilon};\boldsymbol{\epsilon}) = \mathbf{0} + \operatorname{cov}(\boldsymbol{\epsilon}) = \boldsymbol{\Theta} \\
\operatorname{cov}(\boldsymbol{y};\tilde{\boldsymbol{y}}) &= \operatorname{cov}(\tilde{\boldsymbol{y}}+\boldsymbol{\epsilon};\tilde{\boldsymbol{y}}) = \operatorname{cov}(\tilde{\boldsymbol{y}}) + \operatorname{cov}(\boldsymbol{\epsilon};\tilde{\boldsymbol{y}}) = \tilde{\boldsymbol{\Sigma}} + \mathbf{0}.
\end{aligned}$$

If, at variance to the prerequisites of Eq.3.216, not $\boldsymbol{\Theta}, \boldsymbol{\rho}, \boldsymbol{\pi}$ are treated as known but $\boldsymbol{\Theta}, \boldsymbol{\rho}, \boldsymbol{\tau}$, it is also possible to derive the other unknowns:

$$\boldsymbol{\pi} = \boldsymbol{\tau\rho}^{-1} \tag{3.217}$$

$$\tilde{\boldsymbol{\Sigma}} = \boldsymbol{\pi\rho\pi}' = (\boldsymbol{\tau\rho}^{-1})\boldsymbol{\rho}(\boldsymbol{\tau\rho}^{-1})' = \boldsymbol{\tau\rho}^{-1}\boldsymbol{\tau}' \tag{3.218}$$

For most of the triples in $\{\boldsymbol{\Theta}, \boldsymbol{\rho}, \boldsymbol{\Sigma}, \boldsymbol{\pi}, \boldsymbol{\tau}, \tilde{\boldsymbol{\Sigma}}\}$ the remaining triple can be reconstructed.

Remark 3⋆5 A implied covariance matrix like Eq.3.216 will be called a deductive expression. It starts with "known" LVs and works, from top to bottom, down to the covariance of the observed variables.

Remark 3⋆6 If $\boldsymbol{\eta}$ is an unobserved or latent variable, the model Eq.3.215 can be called a factor model, with $\boldsymbol{\eta}$ as one of the factors of the observed variables $\boldsymbol{y}$. Nothing similar to Eq.3.212 is defined in this factor model, and no properties similar to Eq.3.213 can be stated on it.

Remark 3⋆7 The factor model as specified in Eq.3.215 does not imply that the residuals are uncorrelated with each other, i.e. $\boldsymbol{\Theta} = \operatorname{cov}(\boldsymbol{\epsilon})$ is not assumed to be diagonal. Otherwise one would call $\boldsymbol{\eta}$ the "common factor" of $\boldsymbol{y}$, and Eq.3.215 would be the classical Common Factor model. No statement about the rank of $\boldsymbol{\Theta}$ is possible, and any number of factors may be responsible for the variance and covariance of the $\boldsymbol{y}$-variables.

The discussion of factors is continued in Sc.3.3.

3.2.4 Components

The component model. The component model, as shown in Tb.3.2c, is a system of two relations, the composite relation Eq.3.212 and the "factor" relation Eq.3.215. The weights $\boldsymbol{\omega}$ are treated as fixed and known, the loadings $\boldsymbol{\pi}$ are unknown constants.

$$\begin{aligned} \tilde{\boldsymbol{\eta}} &= \boldsymbol{\omega}'\boldsymbol{y} \\ \boldsymbol{\tau} \equiv \operatorname{cov}(\boldsymbol{y}; \tilde{\boldsymbol{\eta}}) &= \boldsymbol{\Sigma\omega} \\ \boldsymbol{\rho} \equiv \operatorname{cov}(\tilde{\boldsymbol{\eta}}) &= \boldsymbol{\omega}'\boldsymbol{\Sigma\omega} \end{aligned} \tag{3.219}$$

$$\begin{aligned} \boldsymbol{y} = \boldsymbol{\pi}\tilde{\boldsymbol{\eta}} + \boldsymbol{\epsilon}, \qquad & \tilde{\boldsymbol{y}} \equiv \mathrm{E}(\boldsymbol{y}|\tilde{\boldsymbol{\eta}}) = \boldsymbol{\pi}\tilde{\boldsymbol{\eta}} \\ & \Longrightarrow \mathrm{E}(\boldsymbol{\epsilon}) = \mathbf{0} \\ & \Longrightarrow \operatorname{cov}(\boldsymbol{\epsilon}; \tilde{\boldsymbol{\eta}}) = \operatorname{cov}(\boldsymbol{\epsilon}; \tilde{\boldsymbol{y}}) = \mathbf{0} \\ & \Longrightarrow \operatorname{cov}(\boldsymbol{y}; \tilde{\boldsymbol{\eta}}) = \operatorname{cov}(\tilde{\boldsymbol{y}}; \tilde{\boldsymbol{\eta}}) = \boldsymbol{\pi} \operatorname{var}(\tilde{\boldsymbol{\eta}}). \end{aligned} \tag{3.220}$$

For this model, all of the composite properties Eqs.3.212, 3.213 and all of the factor properties Eqs.3.216,..., 3.218 hold. But some more properties emerge due to the joint specification of Eqs.3.219, 3.220.

Weights and loadings. There are two expressions for $\boldsymbol{\tau}$, the covariance of predictands and predictors, the inductive expression in Eq.3.212 and the deductive expression in Eq.3.216 which can be equated,

$$\boldsymbol{\tau} = \boldsymbol{\Sigma\omega} = \boldsymbol{\pi\rho}, \tag{3.221}$$

and solved for the loadings (using Eq.3.213),

$$\boldsymbol{\pi} = \boldsymbol{\tau\rho}^{-1} = \boldsymbol{\Sigma\omega\rho}^{-1} = \boldsymbol{\Sigma\omega}(\boldsymbol{\omega}'\boldsymbol{\Sigma\omega})^{-1}, \tag{3.222}$$

This equation implies that the inner product $\boldsymbol{\omega}'\boldsymbol{\pi}$ is unity,

$$\boldsymbol{\omega}'\boldsymbol{\pi} = \boldsymbol{\omega}'[\boldsymbol{\Sigma\omega}(\boldsymbol{\omega}'\boldsymbol{\Sigma\omega})^{-1}] = \mathbf{I}, \qquad \text{by Eq.3.222} \tag{3.223}$$

and that the outer product $\boldsymbol{\pi\omega}'$ is idempotent:

$$\boldsymbol{\pi\omega}' = (\boldsymbol{\pi\omega}')(\boldsymbol{\pi\omega}') = (\boldsymbol{\pi\omega}')^n, n \in \mathbb{N}, \tag{3.224}$$

because, by use of Eq.3.223, $(\boldsymbol{\pi\omega}')(\boldsymbol{\pi\omega}') = \boldsymbol{\pi}(\boldsymbol{\omega}'\boldsymbol{\pi})\boldsymbol{\omega}' = \boldsymbol{\pi}(\mathbf{I})\boldsymbol{\omega}' = \boldsymbol{\pi\omega}'$. In passage we note that:

$$\begin{aligned} \boldsymbol{\omega}'(\mathbf{I} - \boldsymbol{\pi\omega}') &= \boldsymbol{\omega}' - \boldsymbol{\omega}'\boldsymbol{\pi\omega}' = \boldsymbol{\omega}' - \mathbf{I}\boldsymbol{\omega}' \\ &= \mathbf{0} \end{aligned} \tag{3.225}$$

Residuals. The residuals $\boldsymbol{\epsilon}$ and their covariance matrix $\boldsymbol{\Theta}$ are given by

$$\begin{aligned}
\boldsymbol{\epsilon} &= \boldsymbol{y} - \boldsymbol{\pi}\tilde{\boldsymbol{\eta}} = \boldsymbol{y} - \boldsymbol{\pi}\boldsymbol{\omega}'\boldsymbol{y} \\
&= (\mathbf{I} - \boldsymbol{\pi}\boldsymbol{\omega}')\boldsymbol{y} && (3.226) \\
\boldsymbol{\Theta} \equiv \mathrm{cov}(\boldsymbol{\epsilon}) &= (\mathbf{I} - \boldsymbol{\pi}\boldsymbol{\omega}')\boldsymbol{\Sigma}(\mathbf{I} - \boldsymbol{\pi}\boldsymbol{\omega}')' \\
&= \boldsymbol{\Sigma}(\mathbf{I} - \boldsymbol{\pi}\boldsymbol{\omega}')' - \boldsymbol{\pi}\boldsymbol{\omega}'\boldsymbol{\Sigma}(\mathbf{I} - \boldsymbol{\pi}\boldsymbol{\omega}')' \\
&= \boldsymbol{\Sigma}(\mathbf{I} - \boldsymbol{\pi}\boldsymbol{\omega}')' - \boldsymbol{\pi}\boldsymbol{\omega}'\boldsymbol{\Sigma} + \boldsymbol{\pi}\boldsymbol{\omega}'\boldsymbol{\Sigma}\boldsymbol{\omega}\boldsymbol{\pi}' \\
&= \boldsymbol{\Sigma}(\mathbf{I} - \boldsymbol{\pi}\boldsymbol{\omega}')' - \boldsymbol{\pi}\boldsymbol{\tau}' + \boldsymbol{\pi}\boldsymbol{\tau}' \\
&= \boldsymbol{\Sigma}(\mathbf{I} - \boldsymbol{\pi}\boldsymbol{\omega}')' && (3.227) \\
&= \boldsymbol{\Sigma} - \boldsymbol{\Sigma}\boldsymbol{\omega}\boldsymbol{\pi}' = \boldsymbol{\Sigma} - \boldsymbol{\pi}\boldsymbol{\rho}\boldsymbol{\pi}' \\
&= \boldsymbol{\Sigma} - \tilde{\boldsymbol{\Sigma}} && (3.228) \\
\boldsymbol{\omega}'\boldsymbol{\epsilon} &= \boldsymbol{\omega}'(\mathbf{I} - \boldsymbol{\pi}\boldsymbol{\omega}')\boldsymbol{y} = \mathbf{0}\boldsymbol{y} = \mathbf{0} && \text{by Eq.3.225} \\
\boldsymbol{\omega}'\boldsymbol{\Theta}\boldsymbol{\omega} &= \mathbf{0} && \text{by Eq.3.229} \quad (3.229) \\
\boldsymbol{\omega}'\boldsymbol{y} &= \boldsymbol{\omega}'(\tilde{\boldsymbol{y}} + \boldsymbol{\epsilon}) = \boldsymbol{\omega}'\tilde{\boldsymbol{y}} + \boldsymbol{\omega}\boldsymbol{\epsilon} = \boldsymbol{\omega}'\tilde{\boldsymbol{y}} + \mathbf{0} = \boldsymbol{\omega}'\tilde{\boldsymbol{y}}. && (3.230)
\end{aligned}$$

Due to $\boldsymbol{\omega}'\boldsymbol{y} = \boldsymbol{\omega}'\tilde{\boldsymbol{y}}$ the weight relation Eq.3.219 can be replaced by $\tilde{\boldsymbol{\eta}} = \boldsymbol{\omega}'\tilde{\boldsymbol{y}}$.

Covariance deduction. Due to Eq.3.222 the inductive properties of the component model can be extended beyond the properties Eq.3.213 of the aggregate model. Taking $\boldsymbol{y}$ and $\boldsymbol{\omega}$ as given, the covariance matrix of the joint vector of variables can be expressed in terms of $\boldsymbol{\Sigma}$ and $\boldsymbol{\omega}$:

$$\mathrm{cov}\begin{bmatrix} \boldsymbol{\epsilon} \\ \tilde{\boldsymbol{\eta}} \\ \tilde{\boldsymbol{y}} \\ \boldsymbol{y} \end{bmatrix} = \begin{bmatrix} \boldsymbol{\Theta} \equiv \boldsymbol{\Sigma} - \tilde{\boldsymbol{\Sigma}} & . & . & \mathrm{sym.} \\ \mathbf{0} & \boldsymbol{\rho} \equiv \boldsymbol{\omega}'\boldsymbol{\Sigma}\boldsymbol{\omega} & . & . \\ \mathbf{0} & \boldsymbol{\Sigma}\boldsymbol{\omega} & \tilde{\boldsymbol{\Sigma}} \equiv \boldsymbol{\Sigma}\boldsymbol{\omega}\boldsymbol{\rho}^{-1}\boldsymbol{\omega}'\boldsymbol{\Sigma} & . \\ \boldsymbol{\Theta} & \boldsymbol{\Sigma}\boldsymbol{\omega} & \tilde{\boldsymbol{\Sigma}} & \boldsymbol{\Sigma} \end{bmatrix} \qquad (3.231)$$

which is proved by Eq.3.213 and

$$\begin{aligned}
\tilde{\boldsymbol{\Sigma}} \equiv \mathrm{cov}(\tilde{\boldsymbol{y}}) &= \mathrm{cov}(\boldsymbol{\pi}\boldsymbol{\omega}'\boldsymbol{y}) = \boldsymbol{\pi}\boldsymbol{\omega}'\boldsymbol{\Sigma}\boldsymbol{\omega}\boldsymbol{\pi}' \\
&= [\boldsymbol{\Sigma}\boldsymbol{\omega}(\boldsymbol{\omega}'\boldsymbol{\Sigma}\boldsymbol{\omega})^{-1}]\boldsymbol{\omega}'\boldsymbol{\Sigma}\boldsymbol{\omega}[\boldsymbol{\Sigma}\boldsymbol{\omega}(\boldsymbol{\omega}'\boldsymbol{\Sigma}\boldsymbol{\omega})^{-1}]' \\
&= \boldsymbol{\Sigma}\boldsymbol{\omega}(\boldsymbol{\omega}'\boldsymbol{\Sigma}\boldsymbol{\omega})^{-1}\boldsymbol{\omega}'\boldsymbol{\Sigma} \\
&= \boldsymbol{\tau}\boldsymbol{\rho}^{-1}\boldsymbol{\tau}' = \boldsymbol{\pi}\boldsymbol{\rho}\boldsymbol{\pi} = \boldsymbol{\pi}\boldsymbol{\tau}' = \boldsymbol{\tau}\boldsymbol{\pi}' \\
\mathrm{cov}(\tilde{\boldsymbol{y}}; \tilde{\boldsymbol{\eta}}) &= \mathrm{cov}(\boldsymbol{\pi}\boldsymbol{\omega}'\boldsymbol{y}; \boldsymbol{\omega}'\boldsymbol{y}) = \boldsymbol{\pi}\boldsymbol{\omega}'\boldsymbol{\Sigma}\boldsymbol{\omega} = [\boldsymbol{\Sigma}\boldsymbol{\omega}(\boldsymbol{\omega}'\boldsymbol{\Sigma}\boldsymbol{\omega})^{-1}]\boldsymbol{\omega}'\boldsymbol{\Sigma}\boldsymbol{\omega} \\
&= \boldsymbol{\Sigma}\boldsymbol{\omega}.
\end{aligned}$$

Decomposition of $\boldsymbol{y}$. If $\tilde{\boldsymbol{\eta}}$ in Eq.3.220 is substituted from Eq.3.219 the combined equations read as:

$$\boldsymbol{y} = \boldsymbol{\pi}\boldsymbol{\omega}'\boldsymbol{y} + \boldsymbol{\epsilon} = \boldsymbol{\pi}\boldsymbol{\omega}'\boldsymbol{y} + (\mathbf{I} - \boldsymbol{\pi}\boldsymbol{\omega}')\boldsymbol{y} = \tilde{\boldsymbol{y}} + \boldsymbol{\epsilon}. \qquad (3.232)$$

Here $\boldsymbol{y}$ is decomposed into two parts, the systematic part $\tilde{\boldsymbol{y}} = \boldsymbol{\pi}\boldsymbol{\omega}'\boldsymbol{y}$ and the residual part. Eq.3.232 is different from Eq.3.203 which contains as third part of $\boldsymbol{y}$ the expression $\boldsymbol{\pi}\boldsymbol{\delta}$.

Let $K = \text{rank}(\boldsymbol{y})$ and $J = \text{rank}(\tilde{\boldsymbol{\eta}})$, so that $\boldsymbol{\pi}$ and $\boldsymbol{\omega}$ are $K \times J$ matrices. The predictor specification Eq.3.220 guarantees that $\boldsymbol{\epsilon}$ and $\tilde{\boldsymbol{y}}$ as well as $\boldsymbol{\epsilon}$ and $\tilde{\boldsymbol{\eta}}$ are uncorrelated. Hence the K-dimensional space of $\boldsymbol{y}$ is split up into two orthogonal spaces, the J-dimensional space of the components $\tilde{\boldsymbol{y}}$ and the $(K-J)$-dimensional residual space. The way in which the K-dimensional space is split up between components and residual space is determined by the aprioristic weights $\boldsymbol{\omega}$.

Unbiased prediction. By taking the expectations of both sides of Eq.3.232, and making use of $\text{E}(\boldsymbol{\epsilon}) = \mathbf{0}$ which is implied by predictor specification (Eq.3.220) one can establish that the observed variables $\boldsymbol{y}$ and the systematic part $\tilde{\boldsymbol{y}}$ are equal in expectation, i.e. $\tilde{\boldsymbol{y}}$ is a consistent, unbiased predictor of $\boldsymbol{y}$:

$$\begin{aligned} \text{E}(\boldsymbol{y}) &= \text{E}(\tilde{\boldsymbol{y}}) + \text{E}(\boldsymbol{\epsilon}) \\ &= \text{E}[(\boldsymbol{\pi}\boldsymbol{\omega}')\boldsymbol{y}] + \text{E}[(\mathbf{I} - \boldsymbol{\pi}\boldsymbol{\omega}')\boldsymbol{y}] = \text{E}(\boldsymbol{\pi}\boldsymbol{\omega}'\boldsymbol{y}) + \mathbf{0} \\ &= \text{E}(\tilde{\boldsymbol{y}}). \end{aligned} \tag{3.233}$$

Rank reduction. The rank of $\boldsymbol{\Theta}$ is less than the rank of $\boldsymbol{\Sigma}$ (Horst's (1965) rank reduction theorem).

$$\text{rank}(\boldsymbol{\Theta}) = \text{rank}(\boldsymbol{\Sigma}) - \text{rank}(\boldsymbol{\rho}) \tag{3.234}$$

$$\text{rank}(\boldsymbol{\epsilon}) = \text{rank}(\boldsymbol{y}) - \text{rank}(\tilde{\boldsymbol{\eta}}) \tag{3.235}$$

The rank reduction theorem implies the important consequence that the number of components is limited: Not more than $K = \text{rank}(\boldsymbol{\Sigma})$ components can be found. Furthermore, only if $\boldsymbol{\omega}_j'\boldsymbol{\Theta}_i\boldsymbol{\omega}_j > 0$, is the composite $\tilde{\boldsymbol{\eta}}_j = \boldsymbol{\omega}_j'\boldsymbol{y}$ a new component which is different from the earlier components $\tilde{\boldsymbol{\eta}}_i, i < j$.

Scale ambiguity. The component model Eq.3.219 takes the weights as known and fixed. The essential properties Eq.3.223 of the component model, however, holds for the family of weight vectors which are proportional to the original weights, $\boldsymbol{\omega}_f = f \cdot \boldsymbol{\omega}$, $\forall f \neq 0$. The proportionality constant f changes the scale (the standard deviation) of $\tilde{\boldsymbol{\eta}}$ and, hence, of $\boldsymbol{\rho}, \boldsymbol{\tau}, \boldsymbol{\pi}$.

Variants. The component model Eqs.3.219, 3.220 has two sets of coefficients, the weights $\boldsymbol{\omega}$, and the loadings $\boldsymbol{\pi}$ together with the variance $\boldsymbol{\rho}$. But only one set is "basic", and the other set can be derived immediately. More specifically, the composite model Eq.3.212 implies

$$\begin{aligned} (\boldsymbol{y}, \boldsymbol{\omega}) &\rightarrow (\tilde{\boldsymbol{\eta}}) \quad \text{and} \quad (\boldsymbol{\Sigma}, \boldsymbol{\omega}) \rightarrow (\boldsymbol{\tau}, \boldsymbol{\rho}) \\ (\boldsymbol{y}, \tilde{\boldsymbol{\eta}}) &\rightarrow (\boldsymbol{\omega}) \quad \text{and} \quad (\boldsymbol{\Sigma}, \boldsymbol{\tau}) \rightarrow (\boldsymbol{\omega}, \boldsymbol{\rho}), \end{aligned}$$

the factor model Eq.3.215 implies

$$(\boldsymbol{\eta}, \boldsymbol{\epsilon}, \boldsymbol{\pi}) \rightarrow (\boldsymbol{y}, \tilde{\boldsymbol{y}}) \quad \text{and} \quad (\boldsymbol{\pi}, \boldsymbol{\rho}, \boldsymbol{\Theta}) \rightarrow (\boldsymbol{\tau}, \tilde{\boldsymbol{\Sigma}}, \boldsymbol{\Sigma})$$

and other conclusions, and the component model which compromises both relations allows the inductive conclusions

$$(\boldsymbol{y},\boldsymbol{\omega}) \rightarrow (\tilde{\boldsymbol{\eta}},\boldsymbol{\epsilon}) \quad \text{and} \quad (\boldsymbol{\Sigma},\boldsymbol{\omega}) \rightarrow (\boldsymbol{\pi},\boldsymbol{\tau},\boldsymbol{\rho},\boldsymbol{\Theta},\tilde{\boldsymbol{\Sigma}})$$

and the deductive conclusions

$$(\tilde{\boldsymbol{\eta}},\boldsymbol{\epsilon},\boldsymbol{\pi}) \rightarrow (\boldsymbol{y},\tilde{\boldsymbol{y}},\boldsymbol{\omega}) \quad \text{and} \quad (\boldsymbol{\pi},\boldsymbol{\rho},\boldsymbol{\Theta}) \rightarrow (\boldsymbol{\tau},\tilde{\boldsymbol{\Sigma}},\boldsymbol{\Sigma},\boldsymbol{\omega}).$$

However, the assumptions incorporated into the model Eqs.3.219, 3.220 are not sufficient to facilitate the conclusions of principal component analysis:

$$(\boldsymbol{y},\boldsymbol{\rho}) \rightarrow (\tilde{\boldsymbol{\eta}},\boldsymbol{\epsilon}) \quad \text{and} \quad (\boldsymbol{\Sigma},\boldsymbol{\rho}) \rightarrow (\boldsymbol{\omega},\boldsymbol{\pi},\boldsymbol{\tau},\boldsymbol{\rho},\boldsymbol{\Theta},\tilde{\boldsymbol{\Sigma}}).$$

Summary. $\tilde{\boldsymbol{\eta}}$ is called a component of $\boldsymbol{y}$, if it is at the same time a linear aggregate of $\boldsymbol{y}$, ($\tilde{\boldsymbol{\eta}} = \boldsymbol{\omega}'\boldsymbol{y}$) and the predictor for its aggregands ($\boldsymbol{y} = \boldsymbol{\pi}\tilde{\boldsymbol{\eta}} + \boldsymbol{\epsilon}$). The most remarkable properties of components are:

$$\begin{aligned} \boldsymbol{\omega}'\boldsymbol{\pi} &= \mathbf{I}, \\ \tilde{\boldsymbol{y}} &= \boldsymbol{\pi}\boldsymbol{\omega}'\boldsymbol{y}, \\ \boldsymbol{\omega}'\boldsymbol{\epsilon} &= \mathbf{0}, \\ \boldsymbol{\omega}'\boldsymbol{y} &= \boldsymbol{\omega}'\tilde{\boldsymbol{y}} = \tilde{\boldsymbol{\eta}}. \end{aligned}$$

Components separate by means of aprioristic weights $\boldsymbol{\omega}$ the K-dimensional space of $\boldsymbol{y}$ into the J-dimensional space of the components and the $(K-J)$-dimensional residual space. The principal component model differs from the (ordinary) component model in so far as the weights are not considered as known; the directions of the principal components are not determined by aprioristic weights but by the postulate that they have maximum variance (or by equivalent postulates).

Components in the sense of this subsection are the variates created by principal components analysis, canonical correlation analysis, correspondence analysis, discriminant analysis, estimated factor scores in common factor analysis, and redundancy analysis.

3.2.5 Principal Components

Orthogonal transformation $\boldsymbol{\Omega}$. Given a set of random variables $\boldsymbol{y}_{[K\times1]} \sim \mathrm{M}(0,\boldsymbol{\Sigma},\ldots)$ and a orthonormal square matrix of weights $\boldsymbol{\Omega} \equiv [\omega_{kj}]$,

$$\boldsymbol{\Omega} = \boldsymbol{\Omega}^{-1\prime} \qquad \boldsymbol{\Omega}\boldsymbol{\Omega}' = \boldsymbol{\Omega}'\boldsymbol{\Omega} = \mathbf{I}, \tag{3.236}$$

a new set of random variables $\boldsymbol{\eta}_{[J\times1]} \equiv [\eta_j]$ is formed as linear aggregates:

$$\boldsymbol{\eta} = \boldsymbol{\Omega}'\boldsymbol{y}. \tag{3.237}$$

The joint product matrix of original, and aggregate, variables is

$$\mathrm{E}\left(\begin{bmatrix} \boldsymbol{\eta} \\ \boldsymbol{y} \end{bmatrix} \begin{bmatrix} \boldsymbol{\eta}' & \boldsymbol{y}' \end{bmatrix}\right) = \begin{bmatrix} \mathbf{P} & \mathbf{T}' \\ \mathbf{T} & \boldsymbol{\Sigma} \end{bmatrix} = \begin{bmatrix} \boldsymbol{\Omega}'\boldsymbol{\Sigma}\boldsymbol{\Omega} & \boldsymbol{\Sigma}\boldsymbol{\Omega} \\ \boldsymbol{\Omega}'\boldsymbol{\Sigma} & \boldsymbol{\Sigma} \end{bmatrix}. \tag{3.238}$$

Backtransformation. Premultiplication by a matrix $\mathbf{\Pi}$ transforms the aggregates $\boldsymbol{\eta}$ back to their originals $\boldsymbol{y}$:

$$\boldsymbol{y} = \mathbf{\Pi}\boldsymbol{\eta}. \tag{3.239}$$

It turns out that this "loading" matrix $\mathbf{\Pi}$ must be identical to $\mathbf{\Omega}$,

$$\mathbf{\Pi} = \mathbf{\Omega},$$

in order to get a complete backtransformation: $\boldsymbol{y} = \mathbf{\Pi}\boldsymbol{\eta} = \mathbf{\Omega}\boldsymbol{\eta} = \mathbf{\Omega}\mathbf{\Omega}'\boldsymbol{y} = \mathbf{I}\boldsymbol{y} = \boldsymbol{y}$.

Principal components. If simultaneously $\mathbf{\Omega}'\mathbf{\Omega} = \mathbf{I}$ and $\mathbf{\Omega}'\mathbf{\Sigma}\mathbf{\Omega}$ is diagonal,

$$\mathbf{\Omega}'\mathbf{\Sigma}\mathbf{\Omega} \equiv \mathbf{\Lambda}^2 = \mathbf{I} * \mathbf{\Lambda}^2, \tag{3.240}$$

then $\mathbf{\Omega}$ is the matrix of the eigenvectors of $\boldsymbol{y}$, and $\mathbf{\Omega}'\mathbf{\Sigma}\mathbf{\Omega}$ contains the eigenvalues λ^2, and the variables $\boldsymbol{\eta}$ are called the principal components of $\boldsymbol{y}$. The two conditions that $\mathbf{\Omega}'\mathbf{\Omega} = \mathbf{I}$ and that $\mathbf{\Omega}'\mathbf{\Sigma}\mathbf{\Omega}$ be diagonal *uniquely* define the weight matrix $\mathbf{\Omega}$. The joint product matrix of original, and aggregate, variables can now be expressed in terms of $\mathbf{\Omega}$ and $\mathbf{\Lambda}^2$:

$$\mathrm{E}\left(\begin{bmatrix}\boldsymbol{\eta}\\ \boldsymbol{y}\end{bmatrix}\begin{bmatrix}\boldsymbol{\eta}' & \boldsymbol{y}'\end{bmatrix}\right) = \begin{bmatrix}\mathbf{P} & \mathbf{T}'\\ \mathbf{T} & \mathbf{\Sigma}\end{bmatrix} = \begin{bmatrix}\mathbf{\Lambda}^2 & \mathbf{\Lambda}^2\mathbf{\Omega}'\\ \mathbf{\Omega}\mathbf{\Lambda}^2 & \mathbf{\Omega}\mathbf{\Lambda}^2\mathbf{\Omega}'\end{bmatrix}. \tag{3.241}$$

Recapitulation. To sum up, the weight matrix $\mathbf{\Omega}$ is identical to loading matrix $\mathbf{\Pi}$, Eq.3.239, which is due to the assumption $\mathbf{\Omega}'\mathbf{\Omega} = \mathbf{\Omega}\mathbf{\Omega}'$, and proportional to $\mathbf{T}$ (the loading structure matrix, Eq.3.238), which is due to the diagonality of $\mathbf{\Lambda}^2$, the proportionality constants being the reciprocals of the eigenvalues $\mathbf{\Lambda}^2$.

$$\mathbf{\Omega} = \mathbf{\Pi} = \mathbf{T}\mathbf{\Lambda}^{-2}. \tag{3.242}$$

Remark 3⋆8 Eqs.3.236, 3.242 present a special case of the definition of principal components. The general definition contains two essential specifications,
(a) $\mathbf{\Omega}'\mathbf{\Omega}$=diagonal,
(b) $\mathbf{\Omega}'\mathbf{\Sigma}\mathbf{\Omega}$=diagonal,
and the additional specifications
(c) a restriction to obtain an unambiguous ordering of the eigenvalues, the columns of $\mathbf{\Omega}$ and the components $\boldsymbol{\eta}_j$: It is common practice that the eigenvalues λ_j^2 are ordered in decreasing order of magnitude. (Here we do not cover the special cases that some eigenvalues may be equal to each other or equal to zero.)
(d) a restriction to obtain unambiguous scale units of $\boldsymbol{\eta}$, e.g. $\mathbf{\Omega}'\mathbf{\Omega} = \mathbf{I}$ or $\mathbf{\Omega}'\mathbf{\Sigma}\mathbf{\Omega} = \mathbf{I}$ (restriction for scale unambiguity (RSU)).
(e) a restriction to obtain unambiguous scale directions of $\boldsymbol{\eta}$, e.g. $\sum_k \omega_{kj} > 0$ or $\sum_k \omega_{kj}^3 > 0$

Conditional expectation. The notion of conditional expectation can be applied to equations Eq.3.237 and Eq.3.239:

$$\mathrm{E}(\boldsymbol{\eta}|\boldsymbol{y}) = \boldsymbol{\Omega}'\boldsymbol{y} \tag{3.243}$$

$$\mathrm{E}(\boldsymbol{y}|\boldsymbol{\eta}) = \boldsymbol{\Omega}\boldsymbol{\eta} \equiv \boldsymbol{\Pi}\boldsymbol{\eta}. \tag{3.244}$$

The major PCs. Let the eigenvalues be partitioned into two sets, where the first sets contains the ones large in value, and the other the small eigenvalues. One may speak of the major and the minor eigenvalues, $\boldsymbol{\Lambda}^2 = [\boldsymbol{\Lambda}^2_{\mathrm{major}} \oplus \boldsymbol{\Lambda}^2_{\mathrm{minor}}]$. The principal components are split up accordingly into major and minor components, $\boldsymbol{\eta} = [\boldsymbol{\eta}'_{\mathrm{major}}, \boldsymbol{\eta}'_{\mathrm{minor}}]'$, and the weight matrix is also split up accordingly into the major and minor vectors, $\boldsymbol{\Omega} = [\boldsymbol{\Omega}_{\mathrm{major}}, \boldsymbol{\Omega}_{\mathrm{minor}}]$. With this partition, Eq.3.237 is rewritten,

$$\begin{bmatrix} \boldsymbol{\eta}_{\mathrm{major}} \\ \boldsymbol{\eta}_{\mathrm{minor}} \end{bmatrix} = \begin{bmatrix} \boldsymbol{\Omega}_{\mathrm{major}} & \boldsymbol{\Omega}_{\mathrm{minor}} \end{bmatrix}' \boldsymbol{y}, \tag{3.245}$$

or

$$\begin{aligned} \boldsymbol{\eta}_{\mathrm{major}} &= \boldsymbol{\Omega}'_{\mathrm{major}}\boldsymbol{y} \\ \boldsymbol{\eta}_{\mathrm{minor}} &= \boldsymbol{\Omega}'_{\mathrm{minor}}\boldsymbol{y}, \end{aligned}$$

and Eq.3.239 is rewritten:

$$\begin{aligned} \boldsymbol{y} &= \begin{bmatrix} \boldsymbol{\Omega}_{\mathrm{major}} & \boldsymbol{\Omega}_{\mathrm{minor}} \end{bmatrix} \begin{bmatrix} \boldsymbol{\eta}_{\mathrm{major}} \\ \boldsymbol{\eta}_{\mathrm{minor}} \end{bmatrix} \\ &= \boldsymbol{\Omega}_{\mathrm{major}}\boldsymbol{\eta}_{\mathrm{major}} + \boldsymbol{\Omega}_{\mathrm{minor}}\boldsymbol{\eta}_{\mathrm{minor}} \\ &\equiv \tilde{\boldsymbol{y}} + e. \end{aligned} \tag{3.246}$$

The conditional expectation Eq.3.243 is not touched by the partition of the principal components in Eq.3.245. In Eq.3.244, however, the condition *can* be altered: Instead of taking all dimensions of $\boldsymbol{\eta}$ as condition, only the major components can be taken.

$$\mathrm{E}(\boldsymbol{y}|\boldsymbol{\eta}_{\mathrm{major}}) = \boldsymbol{\Omega}_{\mathrm{major}}\boldsymbol{\eta}_{\mathrm{major}}. \tag{3.247}$$

That the coefficients $\boldsymbol{\Omega}_{\mathrm{major}}$ in Eq.3.247 are identical to the corresponding columns of the coefficient matrix $\boldsymbol{\Omega}$ in Eq.3.244 is not trivial. (For example, the omission of some predictors in the passage from Eq.3.123 to Eq.3.124 leads to different coefficients.) That the parameters do not change in the passage from the full model Eqs.3.239, 3.244 to the restricted model Eqs.3.246, 3.247 is caused by the orthogonality of the used predictors $\boldsymbol{\eta}_{\mathrm{major}}$ and the omitted predictors $\boldsymbol{\eta}_{\mathrm{minor}}$.

To sum up, even if only one, or a few, of the principal components $\boldsymbol{\eta}$ are taken as predictors for the original variables $\boldsymbol{y}$, it remains that weight and loading matrix are identical to each other and proportional to the loading structure matrix. (If the condition $\boldsymbol{\Omega}'\boldsymbol{\Omega} = \mathbf{I}$ is changed into $\boldsymbol{\Omega}'\boldsymbol{\Sigma}\boldsymbol{\Omega} = \mathbf{I}$, then $\boldsymbol{\Pi} = \mathbf{T} = \boldsymbol{\Omega}\boldsymbol{\Lambda}^2$.) The proportionality of weights and loadings also characterizes uniquely the principal components. The results are assembled in the following theorem.

Theorem: PCs and conditional expectation. If the vector ω satisfies the two conditions

$$\mathrm{E}(\boldsymbol{y}|\omega'\boldsymbol{y}) = \omega\omega'\boldsymbol{y}\ , \qquad \omega'\omega = 1, \tag{3.248}$$

then:
(i) is ω an eigenvector of $\boldsymbol{\Sigma} = \mathrm{E}(\boldsymbol{y}\boldsymbol{y}')$,
(ii) is $\lambda^2 = \omega'\boldsymbol{\Sigma}\omega$ the corresponding eigenvalue, and
(iii) is $\eta = \omega'\boldsymbol{y}$ a principal component of $\boldsymbol{y}$.

Properties of PCs. The first principal component has several optimum properties that make it suitable to be used as a predictor.

$$\mathrm{E}(\boldsymbol{e}) = \mathbf{0}, \qquad \mathrm{E}(\eta\boldsymbol{e}') = \mathbf{0} \tag{3.249}$$

$$\lambda^2 = \mathrm{var}(\eta) = \mathrm{var}(\omega'\boldsymbol{y}) = \max_{\forall\omega\in R^K} \tag{3.250}$$

$$\sum_k \mathrm{var}(\tilde{y}_k) = \mathrm{trace}(\tilde{\boldsymbol{\Sigma}}) = \omega'\boldsymbol{\Sigma}\omega = \lambda^2 = \max_{\omega} \tag{3.251}$$

$$\sum_k \mathrm{var}(\epsilon_k) = \mathrm{trace}(\boldsymbol{\Theta}) = \mathrm{trace}(\boldsymbol{\Sigma} - \tilde{\boldsymbol{\Sigma}}) = \mathrm{trace}(\boldsymbol{\Sigma}) - \lambda^2 = \min_{\omega}. \tag{3.252}$$

Definition: Predictor specification for PCs. If the conditional expectation Eq.3.248 is specified to be a predictor, this has the formal implications
(i) that ω is the eigenvector of $\boldsymbol{\Sigma}$,
(ii) that ω is the eigenvector corresponding to the largest eigenvalue λ^2_{max},
and the operational consequences
(iii) that $\tilde{y} = \boldsymbol{\pi}\eta = \boldsymbol{\pi}\omega'\boldsymbol{y}$ is a least squares approximation to the data $\boldsymbol{y}$.

3.2.6 The Principal Components Model

Principal Axes analysis, the name as well as the method, dates back to the time when the main problems in mathematics were posed by astronomy and the mechanics of the heavenly bodies. The original problem was to determine the whole elliptical orbit of heavenly bodies while still only some points on the orbit were measured. The ellipsis is known when the origin, the directions (the eigenvectors) and the extensions (the eigenvalues) of the largest, second largest and smallest axis of the three-dimensional ellipsoid are known. The representation of a point in terms of the principal axes is called the canonical (orderly, proper, *eigentümliche*) representation. The problem of describing the factual orbit when the "secular" orbit is distorted by the gravitation of other heavenly bodies (*die Theorie der Säkularstörungen*) becomes more than three-dimensional. The first iterative method for solving these equations was developed by C. G. J. Jacobi in 1846.

Principal Components, or more generally, eigenvalue/eigenvector representations of matrices possess an abundance of mathematical properties which can be detected and proved in a lot of different ways. So it is no wonder that the PC method was discovered so early and rediscovered so often. Many of these properties can be used to numerically determine the PCs. For example, the power method was discovered independently by von Mises and Pollaczek-Geiringer (1929), Hotelling (1933) and also Wold (1966a, 1966b).

The PC model can be exposed and developed in very different ways. In the following we shall use a five-step procedure based on predictor specification.

Step 1, variables and relations. Let be give a set of MVs $\boldsymbol{y}_{[K\times 1]} \sim \mathrm{M}(0, \boldsymbol{\Sigma}, \ldots)$ and let the normalized weight vector $\boldsymbol{\omega}$ define a linear composite η with zero mean and variance ρ,

$$\eta = \boldsymbol{\omega}'\boldsymbol{y}, \qquad \boldsymbol{\omega}'\boldsymbol{\omega} = 1, \qquad \mathrm{cov}(\eta) = \boldsymbol{\omega}'\boldsymbol{\Sigma}\boldsymbol{\omega} = \rho. \tag{3.253}$$

From Eq.3.253 follows immediately that

$$\boldsymbol{\tau} \equiv \mathrm{cov}(\boldsymbol{y}; \eta) = \mathrm{cov}(\boldsymbol{y}; \boldsymbol{\omega}'\boldsymbol{y}) = \mathrm{E}(\boldsymbol{y}\boldsymbol{y}'\boldsymbol{\omega}) = \boldsymbol{\Sigma}\boldsymbol{\omega}. \tag{3.254}$$

The model is defined by the equation

$$\boldsymbol{y} = \boldsymbol{\pi}\boldsymbol{\eta} + \boldsymbol{\epsilon}, \tag{3.255}$$

where $\boldsymbol{\pi}$ is the loadings pattern vector. Both $\boldsymbol{\omega}$ and $\boldsymbol{\pi}$ are unknown but required to be proportional to each other:

$$\boldsymbol{\omega} \propto \boldsymbol{\pi}. \tag{3.256}$$

That is, each indicator y_k enters the linear aggregate η by a weight ω_k which is proportional to the "reliability" π_k.

Step 2, predictor specification. The conditional expectation of the observed variables $\boldsymbol{y}$ is specified to be

$$\begin{aligned} \boldsymbol{y} = \boldsymbol{\pi}\eta + \boldsymbol{\epsilon}, \qquad & \tilde{\boldsymbol{y}} \equiv \mathrm{E}(\boldsymbol{y}|\eta) = \boldsymbol{\pi}\eta \\ & \Longrightarrow \mathrm{E}(\boldsymbol{\epsilon}) = \mathbf{0} \\ & \Longrightarrow \mathrm{cov}(\boldsymbol{\epsilon}; \eta) = \mathrm{cov}(\boldsymbol{\epsilon}; \tilde{\boldsymbol{y}}) = \mathbf{0} \\ & \Longrightarrow \mathrm{cov}(\boldsymbol{y}; \eta) = \mathrm{cov}(\tilde{\boldsymbol{y}}; \eta) = \boldsymbol{\pi}\,\mathrm{var}(\eta). \end{aligned} \tag{3.257}$$

which implies that the residuals $\boldsymbol{\epsilon}$ are uncorrrelated with the predictor η and with the prediction $\tilde{\boldsymbol{y}}$. That is to say that the linear composite η is the best predictor for the observed variables. The predictor specification determines not only the loadings $\boldsymbol{\pi}$ such that the residual variables $\boldsymbol{\epsilon}$ are uncorrelated with the components η and with the systematic part $\tilde{\boldsymbol{y}}$ of the observed variables $\boldsymbol{y}$, and such that the variances of the residual variables are minimal; it determines also the weights $\boldsymbol{\omega}$, so that no linear combination other than $\boldsymbol{\omega}'\boldsymbol{y}$ gives a better fit to the observed variables $\boldsymbol{y}$.

Step 3, covariance deduction. The covariances between predictors $\boldsymbol{y}$ and predictand η are deduced from Eqs.3.253, 3.255, whereby the zero correlation Eq.3.257 implied by predictor specification is used:

$$\boldsymbol{\tau} \equiv \mathrm{cov}(\boldsymbol{y}; \eta) = \boldsymbol{\pi}\boldsymbol{\rho}. \tag{3.258}$$

Step 4, stationary relation. We recall Eq.3.223 which holds for any component model,

$$\boldsymbol{\omega}'\boldsymbol{\pi} = 1. \tag{3.259}$$

The two expressions for the predictor-predictand covariance $\text{cov}(\boldsymbol{y};\eta)$, the inductive expression Eq.3.254 and the deductive expression Eq.3.258 are equated to get the relation:

$$\boldsymbol{\pi}\rho = \boldsymbol{\Sigma}\boldsymbol{\omega} \tag{3.260}$$

The only weight vector $\boldsymbol{\omega}$ which fulfills the three conditions $\boldsymbol{\omega}'\boldsymbol{\omega} = 1$ (Eq.3.253), $\boldsymbol{\omega}'\boldsymbol{\pi} = 1$ (Eq.3.259) and $\boldsymbol{\Sigma}\boldsymbol{\omega} \propto \boldsymbol{\pi}$ (Eq.3.260) is the loading vector $\boldsymbol{\pi}$ itself:

$$\boldsymbol{\omega} = \boldsymbol{\pi} \tag{3.261}$$

If $\boldsymbol{\pi}$ in Eq.3.260 is replaced from Eq.3.261,

$$\boldsymbol{\Sigma}\boldsymbol{\omega} = \rho\boldsymbol{\omega}, \tag{3.262}$$

one gets a relation called the "stationary relation" (normal equation system), because the vector $\boldsymbol{\omega}$ has to be determined so as to balance and to hold in an equilibrium both sides of the equation. The stationary relation has exactly the form of a symmetric eigenvalues problem (Eq.3.165), with eigenvalue $\lambda^2 = \rho$ and eigenvector $\boldsymbol{\omega}$.

Note on Eq.3.259: The vector $\boldsymbol{\omega}$ is not the only one to satisfy the relation $\boldsymbol{\omega}'\boldsymbol{\pi} = 1$. A vector $\boldsymbol{\omega}$ with one element equal to the reciprocal of the corresponding element of $\boldsymbol{\pi}$, $\omega_k = 1/p_k$, and all elements of $\boldsymbol{\omega}$ equal to zero, would satisfy $\boldsymbol{\omega}'\boldsymbol{\pi} = 1$ as well. However, no other vector than the left inverse of $\boldsymbol{\pi}$ can be found that simultaneously satisfies both conditions $\boldsymbol{\omega}'\boldsymbol{\pi} = 1$, and $\boldsymbol{\pi} \propto \boldsymbol{\Sigma}\boldsymbol{\omega}$.

Step 5a, PLS algorithm for raw data. The PLS algorithm (Table 3.3) proceeds in terms of the observed values y_{kn} and a standardized linear aggregate $\hat{\eta}_n$. It starts with arbitrary starting values for the weights ${}_0w_k$ and arrives at the final weight w_k.

Step 5b, PLS algorithm for covariance matrix. The raw data are not required for the iterative estimation of the weights. The computation of the linear aggregates are totally omitted from the algorithm in Table 3.4 which uses the covariance matrix (or correlation matrix) as input. If Steps a and c are combined into one equation,

$$\lambda^2\boldsymbol{w} := \boldsymbol{S}\boldsymbol{w}, \tag{3.263}$$

the algorithm emerges to be identical to the power algorithm, Eq.3.168. Hence the PLS algorithm will converge under the same conditions as the power algorithm.

Remark 3⋆9 The algorithm can be extended to apply to several dimensions of the principal component simultaneously. Given an arbitrary matrix of starting values for the weights $\boldsymbol{W}$, repeat the two steps

$$\boldsymbol{P} := \boldsymbol{S}\boldsymbol{W} \tag{3.264}$$

Table 3.3: Principal Component model, PLS algorithm for raw data

$\boldsymbol{y}_{[K\times N]} = [\boldsymbol{y}_{[K\times 1]n}] = [y_{kn}]$	Rectangular data matrix for $k = 1\dots K$ variables, $n = 1\dots N$ observations, with zero mean and covariance $\boldsymbol{S} = \frac{1}{N}\boldsymbol{y}\boldsymbol{y}'$.
${}_0\boldsymbol{w}$	Starting values are (nearly) arbitrary.
Repeat Steps a...d, beginning at Step d, for cycles $c = 0, 1, 2, \dots$	
a	Compute the MV-LV covariances, which are proportional to the loadings and to the weights
$\breve{w}_k := \bar{\sum}_n y_{kn}\hat{\eta}_n$ $\breve{\boldsymbol{w}} \equiv \boldsymbol{p} := \mathrm{cov}(\boldsymbol{y}; \hat{\eta})$ $= \mathrm{cov}(\boldsymbol{y}; \boldsymbol{w}'\boldsymbol{y}) = \boldsymbol{S}\boldsymbol{w}$	
b	Use the covariances as provisional weights and make provisional linear aggregate $\breve{\eta}$
$\breve{\eta}_n := \sum_k \breve{w}_k y_{kn}$ $\lambda^4 \equiv \mathrm{var}(\breve{\eta}) := \bar{\sum}_n \breve{\eta}_n^2$ $= \breve{\boldsymbol{w}}'\boldsymbol{S}\breve{\boldsymbol{w}}$	
c	Divide provisional weights by standard deviation of the provisional aggregate, to get correctly scaled weights.
$w_k := \breve{w}_k/\lambda^2$ $\boldsymbol{w} := \breve{\boldsymbol{w}}/\lambda^2$	
d	Make standardized linear aggregate.
$\hat{\eta}_n := \sum_k w_k y_{kn} = \boldsymbol{w}'\boldsymbol{y}_n$ $\hat{\eta} := \boldsymbol{w}'\boldsymbol{y}$	now $\mathrm{var}(\hat{\eta}) = \boldsymbol{w}'\boldsymbol{S}\boldsymbol{w} = 1$
until convergence occurs.	Note: Iteration index on $\boldsymbol{w}, \boldsymbol{p}, \hat{\eta}, \breve{\eta}, \breve{\boldsymbol{w}}, \lambda$ is omitted.

$$\boldsymbol{W} := \boldsymbol{P}(\boldsymbol{P}'\boldsymbol{P})^{-1}. \tag{3.265}$$

until convergence is observed. Then $\boldsymbol{\Lambda}^2 = \boldsymbol{P}'\boldsymbol{P}$ will be the diagonal matrix with the largest eigenvalues. If, however, all columns of the starting matrix $\boldsymbol{W}$ are orthogonal to an eigenvector that corresponds to one of the largest eigenvalues, this eigenvalue/eigenvector will not be contained in the solution.

Remark 3⋆10 Two algorithms similar to the one proposed in Eqs.3.264, 3.265 have been proposed by Horst (1965:ch.8). His rank reduction method makes use of the uncorrelatedness of PCs, so, that in each iteration step he transforms the weights so as to give uncorrelated PCs. Letting $\boldsymbol{Q}$ denote the triangular (Cholesky) factor of $\boldsymbol{R} = \boldsymbol{W}'\boldsymbol{S}\boldsymbol{W}$, the rank reduction algorithm iterates these

Table 3.4: Principal Component model, PLS algorithm for covariance data

	$\boldsymbol{S} = \frac{1}{N}\boldsymbol{y}\boldsymbol{y}'$	Covariance matrix
	${}_0\boldsymbol{w}$	Starting values are (nearly) arbitrary.
Repeat Steps a...c for cycles $c = 0, 1, 2, \ldots$		
a	$\breve{\boldsymbol{w}} := \boldsymbol{S}\boldsymbol{w}$	
b	$\lambda^4 := \breve{\boldsymbol{w}}'\boldsymbol{S}\breve{\boldsymbol{w}}$	
c	$\boldsymbol{w} := \breve{\boldsymbol{w}}/\lambda^2$	
until convergence occurs.		
		Note : Iteration index on $\boldsymbol{w}, \breve{\boldsymbol{w}}, \lambda$ is omitted

two steps:

$$\boldsymbol{Q}\boldsymbol{Q}' := \boldsymbol{R} := \boldsymbol{W}'\boldsymbol{S}\boldsymbol{W} \tag{3.266}$$

$$\boldsymbol{W} := \boldsymbol{S}\boldsymbol{W}\boldsymbol{Q}^{-1}. \tag{3.267}$$

Remark 3⋆11 Horst's (1965) orthogonalization algorithm makes use of the orthogonality of eigenvectors, PC loading matrices and PC weight matrices; so in each iteration step (instead of Eq.3.266) the loading matrix $\boldsymbol{P}$ is purified by a Gram-Schmidt process to give an upper triangle matrix $\boldsymbol{U}$ and an orthonormal matrix $\boldsymbol{W}$ which then is used as a new approximation to the PC weight matrix.

$$\boldsymbol{P} := \boldsymbol{S}\boldsymbol{W} \tag{3.268}$$

$$\boldsymbol{W}\boldsymbol{U} := \boldsymbol{P} \tag{3.269}$$

Horst has proved the convergence of both methods. It would not be too difficult to modify his proofs to apply to the algorithm Eqs.3.264, 3.265. For the one-dimensional case, cf. Lyttkens, Areskoug and Wold (1975), Lyttkens (1966, 1973, 1982).

Remark 3⋆12 Notice that the presentation of the PC model defines no estimation criterion like "maximum variance" and that the derivation of the stationary condition (to be solved for the parameters) includes no differentiation of the criterion with respect to the parameters. This is due to the predictor specification: The optimum property follows as a corollary from the conditional expectation, while the stationary conditions are found by using the uncorrelatedness of predictors and residuals.

Table 3.5: Factor score estimation

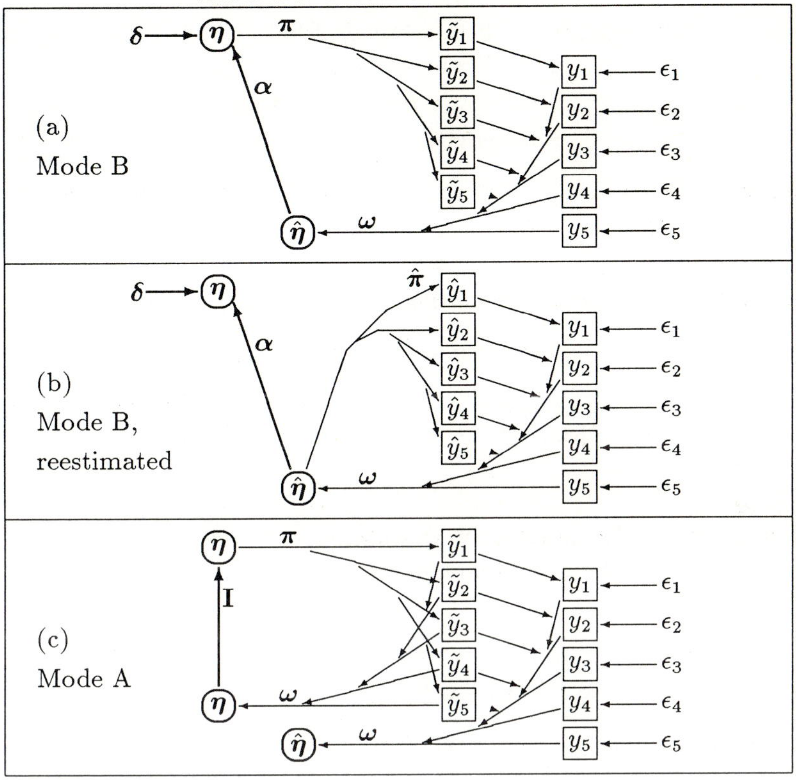

3.3 Factor Score Estimation

The description of the PLS algorithm in Sc.2 introduced two types of approximations to the LV, the outside approximation $\hat{\boldsymbol{\eta}}$ and the inside approximation $\tilde{\boldsymbol{\eta}}$. The outside approximation is a linear function of the MVs of the same block, $\hat{\boldsymbol{\eta}} = \boldsymbol{w}'\boldsymbol{y}$. The inside approximation is a linear function of the adjacent LVs (and thence of their MVs), but it is not a linear function of the block $\boldsymbol{y}$ under consideration. Thus it can be treated like a factor, like a dimension beyond the space of $\boldsymbol{y}$. In the course of the PLS iterations, the inside approximation is replaced by the outside approximation, the factor by a component. The concept of latent variable, as used in Sc.3.3, implies that the LV is not a linear function of the MVs. The three approaches of factor score estimation to be discussed in Sc.3.3 are sketched out in Table 3.5.

3.3.1 Properties of Factor Scores

Factor model. Let be given the factor model, as specified in Sc.3.2.3 and repeated here, with factor variables $\boldsymbol{\eta}_{[J\times 1]} \sim (\mathbf{0}, \boldsymbol{\rho}, \ldots)$, residual variables $\boldsymbol{\epsilon}_{[K\times 1]} \sim$

$(\mathbf{0}, \mathbf{\Theta}, \ldots)$ and a loading matrix $\boldsymbol{\pi}_{[K\times J]}$:

$$\begin{aligned} \boldsymbol{y} = \boldsymbol{\pi}\boldsymbol{\eta} + \boldsymbol{\epsilon}, \qquad & \tilde{\boldsymbol{y}} \equiv \mathrm{E}(\boldsymbol{y}|\boldsymbol{\eta}) = \boldsymbol{\pi}\boldsymbol{\eta} \\ & \Longrightarrow \mathrm{E}(\boldsymbol{\epsilon}) = \mathbf{0} \\ & \Longrightarrow \mathrm{cov}(\boldsymbol{\epsilon}; \boldsymbol{\eta}) = \mathrm{cov}(\boldsymbol{\epsilon}; \tilde{\boldsymbol{y}}) = \mathbf{0} \\ & \Longrightarrow \mathrm{cov}(\boldsymbol{y}; \boldsymbol{\eta}) = \mathrm{cov}(\tilde{\boldsymbol{y}}; \boldsymbol{\eta}) = \boldsymbol{\pi}\,\mathrm{var}(\boldsymbol{\eta}). \end{aligned} \tag{3.301}$$

We introduce a matrix $\boldsymbol{\omega}_{[K\times J]}$ of unknown weight coefficients, which form the linear aggregates $\hat{\boldsymbol{\eta}}_{[J\times 1]}$:

$$\hat{\boldsymbol{\eta}} = \boldsymbol{\omega}'\boldsymbol{y}. \tag{3.302}$$

If $\boldsymbol{y}$ in Eq.3.302 is substituted from the factor relation Eq.3.301, one gets:

$$\hat{\boldsymbol{\eta}} = \boldsymbol{\omega}'(\boldsymbol{\pi}\boldsymbol{\eta} + \boldsymbol{\epsilon}) = \boldsymbol{\omega}'\boldsymbol{\pi}\boldsymbol{\eta} + \boldsymbol{\omega}'\boldsymbol{\epsilon}. \tag{3.303}$$

Because $\hat{\boldsymbol{\eta}}$ is intended to replace $\boldsymbol{\eta}$, its relations to $\boldsymbol{\eta}$, $\boldsymbol{y}$ and $\boldsymbol{\epsilon}$ will be explored, in order to make an appropriate choice of postulates which make $\hat{\boldsymbol{\eta}}$ a predictor of $\boldsymbol{\eta}$ or $\boldsymbol{y}$.

Unbiasedness. A factor estimate $\hat{\boldsymbol{\eta}}$ is called unbiased if the expectation of $\hat{\boldsymbol{\eta}}$ conditional on $\boldsymbol{\eta}$ is identical to the expectation of $\boldsymbol{\eta}$. This conditional expectation is

$$\begin{aligned} \mathrm{E}(\hat{\boldsymbol{\eta}}|\boldsymbol{\eta}) &= \mathrm{E}(\boldsymbol{\omega}'\boldsymbol{\pi}\boldsymbol{\eta} + \boldsymbol{\omega}'\boldsymbol{\epsilon}|\boldsymbol{\eta}) = \boldsymbol{\omega}'\boldsymbol{\pi}\,\mathrm{E}(\boldsymbol{\eta}|\boldsymbol{\eta}) + \boldsymbol{\omega}'\,\mathrm{E}(\boldsymbol{\epsilon}|\boldsymbol{\eta}) \\ &= \boldsymbol{\omega}'\boldsymbol{\pi}\,\mathrm{E}(\boldsymbol{\eta}) + \boldsymbol{\omega}'\mathbf{0}, \end{aligned} \tag{3.304}$$

by the use of Eq.3.303 the fact that $\mathrm{E}(\boldsymbol{\epsilon}) = \mathrm{E}(\boldsymbol{\epsilon}|\boldsymbol{\eta}) = \mathbf{0}$, which is implied by Eq.3.301 and is discussed in Sc.3.2.3. For $\hat{\boldsymbol{\eta}}$ to be an unbiased estimator of $\boldsymbol{\eta}$ we must postulate that the inner product of weight and loading matrix a unit matrix:

$$\begin{aligned} \hat{\boldsymbol{\eta}} \text{ is unbiased} \quad &\equiv \quad \mathrm{E}(\hat{\boldsymbol{\eta}}|\boldsymbol{\eta}) = \mathrm{E}(\boldsymbol{\eta}) \\ &\Leftarrow \quad \boldsymbol{\omega}'\boldsymbol{\pi} = \mathbf{I} \end{aligned} \tag{3.305}$$

As long as $\boldsymbol{\pi}$ has more rows than columns, there is an infinite number of matrices $\boldsymbol{\omega}$ which fulfil $\boldsymbol{\omega}'\boldsymbol{\pi} = \mathbf{I}$. Possible choices are:

$$\begin{aligned} \boldsymbol{\omega} &= \boldsymbol{\pi}(\boldsymbol{\pi}'\boldsymbol{\pi})^{-1} \\ \boldsymbol{\omega} &= \mathbf{\Theta}^{-1}\boldsymbol{\pi}(\boldsymbol{\pi}'\mathbf{\Theta}^{-1}\boldsymbol{\pi})^{-1} \\ \boldsymbol{\omega} &= \mathbf{\Sigma}^{-1}\boldsymbol{\pi}(\boldsymbol{\pi}'\mathbf{\Sigma}^{-1}\boldsymbol{\pi})^{-1} \end{aligned}$$

Univocality. A multidimensional factor estimate $\hat{\boldsymbol{\eta}}$ is called univocal if each dimension η_j is regressed only its own estimated counterpart $\hat{\eta}_j$ and all other regressions are zero.

$$\begin{aligned} \mathrm{E}(\boldsymbol{\eta}|\hat{\boldsymbol{\eta}}) &= \boldsymbol{\alpha}\hat{\boldsymbol{\eta}} \\ \boldsymbol{\alpha} &= \mathrm{E}(\boldsymbol{\eta}\hat{\boldsymbol{\eta}}')\,\mathrm{E}^{-1}(\hat{\boldsymbol{\eta}}\hat{\boldsymbol{\eta}}') = \mathrm{E}(\boldsymbol{\eta}(\boldsymbol{\omega}'\boldsymbol{\pi}\boldsymbol{\eta} + \boldsymbol{\omega}'\boldsymbol{\epsilon})')\hat{\boldsymbol{\rho}}^{-1} \\ &= [\mathrm{E}(\boldsymbol{\eta}\boldsymbol{\eta}')\boldsymbol{\pi}'\boldsymbol{\omega}) + \mathrm{E}(\boldsymbol{\eta}\boldsymbol{\epsilon})\boldsymbol{\omega}]\hat{\boldsymbol{\rho}}^{-1} \\ &= \boldsymbol{\rho}\boldsymbol{\pi}'\boldsymbol{\omega}\hat{\boldsymbol{\rho}}^{-1} = \boldsymbol{\omega}'\boldsymbol{\pi}\boldsymbol{\rho}\hat{\boldsymbol{\rho}}^{-1}, \end{aligned}$$

We use the fact that $\mathrm{E}(\boldsymbol{\epsilon}\boldsymbol{\eta}') = \mathbf{0}$, which is implied by Eq.3.301 and discussed in Sc.3.2.3. The general definition of univocality is now:

$$\begin{aligned} \hat{\boldsymbol{\eta}} \text{ is univocal} \quad &\equiv \quad \mathrm{E}(\boldsymbol{\eta}|\hat{\boldsymbol{\eta}}) = \text{diagonal} \\ &\Leftarrow \quad \boldsymbol{\omega}'\boldsymbol{\pi}\boldsymbol{\rho}\hat{\boldsymbol{\rho}}^{-1} = \text{diagonal} \end{aligned} \tag{3.306}$$

For the more common special case that $\boldsymbol{\omega}'\boldsymbol{\pi}$ and $\boldsymbol{\rho}$ are diagonal, one only has to require that $\hat{\boldsymbol{\rho}} = \text{diagonal}$.

Uncorrelatedness. If the latent variables $\boldsymbol{\eta}$ are uncorrelated, one would like to have also uncorrelated estimates $\hat{\boldsymbol{\eta}}$. We postmultiply Eq.3.302 by $\hat{\boldsymbol{\eta}}'$ and take the expectation of both sides:

$$\begin{aligned} \hat{\boldsymbol{\rho}} \equiv \mathrm{E}(\hat{\boldsymbol{\eta}}\hat{\boldsymbol{\eta}}') = \mathrm{E}(\boldsymbol{\omega}'\boldsymbol{y}\boldsymbol{y}'\boldsymbol{\omega}) &= \boldsymbol{\omega}'(\boldsymbol{\pi}\boldsymbol{\rho}\boldsymbol{\pi}' + \boldsymbol{\Theta})\boldsymbol{\omega} \\ = \boldsymbol{\omega}'\boldsymbol{\pi}\boldsymbol{\rho}\boldsymbol{\pi}'\boldsymbol{\omega} + \boldsymbol{\omega}'\boldsymbol{\Theta}\boldsymbol{\omega} \end{aligned} \tag{3.307}$$

To get uncorrelatedness one firstly has to postulate univocality, and secondly that $\boldsymbol{\omega}'\boldsymbol{\Theta}\boldsymbol{\omega}$ is diagonal (or even zero).

$$\begin{aligned} \hat{\boldsymbol{\eta}} \text{ is uncorrelated} \quad &\equiv \quad \mathrm{E}(\hat{\boldsymbol{\eta}}\hat{\boldsymbol{\eta}}') = \text{diagonal} \\ &\Leftarrow \quad \left\{ \begin{array}{r} \boldsymbol{\rho} = \text{diagonal} \\ \boldsymbol{\omega}'\boldsymbol{\pi} = \text{diagonal} \\ \boldsymbol{\omega}'\boldsymbol{\Theta}\boldsymbol{\omega} = \text{diagonal} \end{array} \right\} \end{aligned} \tag{3.308}$$

Unbiasedness of structure $\boldsymbol{\tau}$. We multiply Eq.3.302 by $\boldsymbol{y}'$ and take the expectation of both sides:

$$\begin{aligned} \hat{\boldsymbol{\tau}} \equiv \mathrm{cov}(\boldsymbol{y}\hat{\boldsymbol{\eta}}') = \mathrm{E}(\boldsymbol{y}\boldsymbol{y}'\boldsymbol{\omega}) &= (\boldsymbol{\pi}\boldsymbol{\rho}\boldsymbol{\pi}' + \boldsymbol{\Theta})\boldsymbol{\omega} = \boldsymbol{\pi}\boldsymbol{\rho}\boldsymbol{\pi}'\boldsymbol{\omega} + \boldsymbol{\Theta}\boldsymbol{\omega} \\ = \boldsymbol{\tau}\boldsymbol{\pi}'\boldsymbol{\omega} + \boldsymbol{\Theta}\boldsymbol{\omega} \end{aligned} \tag{3.309}$$

If one wants the covariance of MVs with the linear aggregates to be identical to the MV-LV covariance (the LV loading structure matrix $\boldsymbol{\tau}$), one has to postulate that $\boldsymbol{\pi}'\boldsymbol{\omega}$ be an identity matrix (cf. Eq.3.305) and that $\boldsymbol{\Theta}\boldsymbol{\omega}$ be a zero matrix — or that the effects of the two summands in Eq.3.309 cancel out each other.

$$\begin{aligned} \hat{\boldsymbol{\tau}} \text{ is unbiased} \quad &\equiv \quad \mathrm{E}(\boldsymbol{y}\boldsymbol{\eta}') = \mathrm{E}(\boldsymbol{y}\hat{\boldsymbol{\eta}}') \\ &\Leftarrow \quad \left\{ \begin{array}{r} \boldsymbol{\omega}'\boldsymbol{\pi} = \mathbf{I} \\ \boldsymbol{\Theta}\boldsymbol{\omega} = \mathbf{0} \end{array} \right\} \end{aligned} \tag{3.310}$$

Re-estimated solutions. The second summand in Eq.3.309 can be null only if the residuals $\boldsymbol{\epsilon}$ and their covariance matrix $\boldsymbol{\Theta}$ do not have full rank. This condition typically holds for component residuals, not for factor residuals. Component residuals (or block residuals, reestimated residuals) are based on a re-specified version of the model Eq.3.301, where the "true" latent variable $\boldsymbol{\eta}$ is replaced by the linear aggregate $\hat{\boldsymbol{\eta}}$:

$$\begin{aligned} \boldsymbol{y} = \hat{\boldsymbol{\pi}}\hat{\boldsymbol{\eta}} + \hat{\boldsymbol{\epsilon}}, \qquad & \hat{\boldsymbol{y}} \equiv \mathrm{E}(\boldsymbol{y}|\hat{\boldsymbol{\eta}}) = \hat{\boldsymbol{\pi}}\hat{\boldsymbol{\eta}} \\ & \Longrightarrow \mathrm{E}(\hat{\boldsymbol{\epsilon}}) = \mathbf{0} \\ & \Longrightarrow \mathrm{cov}(\hat{\boldsymbol{\epsilon}}; \hat{\boldsymbol{\eta}}) = \mathrm{cov}(\hat{\boldsymbol{\epsilon}}; \hat{\boldsymbol{y}}) = \mathbf{0} \\ & \Longrightarrow \mathrm{cov}(\boldsymbol{y}; \hat{\boldsymbol{\eta}}) = \mathrm{cov}(\hat{\boldsymbol{y}}; \hat{\boldsymbol{\eta}}) = \hat{\boldsymbol{\pi}}\, \mathrm{var}(\hat{\boldsymbol{\eta}}). \end{aligned} \tag{3.311}$$

This, however, is a component model, as discussed in Sc.3.2.4.

In the following subsections some well-known methods of factor score estimation will discussed. They will be introduced by predictor specification, and their properties with respect to the ones outlined above will be derived. Besides the LV $\boldsymbol{\eta}$ and the estimate $\hat{\boldsymbol{\eta}}$ two more variables are treated explicitly, which are introduced in Eqs.3.201, 3.202 as $\tilde{\boldsymbol{y}}$ and $\tilde{\boldsymbol{\eta}}$, the systematic parts of MVs and LVs, respectively. These systematic parts are conditional expectations with respect to, and linear aggregates of, some other variables. Depending on which variables are taken as conditions, three different models which are depicted in Tb.3.5 will be treated in the next three subsections. The three models include the residual variable $\boldsymbol{\delta}$ which accounts for all variation of the LV that cannot be predicted by a linear combination of the MVs. Hence it is natural to specify a model which implies that the residual $\boldsymbol{\delta}$ as uncorrelated with the observables $\boldsymbol{y}$.

3.3.2 Factor Estimation ModeB

Model. The first model, which displayed in Tb.3.5a, has two relations, which are given here together with the predictor specifications and the directly implied zero correlations of predictors and residuals:

$$\begin{aligned} \boldsymbol{y} = \boldsymbol{\pi}\boldsymbol{\eta} + \boldsymbol{\epsilon}, \qquad & \tilde{\boldsymbol{y}} \equiv \mathrm{E}(\boldsymbol{y}|\boldsymbol{\eta}) = \boldsymbol{\pi}\boldsymbol{\eta} \\ & \Longrightarrow \mathrm{E}(\boldsymbol{\epsilon}) = \mathbf{0} \\ & \Longrightarrow \mathrm{cov}(\boldsymbol{\epsilon}; \boldsymbol{\eta}) = \mathrm{cov}(\boldsymbol{\epsilon}; \tilde{\boldsymbol{y}}) = \mathbf{0} \\ & \Longrightarrow \mathrm{cov}(\boldsymbol{y}; \boldsymbol{\eta}) = \mathrm{cov}(\tilde{\boldsymbol{y}}; \boldsymbol{\eta}) = \boldsymbol{\pi}\, \mathrm{var}(\boldsymbol{\eta}) \end{aligned} \tag{3.312}$$

$$\begin{aligned} \boldsymbol{\eta} = \boldsymbol{\omega}\boldsymbol{y} + \boldsymbol{\delta}, \qquad & \hat{\boldsymbol{\eta}} \equiv \mathrm{E}(\boldsymbol{\eta}|\boldsymbol{y}) = \boldsymbol{\omega}\boldsymbol{y} \\ & \Longrightarrow \mathrm{E}(\boldsymbol{\delta}) = \mathbf{0} \\ & \Longrightarrow \mathrm{cov}(\boldsymbol{\delta}; \boldsymbol{y}) = \mathrm{cov}(\boldsymbol{\delta}; \hat{\boldsymbol{\eta}}) = \mathbf{0} \\ & \Longrightarrow \mathrm{cov}(\boldsymbol{\eta}; \boldsymbol{y}) = \mathrm{cov}(\hat{\boldsymbol{\eta}}; \boldsymbol{y}) = \boldsymbol{\omega}\, \mathrm{var}(\boldsymbol{y}). \end{aligned} \tag{3.313}$$

Implications. This model implies all of the properties Eqs.3.216,..., 3.218 which follow directly from Eq.3.215. The variances and covariances of all variables, collected into one matrix, are:

$$\mathrm{var}\begin{bmatrix} \boldsymbol{\delta} \\ \boldsymbol{\epsilon} \\ \boldsymbol{\eta} \\ \hat{\boldsymbol{\eta}} \\ \tilde{\boldsymbol{y}} \\ \boldsymbol{y} \end{bmatrix} = \begin{bmatrix} \boldsymbol{\varphi} & . & . & . & . & \mathrm{sym.} \\ -\boldsymbol{\pi}\boldsymbol{\varphi} & \boldsymbol{\Theta} & . & . & . & . \\ \boldsymbol{\varphi} & \mathbf{0} & \boldsymbol{\rho} & . & . & . \\ 0 & \boldsymbol{\varphi}\boldsymbol{\pi}' & \hat{\boldsymbol{\rho}} & \hat{\boldsymbol{\rho}} \equiv \boldsymbol{\rho} - \boldsymbol{\varphi} & . & . \\ \boldsymbol{\pi}\boldsymbol{\varphi} & \mathbf{0} & \boldsymbol{\pi}\boldsymbol{\rho} & \boldsymbol{\pi}\hat{\boldsymbol{\rho}} & \tilde{\boldsymbol{\Sigma}} \equiv \boldsymbol{\pi}\boldsymbol{\rho}\boldsymbol{\pi}' & . \\ 0 & \boldsymbol{\Theta} & \boldsymbol{\pi}\boldsymbol{\rho} & \boldsymbol{\pi}\boldsymbol{\rho} & \tilde{\boldsymbol{\Sigma}} & \tilde{\boldsymbol{\Sigma}} + \boldsymbol{\Theta} \end{bmatrix}, \tag{3.314}$$

which is proved by:

$$\begin{aligned}
\operatorname{cov}(\boldsymbol{\eta};\boldsymbol{\delta}) &= \operatorname{cov}(\hat{\boldsymbol{\eta}}+\boldsymbol{\delta};\boldsymbol{\delta}) = \operatorname{cov}(\hat{\boldsymbol{\eta}};\boldsymbol{\delta}) + \operatorname{cov}(\boldsymbol{\delta};\boldsymbol{\delta}) = 0 + \boldsymbol{\varphi} \\
\operatorname{cov}(\boldsymbol{\epsilon};\boldsymbol{\delta}) &= \operatorname{cov}(\boldsymbol{y}-\boldsymbol{\pi\eta};\boldsymbol{\delta}) = \operatorname{cov}(\boldsymbol{y};\boldsymbol{\delta}) - \operatorname{cov}(\boldsymbol{\pi\eta};\boldsymbol{\delta}) = \mathbf{0} - \boldsymbol{\pi\varphi} \\
\operatorname{cov}(\hat{\boldsymbol{\eta}};\boldsymbol{\epsilon}) &= \operatorname{cov}(\boldsymbol{\eta}-\boldsymbol{\delta};\boldsymbol{\epsilon}) = \operatorname{cov}(\boldsymbol{\eta};\boldsymbol{\epsilon}) - \operatorname{cov}(\boldsymbol{\delta};\boldsymbol{\epsilon}) = \mathbf{0} + \boldsymbol{\varphi\pi}' \\
\operatorname{cov}(\tilde{\boldsymbol{y}};\boldsymbol{\delta}) &= \operatorname{cov}(\boldsymbol{\pi\eta};\boldsymbol{\delta}) = \boldsymbol{\pi\varphi} \\
\hat{\boldsymbol{\rho}} \equiv \operatorname{var}(\hat{\boldsymbol{\eta}}) &= \operatorname{var}(\boldsymbol{\eta}-\boldsymbol{\delta}) = \operatorname{var}(\boldsymbol{\eta}) + \operatorname{var}(\boldsymbol{\delta}) + 2\operatorname{cov}(\boldsymbol{\eta};-\boldsymbol{\delta}) \\
&= \boldsymbol{\rho} + \boldsymbol{\varphi} - 2\boldsymbol{\varphi} = \boldsymbol{\rho} - \boldsymbol{\varphi} \\
\operatorname{cov}(\boldsymbol{\eta};\hat{\boldsymbol{\eta}}) &= \operatorname{cov}(\hat{\boldsymbol{\eta}}+\boldsymbol{\delta};\hat{\boldsymbol{\eta}}) = \operatorname{cov}(\hat{\boldsymbol{\eta}};\hat{\boldsymbol{\eta}}) + \operatorname{cov}(\boldsymbol{\delta};\hat{\boldsymbol{\eta}}) = \hat{\boldsymbol{\rho}} + \mathbf{0} \\
\hat{\boldsymbol{\tau}} \equiv \operatorname{cov}(\boldsymbol{y};\hat{\boldsymbol{\eta}}) &= \operatorname{cov}(\boldsymbol{\pi\eta}+\boldsymbol{\epsilon};\hat{\boldsymbol{\eta}}) = \boldsymbol{\pi}\hat{\boldsymbol{\rho}} + \boldsymbol{\pi\varphi} = \boldsymbol{\pi\rho} \equiv \boldsymbol{\tau} \\
\operatorname{cov}(\tilde{\boldsymbol{y}};\hat{\boldsymbol{\eta}}) &= \operatorname{cov}(\boldsymbol{\pi\eta};\boldsymbol{\eta}-\boldsymbol{\delta}) = \boldsymbol{\pi}(\boldsymbol{\rho}-\boldsymbol{\varphi}) = \boldsymbol{\pi}\hat{\boldsymbol{\rho}}.
\end{aligned}$$

Induction in terms of $\boldsymbol{\Sigma}, \boldsymbol{\omega}$. The (pure) inductive expressions for the covariances use only $\boldsymbol{\Sigma}$ and $\boldsymbol{\omega}$. It turns out that most of the covariances, which are marked by an asterisk in Eq.3.315, have no inductive form, as $\boldsymbol{\rho}, \boldsymbol{\varphi}, \boldsymbol{\Theta}$ cannot be expressed in terms of $\boldsymbol{\Sigma}$ and $\boldsymbol{\omega}$:

$$\operatorname{var}\begin{bmatrix} \boldsymbol{\delta} \\ \boldsymbol{\epsilon} \\ \boldsymbol{\eta} \\ \hat{\boldsymbol{\eta}} \\ \tilde{\boldsymbol{y}} \\ \boldsymbol{y} \end{bmatrix} = \begin{bmatrix} * & . & . & . & . & \text{sym.} \\ * & * & . & . & . & . \\ * & 0 & * & . & . & . \\ 0 & * & * & \hat{\boldsymbol{\rho}} \equiv \boldsymbol{\omega}'\boldsymbol{\Sigma\omega} & . & . \\ * & 0 & * & * & * & . \\ 0 & * & \boldsymbol{\tau} \equiv \boldsymbol{\Sigma\omega} & \hat{\boldsymbol{\tau}} \equiv \boldsymbol{\Sigma\omega} & * & \boldsymbol{\Sigma} \end{bmatrix} \tag{3.315}$$

This is proved by:

$$\begin{aligned}
\hat{\boldsymbol{\tau}} \equiv \operatorname{cov}(\boldsymbol{y};\hat{\boldsymbol{\eta}}) &= \operatorname{cov}(\boldsymbol{y};\boldsymbol{\omega}'\boldsymbol{y}) = \boldsymbol{\Sigma\omega} \\
\hat{\boldsymbol{\rho}} \equiv \operatorname{var}(\hat{\boldsymbol{\eta}}) &= \operatorname{var}(\boldsymbol{\omega}'\boldsymbol{y}) = \boldsymbol{\omega}'\boldsymbol{\Sigma\omega}
\end{aligned}$$

Induction in terms of $\boldsymbol{\Sigma}, \boldsymbol{\tau}, \boldsymbol{\rho}$. The two expressions for the MV-LV covariance $\boldsymbol{\tau}$ and $\hat{\boldsymbol{\tau}}$, which are derived in Eqs.3.314, 3.315, can be equated,

$$\boldsymbol{\tau} = \hat{\boldsymbol{\tau}} = \boldsymbol{\Sigma\omega} = \boldsymbol{\pi\rho}, \tag{3.316}$$

and solved for the weights, which gives several expressions for the $\boldsymbol{\omega}$ in terms of $\boldsymbol{\tau}$, $\hat{\boldsymbol{\tau}}$ and $\boldsymbol{\pi\rho}$.

$$\boldsymbol{\omega} = \boldsymbol{\Sigma}^{-1}\hat{\boldsymbol{\tau}} = \boldsymbol{\Sigma}^{-1}\boldsymbol{\tau} = \boldsymbol{\Sigma}^{-1}\boldsymbol{\pi\rho} \tag{3.317}$$

If, besides $\boldsymbol{\Sigma}$ and $\boldsymbol{\tau}$, also $\boldsymbol{\rho}$ is known, all coefficients and covariances of the model can be derived: $\boldsymbol{\varphi} := \boldsymbol{\rho} - \boldsymbol{\omega}'\boldsymbol{\Sigma\omega}$, $\boldsymbol{\pi} := \boldsymbol{\tau\rho}^{-1}$, $\hat{\boldsymbol{\Sigma}} := \boldsymbol{\pi\rho\pi}'$, $\boldsymbol{\Theta} := \boldsymbol{\Sigma} - \hat{\boldsymbol{\Sigma}}$, $\boldsymbol{\Gamma} := -\boldsymbol{\pi\varphi}$, and all asterisks in Eq.3.315 can be replaced by valid expressions.

Properties. The factor estimate $\hat{\boldsymbol{\eta}}$ is biased, according to the definition Eq.3.305:

$$\boldsymbol{\omega}'\boldsymbol{\pi} = \boldsymbol{\rho\pi}'\boldsymbol{\Sigma}^{-1}\boldsymbol{\pi} \neq \mathbf{I} \tag{3.318}$$

Univocality (Eq.3.306) is assured:

$$\boldsymbol{\alpha} = \operatorname{cov}(\boldsymbol{\eta}\hat{\boldsymbol{\eta}}) \operatorname{var}^{-1}(\hat{\boldsymbol{\eta}}) = \hat{\boldsymbol{\rho}}\hat{\boldsymbol{\rho}}^{-1} = \mathbf{I} \tag{3.319}$$

The model implies that the covariance of the MVs with the "true" LVs $\boldsymbol{\eta}$ is identical to covariance with the linear composites $\hat{\boldsymbol{\eta}}$, i.e. $\boldsymbol{\tau} = \hat{\boldsymbol{\tau}}$ (unbiased structure, cf. Eq.3.310. The unfavorable aspect of the model is indicated by the inevitable covariance $\operatorname{cov}(\boldsymbol{\delta};\boldsymbol{\epsilon}) = -\boldsymbol{\pi}\boldsymbol{\varphi}$, which is to say that the lack of validity $\boldsymbol{\delta}$ is the opposite of the measurement error $\boldsymbol{\epsilon}$.

As demonstrated in Eq.3.314, the composite $\hat{\boldsymbol{\eta}}$ and the residuals $\boldsymbol{\epsilon}$ are correlated, so $\hat{\boldsymbol{\eta}}$ can not be used as a predictor for $\boldsymbol{y}$. In the following subsections loadings and residuals are respecified to form a coherent prediction model.

3.3.3 Factor Estimation ModeB, Reestimated Loadings

Model. The model shown in Tb.3.5b is different from Tb.3.5a in so far as the systematic part of $\boldsymbol{y}$ is not based on the LV $\boldsymbol{\eta}$ itself but only on the measurable part $\hat{\boldsymbol{\eta}}$. The model contains the two relations:

$$\begin{aligned}
\boldsymbol{\eta} = \boldsymbol{\omega}\boldsymbol{y} + \boldsymbol{\delta}, \qquad & \hat{\boldsymbol{\eta}} \equiv \mathrm{E}(\boldsymbol{\eta}|\boldsymbol{y}) = \boldsymbol{\omega}\boldsymbol{y} \\
& \Longrightarrow \mathrm{E}(\boldsymbol{\delta}) = \mathbf{0} \\
& \Longrightarrow \operatorname{cov}(\boldsymbol{\delta};\boldsymbol{y}) = \operatorname{cov}(\boldsymbol{\delta};\hat{\boldsymbol{\eta}}) = \mathbf{0} \\
& \Longrightarrow \operatorname{cov}(\boldsymbol{\eta};\boldsymbol{y}) = \operatorname{cov}(\hat{\boldsymbol{\eta}};\boldsymbol{y}) = \boldsymbol{\omega}\operatorname{var}(\boldsymbol{y})
\end{aligned} \tag{3.320}$$

$$\begin{aligned}
\boldsymbol{y} = \hat{\boldsymbol{\pi}}\hat{\boldsymbol{\eta}} + \hat{\boldsymbol{\epsilon}}, \qquad & \hat{\boldsymbol{y}} \equiv \mathrm{E}(\boldsymbol{y}|\hat{\boldsymbol{\eta}}) = \hat{\boldsymbol{\pi}}\hat{\boldsymbol{\eta}} \\
& \Longrightarrow \mathrm{E}(\hat{\boldsymbol{\epsilon}}) = \mathbf{0} \\
& \Longrightarrow \operatorname{cov}(\hat{\boldsymbol{\epsilon}};\hat{\boldsymbol{\eta}}) = \operatorname{cov}(\hat{\boldsymbol{\epsilon}};\hat{\boldsymbol{y}}) = \mathbf{0} \\
& \Longrightarrow \operatorname{cov}(\boldsymbol{y};\hat{\boldsymbol{\eta}}) = \operatorname{cov}(\hat{\boldsymbol{y}};\hat{\boldsymbol{\eta}}) = \hat{\boldsymbol{\pi}}\operatorname{var}(\hat{\boldsymbol{\eta}})
\end{aligned} \tag{3.321}$$

and implies two more zero covariances:

$$\begin{aligned}
\operatorname{cov}(\boldsymbol{\delta};\hat{\boldsymbol{y}}) &= \operatorname{cov}(\boldsymbol{\delta};\hat{\boldsymbol{\pi}}\boldsymbol{\omega}'\boldsymbol{y}) = \mathbf{0} \\
\operatorname{cov}(\boldsymbol{\delta};\hat{\boldsymbol{\epsilon}}) &= \operatorname{cov}(\boldsymbol{\delta};(\mathbf{I}-\hat{\boldsymbol{\pi}}\boldsymbol{\omega}')\boldsymbol{y}) = \mathbf{0}.
\end{aligned}$$

Deductive expression for covariances. This model implies all of the properties Eqs.3.216,..., 3.218 which follow directly from Eq.3.215. The covariances of all variables, collected into one matrix, are:

$$\operatorname{var}\begin{bmatrix} \boldsymbol{\delta} \\ \hat{\boldsymbol{\epsilon}} \\ \boldsymbol{\eta} \\ \hat{\boldsymbol{\eta}} \\ \hat{\boldsymbol{y}} \\ \boldsymbol{y} \end{bmatrix} = \begin{bmatrix}
\boldsymbol{\varphi} & . & . & . & . & \text{sym.} \\
\mathbf{0} & \boldsymbol{\Theta} & . & . & . & . \\
\boldsymbol{\varphi} & \mathbf{0} & \boldsymbol{\rho} & . & . & . \\
\mathbf{0} & \mathbf{0} & \hat{\boldsymbol{\rho}} & \hat{\boldsymbol{\rho}} \equiv \boldsymbol{\rho} - \boldsymbol{\varphi} & . & . \\
\mathbf{0} & \mathbf{0} & \hat{\boldsymbol{\pi}}\hat{\boldsymbol{\rho}} & \hat{\boldsymbol{\pi}}\hat{\boldsymbol{\rho}} & \hat{\boldsymbol{\Sigma}} \equiv \hat{\boldsymbol{\pi}}\hat{\boldsymbol{\rho}}\hat{\boldsymbol{\pi}}' & . \\
\mathbf{0} & \boldsymbol{\Theta} & \hat{\boldsymbol{\pi}}\hat{\boldsymbol{\rho}} & \hat{\boldsymbol{\pi}}\hat{\boldsymbol{\rho}} & \hat{\boldsymbol{\Sigma}} & \hat{\boldsymbol{\Sigma}} + \boldsymbol{\Theta}
\end{bmatrix}, \tag{3.322}$$

which is proved by:

$$\begin{aligned}
\text{cov}(\boldsymbol{\eta};\hat{\boldsymbol{\eta}}) &= \text{cov}(\hat{\boldsymbol{\eta}}+\boldsymbol{\delta};\hat{\boldsymbol{\eta}}) = \hat{\boldsymbol{\rho}}+\mathbf{0}\\
\text{cov}(\hat{\boldsymbol{y}};\boldsymbol{\eta}) &= \text{cov}(\hat{\boldsymbol{\pi}}\hat{\boldsymbol{\eta}};\hat{\boldsymbol{\eta}}+\boldsymbol{\delta}) = \hat{\boldsymbol{\pi}}\hat{\boldsymbol{\rho}}+\mathbf{0}\\
\text{cov}(\hat{\boldsymbol{y}};\hat{\boldsymbol{\eta}}) &= \text{cov}(\hat{\boldsymbol{\pi}}\hat{\boldsymbol{\eta}};\hat{\boldsymbol{\eta}}) = \hat{\boldsymbol{\pi}}\hat{\boldsymbol{\rho}}\\
\boldsymbol{\tau} \equiv \text{cov}(\boldsymbol{y};\boldsymbol{\eta}) &= \text{cov}(\hat{\boldsymbol{\pi}}\hat{\boldsymbol{\eta}}+\hat{\boldsymbol{\epsilon}};\hat{\boldsymbol{\eta}}+\boldsymbol{\delta}) = \hat{\boldsymbol{\pi}}\hat{\boldsymbol{\rho}}+\mathbf{0}+\mathbf{0}+\mathbf{0}\\
\hat{\boldsymbol{\tau}} \equiv \text{cov}(\boldsymbol{y};\hat{\boldsymbol{\eta}}) &= \text{cov}(\hat{\boldsymbol{\pi}}\hat{\boldsymbol{\eta}}+\hat{\boldsymbol{\epsilon}};\hat{\boldsymbol{\eta}}) = \hat{\boldsymbol{\pi}}\hat{\boldsymbol{\rho}}+\mathbf{0}\\
\hat{\boldsymbol{\Sigma}} \equiv \text{var}(\hat{\boldsymbol{y}}) &= \text{cov}(\hat{\boldsymbol{\pi}}\hat{\boldsymbol{\eta}}) = \hat{\boldsymbol{\pi}}\hat{\boldsymbol{\rho}}\hat{\boldsymbol{\pi}}'.
\end{aligned}$$

The model implies that the residuals $\boldsymbol{\delta}$ and $\hat{\boldsymbol{\epsilon}}$ are uncorrelated. This simplifies the expressions for the covariances of the residuals with the other variables remarkably. The four expression of MV-LV covariance have all the same structure $\hat{\boldsymbol{\pi}}\hat{\boldsymbol{\rho}}$.

Induction in terms of $\boldsymbol{\Sigma}, \boldsymbol{\omega}$. The (pure) inductive expressions for the covariances use only $\boldsymbol{\Sigma}$ and $\boldsymbol{\omega}$.

$$\begin{aligned}
\hat{\boldsymbol{\rho}} \equiv \text{cov}(\hat{\boldsymbol{\eta}}) &= \text{cov}(\boldsymbol{\omega}'\boldsymbol{y}) = \boldsymbol{\omega}'\boldsymbol{\Sigma}\boldsymbol{\omega}\\
\boldsymbol{\tau} \equiv \text{cov}(\boldsymbol{y};\boldsymbol{\eta}) &= \text{cov}(\boldsymbol{y};\boldsymbol{\omega}'\boldsymbol{y}+\boldsymbol{\delta}) = \boldsymbol{\Sigma}\boldsymbol{\omega}+\mathbf{0}\\
\hat{\boldsymbol{\tau}} \equiv \text{cov}(\boldsymbol{y};\hat{\boldsymbol{\eta}}) &= \text{cov}(\boldsymbol{y};\boldsymbol{\omega}'\boldsymbol{y}) = \boldsymbol{\Sigma}\boldsymbol{\omega}\\
\text{cov}(\hat{\boldsymbol{y}};\boldsymbol{\eta}) &= \text{cov}(\boldsymbol{y}-\hat{\boldsymbol{\epsilon}};\boldsymbol{\omega}'\boldsymbol{y}+\boldsymbol{\delta}) = \boldsymbol{\Sigma}\boldsymbol{\omega}+\mathbf{0}+\mathbf{0}+\mathbf{0}\\
\text{cov}(\hat{\boldsymbol{y}};\hat{\boldsymbol{\eta}}) &= \text{cov}(\boldsymbol{y}-\hat{\boldsymbol{\epsilon}};\boldsymbol{\omega}'\boldsymbol{y}) = \boldsymbol{\Sigma}\boldsymbol{\omega}+\mathbf{0}
\end{aligned} \tag{3.323}$$

It turns out that the inductive expressions for the four MV-LV covariances are identical, the same way as the deductive expressions are.

Induction in terms of $\boldsymbol{\Sigma}, \boldsymbol{\tau}, \boldsymbol{\rho}$. The two expressions for the MV-LV covariance $\boldsymbol{\tau}$, which are derived in Eqs.3.322, 3.323, can be equated,

$$\boldsymbol{\tau} = \hat{\boldsymbol{\tau}} = \boldsymbol{\Sigma}\boldsymbol{\omega} = \hat{\boldsymbol{\pi}}\hat{\boldsymbol{\rho}}, \tag{3.324}$$

and solved for the weights:

$$\boldsymbol{\omega} = \boldsymbol{\Sigma}^{-1}\hat{\boldsymbol{\pi}}\hat{\boldsymbol{\rho}} = \boldsymbol{\Sigma}^{-1}\boldsymbol{\tau} \tag{3.325}$$

Example. We illustrate Eq.3.322 by an little example with two observed variables y_1, y_2. The three theoretical variables η, ϵ_1, ϵ_2 span a three-dimensional model space; the observed space is a two-dimensional subspace embedded in the three-dimensional model space. The estimated LV $\hat{\eta}_B$ is the projection of η onto the observed space. The two estimated residuals have a perfect negative correlation, $\text{cor}(\hat{\epsilon}_1;\hat{\epsilon}_2) = -1$, i.e. they have rank one and are one-dimensional. That is to say, the estimated space is identical to the observed space, but has orthogonal axes $\hat{\eta}_B$ and $\hat{\epsilon}$. The "estimated" validity residual $\hat{\delta}$ is perpendicular on the observed plane; it is uncorrelated with the estimated residuals and $\hat{\eta}$, but negatively correlated to the "true" factor residuals ϵ_1, ϵ_2.

$$\operatorname{var}\begin{bmatrix} y_1 \\ y_2 \\ \epsilon_1 \\ \epsilon_2 \\ \eta \\ \hat{\eta}_B \\ \hat{\eta}_{std} \\ \hat{\delta} \\ \hat{\epsilon}_1 \\ \hat{\epsilon}_2 \end{bmatrix} = \begin{bmatrix} 1.0000 & . & . & . & . & . & . & . & . & \text{sym.} \\ .4000 & 1.0000 & . & . & . & . & . & . & . & . \\ .3600 & 0 & .3600 & . & . & . & . & . & . & . \\ 0 & .7500 & 0 & .7500 & . & . & . & . & . & . \\ .8000 & .5000 & 0 & 0 & 1.0000 & . & . & . & . & . \\ .8000 & .5000 & .2571 & .1607 & .6786 & .6786 & . & . & . & . \\ .9712 & .6070 & .3122 & .1951 & .8238 & .8238 & 1.0000 & . & . & . \\ 0 & 0 & -.2571 & -.1607 & .3214 & 0 & 0 & .3214 & . & . \\ .0568 & -.1895 & .0568 & -.1895 & 0 & 0 & 0 & 0 & .0568 & . \\ -.1895 & .6316 & -.1895 & .6316 & 0 & 0 & 0 & 0 & -.1895 & .6316 \end{bmatrix} \tag{3.326}$$

3.3.4 Factor Estimation ModeA

Factor model. The model displayed in Tb.3.5c uses the factors loadings $\boldsymbol{\pi}$ as weights for the factor scores, after a rescaling so that $\boldsymbol{\pi}'\boldsymbol{\omega} = \mathbf{I}$. The choice of the weights is deliberate, but has proven to be useful in the Principal Component model, Sc.3.2.6, and will again exhibit desirable properties in the Principal Predictor model, Sc.3.4.3. The model consists of the predictor specification for a factor model

$$\begin{aligned} \boldsymbol{y} = \boldsymbol{\pi}\boldsymbol{\eta} + \boldsymbol{\epsilon}, \qquad & \tilde{\boldsymbol{y}} \equiv \mathrm{E}(\boldsymbol{y}|\boldsymbol{\eta}) = \boldsymbol{\pi}\boldsymbol{\eta} \\ & \Longrightarrow \mathrm{E}(\boldsymbol{\epsilon}) = \mathbf{0} \\ & \Longrightarrow \operatorname{cov}(\boldsymbol{\epsilon};\boldsymbol{\eta}) = \operatorname{cov}(\boldsymbol{\epsilon};\tilde{\boldsymbol{y}}) = \mathbf{0} \\ & \Longrightarrow \operatorname{cov}(\boldsymbol{y};\boldsymbol{\eta}) = \operatorname{cov}(\tilde{\boldsymbol{y}};\boldsymbol{\eta}) = \boldsymbol{\pi}\operatorname{var}(\boldsymbol{\eta}) \end{aligned} \tag{3.327}$$

and the additional definition

$$\boldsymbol{\omega} := \boldsymbol{\pi}(\boldsymbol{\pi}'\boldsymbol{\pi})^{-1}, \tag{3.328}$$

which gives us $\boldsymbol{\pi}'\boldsymbol{\omega} = \mathbf{I}$ and

$$\begin{aligned} \hat{\boldsymbol{\eta}} &= \boldsymbol{\omega}'\boldsymbol{y} \\ &= \boldsymbol{\omega}'(\boldsymbol{\pi}\boldsymbol{\eta} + \boldsymbol{\epsilon}) = \boldsymbol{\omega}'\boldsymbol{\pi}\boldsymbol{\eta} + \boldsymbol{\omega}'\boldsymbol{\epsilon} = \boldsymbol{\eta} + \boldsymbol{\omega}'\boldsymbol{\epsilon} \\ &\equiv \boldsymbol{\eta} + \boldsymbol{v}. \end{aligned} \tag{3.329}$$

The term $\boldsymbol{v} \equiv \boldsymbol{\omega}'\boldsymbol{\epsilon}$ introduces a distortion into the factor estimate $\hat{\boldsymbol{\eta}}$. $\boldsymbol{v}$ is linear aggregate of the factor residuals $\boldsymbol{\epsilon}$ and is uncorrelated to the LV $\boldsymbol{\eta}$.

Implications. The covariances of all variables of the model, collected into one matrix, are:

$$\operatorname{var}\begin{bmatrix} \boldsymbol{\epsilon} \\ \boldsymbol{v} \\ \boldsymbol{\eta} \\ \hat{\boldsymbol{\eta}} \\ \boldsymbol{y} \end{bmatrix} = \begin{bmatrix} \boldsymbol{\Theta} & . & . & . & \text{sym.} \\ \boldsymbol{\omega}'\boldsymbol{\Theta} & \boldsymbol{\psi} \equiv \boldsymbol{\omega}'\boldsymbol{\Theta}\boldsymbol{\omega} & . & . & . \\ \mathbf{0} & \mathbf{0} & \boldsymbol{\rho} & . & . \\ \boldsymbol{\omega}'\boldsymbol{\Theta} & \boldsymbol{\psi} & \boldsymbol{\rho} & \boldsymbol{\rho} + \boldsymbol{\psi} & . \\ \boldsymbol{\Theta} & \boldsymbol{\Theta}\boldsymbol{\omega} & \boldsymbol{\pi}\boldsymbol{\rho} & \boldsymbol{\pi}\boldsymbol{\rho} + \boldsymbol{\Theta}\boldsymbol{\omega} & \boldsymbol{\Sigma} \end{bmatrix} \tag{3.330}$$

which is proved by:

$$\begin{aligned}
\boldsymbol{\psi} \equiv \operatorname{var}(\boldsymbol{v}) &= \operatorname{var}(\boldsymbol{\omega}'\boldsymbol{\epsilon}) = \boldsymbol{\omega}'\boldsymbol{\Theta}\boldsymbol{\omega} \\
\operatorname{cov}(\boldsymbol{v};\boldsymbol{\epsilon}) &= \operatorname{cov}(\boldsymbol{\omega}'\boldsymbol{\epsilon};\boldsymbol{\epsilon}) = \boldsymbol{\omega}'\boldsymbol{\Theta} \\
\operatorname{cov}(\boldsymbol{y};\boldsymbol{v}) &= \operatorname{cov}(\boldsymbol{\pi}\boldsymbol{\eta}+\boldsymbol{\epsilon};\boldsymbol{\omega}'\boldsymbol{\epsilon}) = \mathbf{0} + \boldsymbol{\Theta}\boldsymbol{\omega} \\
\hat{\boldsymbol{\rho}} \equiv \operatorname{cov}(\hat{\boldsymbol{\eta}};\hat{\boldsymbol{\eta}}) &= \boldsymbol{\omega}'\boldsymbol{\Sigma}\boldsymbol{\omega} = \boldsymbol{\omega}'(\boldsymbol{\pi}\boldsymbol{\rho}\boldsymbol{\pi}'+\boldsymbol{\Theta})\boldsymbol{\omega} = \boldsymbol{\rho} + \boldsymbol{\omega}'\boldsymbol{\Theta}\boldsymbol{\omega} = \boldsymbol{\rho} + \boldsymbol{\psi} \\
\operatorname{cov}(\hat{\boldsymbol{\eta}};\boldsymbol{\epsilon}) &= \operatorname{cov}(\boldsymbol{\eta}+\boldsymbol{\omega}'\boldsymbol{\epsilon};\boldsymbol{\epsilon}) = \boldsymbol{\omega}'\boldsymbol{\Theta} \\
\hat{\boldsymbol{\tau}} \equiv \operatorname{cov}(\boldsymbol{y};\hat{\boldsymbol{\eta}}) &= \operatorname{cov}(\boldsymbol{y};\boldsymbol{\eta}+\boldsymbol{\omega}'\boldsymbol{\epsilon}) = \operatorname{cov}(\boldsymbol{y};\boldsymbol{\eta}) + \operatorname{cov}(\boldsymbol{y};\boldsymbol{\omega}'\boldsymbol{\epsilon}) = \boldsymbol{\pi}\boldsymbol{\rho} + \boldsymbol{\Theta}\boldsymbol{\omega}.
\end{aligned}$$

Properties. The distortion introduced by $\boldsymbol{v} \equiv \boldsymbol{\omega}'\boldsymbol{\epsilon}$ into the factor estimate $\hat{\boldsymbol{\eta}}$ inflates the variance of $\hat{\boldsymbol{\eta}}$ by the term $\boldsymbol{\omega}'\boldsymbol{\Theta}\boldsymbol{\omega}$ and the factor structure by $\boldsymbol{\Theta}\boldsymbol{\omega}$. However, no bias in the sense of Eq.3.305 is introduced. The expectation of $\hat{\boldsymbol{\eta}}$ conditional on $\boldsymbol{\eta}$ is, due to $\mathrm{E}(\boldsymbol{\epsilon}) = \mathrm{E}(\boldsymbol{\epsilon}\boldsymbol{\eta}) = \mathrm{E}(\boldsymbol{\epsilon}|\boldsymbol{\eta}) = \mathbf{0}$, equal to the factor $\boldsymbol{\eta}$ itself:

$$\mathrm{E}(\hat{\boldsymbol{\eta}}|\boldsymbol{\eta}) = \mathrm{E}(\boldsymbol{\eta}+\boldsymbol{\omega}'\boldsymbol{\epsilon}|\boldsymbol{\eta}) = \mathrm{E}(\boldsymbol{\eta}|\boldsymbol{\eta}) + \boldsymbol{\omega}'\,\mathrm{E}(\boldsymbol{\epsilon}|\boldsymbol{\eta}) = \mathrm{E}(\boldsymbol{\eta}) \tag{3.331}$$

Hence $\hat{\boldsymbol{\eta}}$ is an unbiased proxy of $\boldsymbol{\eta}$. Furthermore, due to

$$\mathrm{E}(\boldsymbol{\eta}|\hat{\boldsymbol{\eta}}) = \mathrm{E}(\boldsymbol{\eta}|\boldsymbol{\eta}+\boldsymbol{\omega}'\boldsymbol{\epsilon}) = \mathrm{E}(\boldsymbol{\eta}|\boldsymbol{\eta}) + \boldsymbol{\omega}'\,\mathrm{E}(\boldsymbol{\eta}|\boldsymbol{\epsilon}) = \mathrm{E}(\boldsymbol{\eta}) \tag{3.332}$$

the estimate $\hat{\boldsymbol{\eta}}$ is univocal. Notice that this is not an additional predictor specification, but is implied by the predictor specification Eq.3.327 together with the choice of the weights in Eq.3.328. The conditions for uncorrelatedness (Eq.3.308 are not met because $\boldsymbol{\omega}'\boldsymbol{\Theta}\boldsymbol{\omega}$ is not diagonal, in general. The factor structure is biased, $\boldsymbol{\tau} \neq \hat{\boldsymbol{\tau}}$ (Eq.3.310).

Example. We use again the example of Sc.3.3.3 to illustrate the covariances of all variables, but now with ModeA weights according to Eq.3.328:

$$\operatorname{var}\begin{bmatrix} y_1 \\ y_2 \\ \epsilon_1 \\ \epsilon_2 \\ \eta \\ \hat{\eta}_A \\ v \end{bmatrix} = \begin{bmatrix}
1.0000 & . & . & . & . & . & \text{sym.} \\
.4000 & 1.0000 & . & . & . & . & . \\
.3600 & 0 & .3600 & . & . & . & . \\
0 & .7500 & 0 & .7500 & . & . & . \\
.8000 & .5000 & 0 & 0 & 1.0000 & . & . \\
1.1236 & .9213 & .3236 & .4213 & 1.0000 & 1.5276 & . \\
.3236 & .4213 & .3236 & .4213 & 0 & .5276 & .5276
\end{bmatrix} \tag{3.333}$$

The loadings are $\boldsymbol{\tau} = \boldsymbol{\pi} = (0.8, 0.5)'$, and the weights are $\boldsymbol{\omega} = \boldsymbol{\pi}(\boldsymbol{\pi}'\boldsymbol{\pi})^{-1} = (0.8989, 0.5618)'$. Notice the distortion of the component structure

$$\hat{\boldsymbol{\tau}} = \operatorname{cov}(\boldsymbol{y};\boldsymbol{\eta}) + \operatorname{cov}(\boldsymbol{\epsilon};\boldsymbol{v}) = \begin{bmatrix} .8000 \\ .5000 \end{bmatrix} + \begin{bmatrix} .3236 \\ .4213 \end{bmatrix} = \begin{bmatrix} 1.1236 \\ .9213 \end{bmatrix}$$

and of the variance of the component:

$$\hat{\boldsymbol{\rho}} = \boldsymbol{\rho} + \boldsymbol{\omega}'\boldsymbol{\Theta}\boldsymbol{\omega} = 1.000 + 0.5276 = 1.5276$$

3.3.5 Factor Estimation ModeA, Reestimated Loadings

Component loadings. One can reestimate the model such that the linear aggregate $\hat{\boldsymbol{\eta}} = \boldsymbol{\omega}'\boldsymbol{y}$ serves as predictor for the observed variables $\boldsymbol{y}$ and leaves the residuals $\hat{\boldsymbol{\epsilon}}$. The model Eqs.3.327, 3.328 is augmented by:

$$\begin{aligned} \boldsymbol{y} = \hat{\boldsymbol{\pi}}\boldsymbol{\omega}'\boldsymbol{y} + \hat{\boldsymbol{\epsilon}}, \qquad & \hat{\boldsymbol{y}} \equiv \mathrm{E}(\boldsymbol{y}|\boldsymbol{\omega}'\boldsymbol{y}) = \hat{\boldsymbol{\pi}}\boldsymbol{\omega}'\boldsymbol{y} \\ & \Longrightarrow \mathrm{E}(\hat{\boldsymbol{\epsilon}}) = \mathbf{0} \\ & \Longrightarrow \mathrm{cov}(\hat{\boldsymbol{\epsilon}}; \boldsymbol{\omega}'\boldsymbol{y}) = \mathrm{cov}(\hat{\boldsymbol{\epsilon}}; \hat{\boldsymbol{y}}) = \mathbf{0} \\ & \Longrightarrow \mathrm{cov}(\boldsymbol{y}; \boldsymbol{\omega}'\boldsymbol{y}) = \mathrm{cov}(\hat{\boldsymbol{y}}; \boldsymbol{\omega}'\boldsymbol{y}) = \hat{\boldsymbol{\pi}}\,\mathrm{var}(\boldsymbol{\omega}'\boldsymbol{y}). \end{aligned} \tag{3.334}$$

The reestimated parameters and variables are:

$$\begin{aligned} \hat{\boldsymbol{\pi}} &= \hat{\boldsymbol{\tau}}\hat{\boldsymbol{\rho}}^{-1} = (\boldsymbol{\pi}\boldsymbol{\rho} + \boldsymbol{\Theta}\boldsymbol{\omega})(\boldsymbol{\rho} + \boldsymbol{\omega}'\boldsymbol{\Theta}\boldsymbol{\omega})^{-1} \\ &= \boldsymbol{\Sigma}\boldsymbol{\omega}(\boldsymbol{\omega}'\boldsymbol{\Sigma}\boldsymbol{\omega})^{-1} = \boldsymbol{\Sigma}\boldsymbol{\pi}(\boldsymbol{\pi}'\boldsymbol{\pi})^{-1}[(\boldsymbol{\pi}'\boldsymbol{\pi})^{-1}\boldsymbol{\pi}'\boldsymbol{\Sigma}\boldsymbol{\pi}(\boldsymbol{\pi}'\boldsymbol{\pi})^{-1}]^{-1} \\ &= \boldsymbol{\Sigma}\boldsymbol{\pi}(\boldsymbol{\pi}'\boldsymbol{\Sigma}\boldsymbol{\pi})^{-1}(\boldsymbol{\pi}'\boldsymbol{\pi}) \end{aligned} \tag{3.335}$$

$$\hat{\boldsymbol{y}} \equiv \hat{\boldsymbol{\pi}}\hat{\boldsymbol{\eta}} = \hat{\boldsymbol{\pi}}\boldsymbol{\omega}'\boldsymbol{y} = \hat{\boldsymbol{\pi}}\boldsymbol{\omega}'(\boldsymbol{\pi}\boldsymbol{\eta} + \boldsymbol{\epsilon}) = \hat{\boldsymbol{\pi}}(\boldsymbol{\eta} + \boldsymbol{\omega}'\boldsymbol{\epsilon}). \tag{3.336}$$

$$\hat{\boldsymbol{\epsilon}} \equiv \boldsymbol{y} - \hat{\boldsymbol{y}} = (\boldsymbol{\pi}\boldsymbol{\eta} + \boldsymbol{\epsilon}) - \hat{\boldsymbol{\pi}}(\boldsymbol{\eta} + \boldsymbol{\omega}'\boldsymbol{\epsilon}) = (\boldsymbol{\pi} - \hat{\boldsymbol{\pi}})\boldsymbol{\eta} + (\mathbf{I} - \hat{\boldsymbol{\pi}}\boldsymbol{\omega}')\boldsymbol{\epsilon}, \tag{3.337}$$

$\hat{\boldsymbol{y}}$ is the systematic part of the MVs as predicted from the factors scores, and $\hat{\boldsymbol{\epsilon}}$ are the respective residuals.

Implications. The covariances implied by this model are fairly complicated when expressed in terms of the parameters $\boldsymbol{\rho}$, $\boldsymbol{\pi}$ and $\boldsymbol{\Theta}$:

$$\mathrm{var}\begin{bmatrix} \boldsymbol{\epsilon} \\ \boldsymbol{\eta} \\ \hat{\boldsymbol{\eta}} \\ \hat{\boldsymbol{y}} \\ \boldsymbol{y} \end{bmatrix} = \begin{bmatrix} \boldsymbol{\Theta} & . & . & \text{sym.} & \\ \mathbf{0} & \boldsymbol{\rho} & . & . & . \\ \mathbf{0} & \boldsymbol{\rho} & \hat{\boldsymbol{\rho}} \equiv \boldsymbol{\rho} + \boldsymbol{\omega}'\boldsymbol{\Theta}\boldsymbol{\omega} & . & . \\ \boldsymbol{\omega}'\boldsymbol{\Theta} & \hat{\boldsymbol{\pi}}\boldsymbol{\rho} & \boldsymbol{\pi}\boldsymbol{\rho} + \boldsymbol{\Theta}\boldsymbol{\omega} & \hat{\boldsymbol{\Sigma}} \equiv \hat{\boldsymbol{\pi}}\hat{\boldsymbol{\rho}}\hat{\boldsymbol{\pi}}' & . \\ \boldsymbol{\Theta} & \boldsymbol{\pi}\boldsymbol{\rho} & \boldsymbol{\pi}\boldsymbol{\rho} + \boldsymbol{\Theta}\boldsymbol{\omega} & \boldsymbol{\pi}\boldsymbol{\rho}\hat{\boldsymbol{\pi}} + \boldsymbol{\Theta}\boldsymbol{\omega} & \boldsymbol{\Sigma} \end{bmatrix} \tag{3.338}$$

which is proved by:

$$\begin{aligned} \mathrm{cov}(\hat{\boldsymbol{y}}; \hat{\boldsymbol{\eta}}) &= \mathrm{cov}(\hat{\boldsymbol{\pi}}\hat{\boldsymbol{\eta}}; \hat{\boldsymbol{\eta}}) = \hat{\boldsymbol{\tau}}\hat{\boldsymbol{\rho}}\boldsymbol{\rho}^{-1} = \boldsymbol{\pi}\boldsymbol{\rho} + \boldsymbol{\Theta}\boldsymbol{\omega} = \mathrm{cov}(\boldsymbol{y}\boldsymbol{y}; \hat{\boldsymbol{\eta}}) \\ \mathrm{cov}(\hat{\boldsymbol{y}}; \boldsymbol{\eta}) &= \mathrm{cov}(\hat{\boldsymbol{\pi}}(\boldsymbol{\eta} + \boldsymbol{\omega}'\boldsymbol{\epsilon}); \boldsymbol{\eta}) = \hat{\boldsymbol{\pi}}\,\mathrm{cov}(\boldsymbol{\eta}; \boldsymbol{\eta}) + \hat{\boldsymbol{\pi}}\boldsymbol{\omega}'\,\mathrm{cov}(\boldsymbol{\epsilon}; \boldsymbol{\eta}) = \hat{\boldsymbol{\pi}}\boldsymbol{\rho} + \hat{\boldsymbol{\pi}}\boldsymbol{\omega}'\mathbf{0} \\ \hat{\boldsymbol{\Sigma}} \equiv \mathrm{var}(\hat{\boldsymbol{y}}) &= \mathrm{var}(\hat{\boldsymbol{\pi}}\hat{\boldsymbol{\eta}}) = \hat{\boldsymbol{\pi}}\hat{\boldsymbol{\rho}}\hat{\boldsymbol{\pi}}' = \hat{\boldsymbol{\tau}}\hat{\boldsymbol{\rho}}^{-1}\hat{\boldsymbol{\rho}}\hat{\boldsymbol{\rho}}^{-1}\hat{\boldsymbol{\tau}}' \\ &= (\boldsymbol{\pi}\boldsymbol{\rho} + \boldsymbol{\Theta}\boldsymbol{\omega})(\boldsymbol{\rho} + \boldsymbol{\omega}'\boldsymbol{\Theta}\boldsymbol{\omega})^{-1}(\boldsymbol{\pi}\boldsymbol{\rho} + \boldsymbol{\Theta}\boldsymbol{\omega})^{-1} \\ \mathrm{cov}(\boldsymbol{y}; \hat{\boldsymbol{y}}) &= \mathrm{cov}(\boldsymbol{\pi}\boldsymbol{\eta} + \boldsymbol{\epsilon}; \hat{\boldsymbol{\pi}}(\boldsymbol{\eta} + \boldsymbol{\omega}'\boldsymbol{\epsilon})) = \boldsymbol{\pi}\boldsymbol{\rho}\hat{\boldsymbol{\pi}} + \boldsymbol{\Theta}\boldsymbol{\omega} \\ \mathrm{cov}(\hat{\boldsymbol{\epsilon}}; \boldsymbol{\epsilon}) &= \mathrm{cov}[(\boldsymbol{\pi} - \hat{\boldsymbol{\pi}})\boldsymbol{\eta} + (\mathbf{I} - \hat{\boldsymbol{\pi}}\boldsymbol{\omega}')\boldsymbol{\epsilon}; \boldsymbol{\epsilon}] = (\mathbf{I} - \hat{\boldsymbol{\pi}}\boldsymbol{\omega}')\boldsymbol{\Theta} \\ \mathrm{cov}(\hat{\boldsymbol{\epsilon}}; \boldsymbol{\eta}) &= \mathrm{cov}[(\boldsymbol{\pi} - \hat{\boldsymbol{\pi}})\boldsymbol{\eta} + (\mathbf{I} - \hat{\boldsymbol{\pi}}\boldsymbol{\omega}')\boldsymbol{\epsilon}; \boldsymbol{\eta}] = (\boldsymbol{\pi} - \hat{\boldsymbol{\pi}})\boldsymbol{\rho} \\ \mathrm{cov}(\hat{\boldsymbol{\epsilon}}; \boldsymbol{y}) &= \mathrm{cov}[(\boldsymbol{\pi} - \hat{\boldsymbol{\pi}})\boldsymbol{\eta} + (\mathbf{I} - \hat{\boldsymbol{\pi}}\boldsymbol{\omega}')\boldsymbol{\epsilon}; \boldsymbol{\pi}\boldsymbol{\eta} + \boldsymbol{\epsilon}] = (\boldsymbol{\pi} - \hat{\boldsymbol{\pi}})\boldsymbol{\rho}\hat{\boldsymbol{\pi}}' + (\mathbf{I} - \hat{\boldsymbol{\pi}}\boldsymbol{\omega}')\boldsymbol{\Theta} \\ \hat{\boldsymbol{\Theta}} \equiv \mathrm{var}(\hat{\boldsymbol{\epsilon}}) &= \mathrm{var}[(\boldsymbol{\pi} - \hat{\boldsymbol{\pi}})\boldsymbol{\eta} + (\mathbf{I} - \hat{\boldsymbol{\pi}}\boldsymbol{\omega}')\boldsymbol{\epsilon}] \\ &= (\boldsymbol{\pi} - \hat{\boldsymbol{\pi}})\boldsymbol{\rho}(\boldsymbol{\pi} - \hat{\boldsymbol{\pi}})' + (\mathbf{I} - \hat{\boldsymbol{\pi}}\boldsymbol{\omega}')\boldsymbol{\Theta}(\mathbf{I} - \hat{\boldsymbol{\pi}}\boldsymbol{\omega}')^{-1} \end{aligned}$$

Example. We continue the example Eq.3.333

$$\text{var}\begin{bmatrix} y_1 \\ y_2 \\ \epsilon_1 \\ \epsilon_2 \\ \eta \\ \hat{\eta}_A \\ \hat{\epsilon}_1 \\ \hat{\epsilon}_2 \end{bmatrix} = \begin{bmatrix} 1.0000 & . & . & . & . & . & . & \text{sym.} \\ .4000 & 1.0000 & . & . & . & . & . & . \\ .3600 & 0 & .3600 & . & . & . & . & . \\ 0 & .7500 & 0 & .7500 & . & . & . & . \\ .8000 & .5000 & 0 & 0 & 1.0000 & . & . & . \\ 1.1236 & .9213 & .3236 & .4213 & 1.0000 & 1.5276 & . & . \\ .1736 & -.2777 & .1220 & -.3099 & .0645 & 0 & .1736 & . \\ -.2777 & .4443 & -.1952 & .4959 & -.1031 & 0 & -.2777 & .4443 \end{bmatrix} \quad (3.339)$$

3.3.6 Summary of Factor Estimation

Two ways of replacing factors by components were shown, called the Mode A and Mode B estimation technique. The ModeA technique (outwards directed) creates a component that facilitates an optimal prediction of the MVs. However, the Mode A introduces a "distortion" $\boldsymbol{\omega}'\boldsymbol{\epsilon}$ into the estimate $\hat{\boldsymbol{y}}$. The ModeB estimation technique (inwards directed) regresses the LV on the total of the MVs and minimizes var($\boldsymbol{\delta}$), the "lack of validity".

For the sake of comparison, the factor score estimation methods discussed here in Sc.3.3 are listed together with other methods, taken from the factor-analytic literature (cf. McDonald & Burr 1967, Tucker 1971, Hakstian, Zidek & McDonald 1977, Harman 1967, Harris 1967, Mulaik 1972, Tucker 1971).

$$\boldsymbol{\omega} = \boldsymbol{\Sigma}^{-1}\boldsymbol{\pi} \qquad \text{Mode B} \quad (3.340)$$

$$\boldsymbol{\omega} = \boldsymbol{\pi}(\boldsymbol{\pi}'\boldsymbol{\pi})^{-1} \qquad \text{Mode A} \quad (3.341)$$

$$\boldsymbol{\omega} = \boldsymbol{\Theta}\boldsymbol{\pi}(\boldsymbol{\pi}'\boldsymbol{\Theta}^{-1}\boldsymbol{\pi})^{-1} \qquad \text{Bartlett} \quad (3.342)$$

$$\boldsymbol{\omega} = \boldsymbol{\Theta}^{-1}\boldsymbol{\pi}(\boldsymbol{\pi}'\boldsymbol{\Theta}^{-1}\boldsymbol{\rho}\boldsymbol{\Theta}^{-1}\boldsymbol{\pi})^{-1/2} \qquad \text{Anderson \& Rubin} \quad (3.343)$$

Using the Bartlett or Anderson and Rubin weights inside the PLS algorithm would require to compute the residual covariance $\boldsymbol{\Theta}$ in each iteration cycle. Let us end this section by two quotations: "It would be theoretically attractive to estimate the factor scores, factor loadings, and specific variances all at the same time from the data, using maximum likelihood. However, there are too many parameters for this to be possible" (Mardia, Kent and Bibby 1979:275). The other quotation voices concern about component scores in general: "Because of the indeterminacy of the factor scores, it has been suggested that component, canonical or image analysis is more appropriate if factor scores are desirable. This is true for the original study but only for this study. If a component weight matrix, for example, is used with a new sample, the computed scores will not correlate with the theoretical factor in the new sample much better than would estimated common factor scores since the component weight matrix will have capitalized on chance characteristics of the first sample" (Gorsuch 1974:213).

Table 3.6: Four models for two blocks

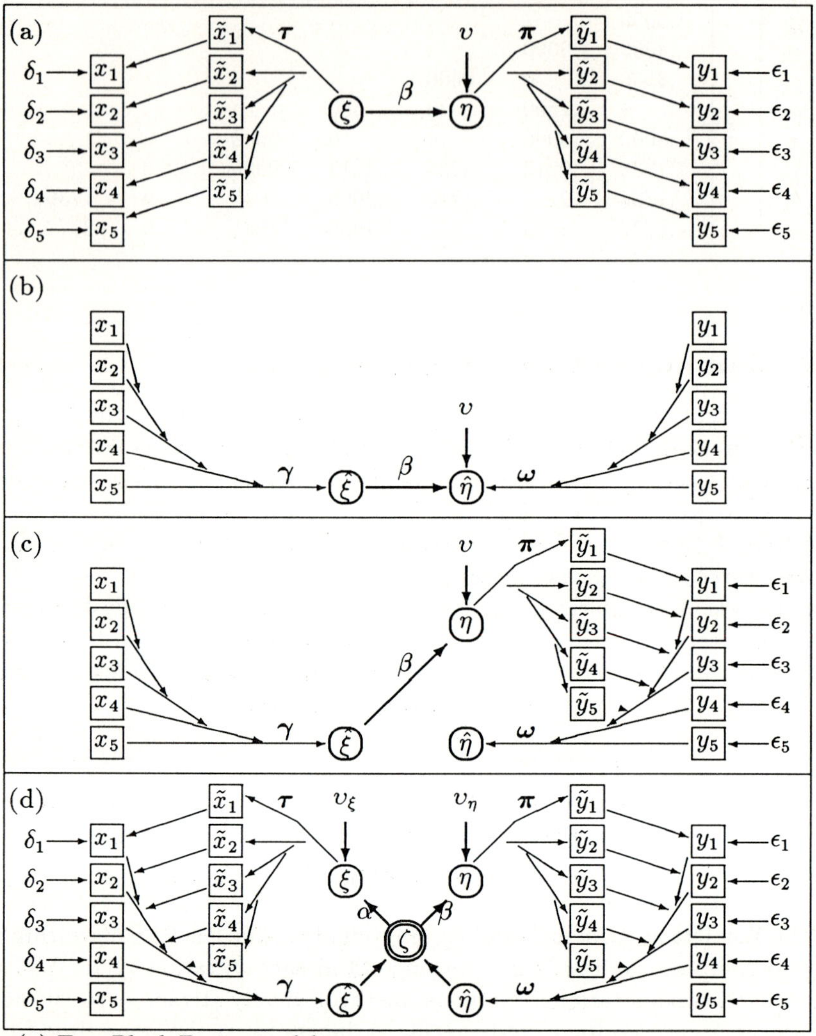

(a) Two-Block Factor model
(b) Canonical Correlation model
(c) Principal Predictor model
(d) Interbattery Factor model

3.4 Predictive Two-Block Models

Hotelling's (1935) canonical correlation (CC) model was the first type of two-block models with one block of predictors and one of predictands (criteria). The CC model has been criticized from the start, beginning with Guttman (1941b). The CC model predicts the "most predictable criterion", which is a purely mathematical criterion and not something that is determined by the researcher to be worth predicting for substantive reasons. In this Sc.3.4 we shall report and discuss several other approaches to prediction in two-block models, namely

1. the Two-Block Factor model
2. Hotelling's 1935 Canonical Correlation model
3. the Principal Predictor model, also known as Rank-Restricted Regression model and Redundancy model (van den Wollenberg)
4. Tucker's 1958 Interbattery Factor model
5. Fortier's 1966 Simultaneous Linear Prediction model and
6. Hauser and Goldberger's 1971 causal model (MIMIC model).

The exposition will use the 5-step deduction of the PLS estimation described in Sc.3.1.5.

3.4.1 The Two-Block Factor Model

Notation. Throughout Sc.3.4, we shall make use the following notation:

$\boldsymbol{x}_{[L\times 1]} = [x_\ell] \sim \mathrm{M}(\mathbf{0}, \boldsymbol{\Sigma}_{xx}, \ldots)$		for predictor MVs
$\boldsymbol{y}_{[K\times 1]} = [y_k] \sim \mathrm{M}(\mathbf{0}, \boldsymbol{\Sigma}_{yy}, \ldots)$		for predictand MVs
$\xi, \tilde{\xi}, \hat{\xi}$		for predictor LVs
	$\eta, \tilde{\eta}, \hat{\eta}$	for predictand LVs
$\boldsymbol{x} = \boldsymbol{\tau}\xi + \boldsymbol{\delta}$	$\boldsymbol{y} = \boldsymbol{\pi}\eta + \boldsymbol{\epsilon}$	the loading relations
$\xi = \boldsymbol{\gamma}'\boldsymbol{x}$	$\eta = \boldsymbol{\omega}'\boldsymbol{y}$	the weight relations
$\xi = \alpha\eta + \zeta$	$\eta = \beta\xi + \upsilon$	the inner relation

The rank of the predictor-predictand covariance matrix is

$$A = \mathrm{rank}(\boldsymbol{\Sigma}_{xy}) \le \min(L, K).$$

The two-block model with LVs. The two-block LV model is shown in Tb.3.6a. Given two sets $\boldsymbol{x}, \boldsymbol{y}$ of observed variables, one introduces two latent variables ξ, η which follow the relations:

$$\begin{aligned} \boldsymbol{x} = \boldsymbol{\tau}\xi + \boldsymbol{\delta}, \qquad & \tilde{\boldsymbol{x}} \equiv \mathrm{E}(\boldsymbol{x}|\xi) = \boldsymbol{\tau}\xi \\ & \Longrightarrow \mathrm{E}(\boldsymbol{\delta}) = \mathbf{0} \\ & \Longrightarrow \mathrm{cov}(\boldsymbol{\delta}; \xi) = \mathrm{cov}(\boldsymbol{\delta}; \tilde{\boldsymbol{x}}) = \mathbf{0} \\ & \Longrightarrow \mathrm{cov}(\boldsymbol{x}; \xi) = \mathrm{cov}(\tilde{\boldsymbol{x}}; \xi) = \boldsymbol{\tau}\,\mathrm{var}(\xi) \end{aligned} \tag{3.401}$$

$$\begin{aligned} \boldsymbol{y} = \boldsymbol{\pi}\eta + \boldsymbol{\epsilon}, \qquad & \tilde{\boldsymbol{y}} \equiv \mathrm{E}(\boldsymbol{y}|\eta) = \boldsymbol{\pi}\eta \\ & \Longrightarrow \mathrm{E}(\boldsymbol{\epsilon}) = \mathbf{0} \\ & \Longrightarrow \mathrm{cov}(\boldsymbol{\epsilon}; \eta) = \mathrm{cov}(\boldsymbol{\epsilon}; \tilde{\boldsymbol{y}}) = \mathbf{0} \\ & \Longrightarrow \mathrm{cov}(\boldsymbol{y}; \eta) = \mathrm{cov}(\tilde{\boldsymbol{y}}; \eta) = \boldsymbol{\pi}\,\mathrm{var}(\eta) \end{aligned} \tag{3.402}$$

$$\begin{aligned} \eta = \beta\xi + \upsilon, \qquad & \tilde{\eta} \equiv \mathrm{E}(\eta|\xi) = \beta\xi \\ & \Longrightarrow \mathrm{E}(\upsilon) = 0 \\ & \Longrightarrow \mathrm{cov}(\upsilon; \xi) = \mathrm{cov}(\upsilon; \tilde{\eta}) = 0 \\ & \Longrightarrow \mathrm{cov}(\eta; \xi) = \mathrm{cov}(\tilde{\eta}; \xi) = \beta\,\mathrm{var}(\xi), \end{aligned} \tag{3.403}$$

where $\boldsymbol{\tau}, \boldsymbol{\pi}, \beta$ are regression coefficients and $\boldsymbol{\delta}, \boldsymbol{\epsilon}, v$ are residual variables. If the latent variables ξ, η are known, the predictor specifications lead immediately to the estimation of the regression coefficients, by the steps presented in Sc.3.1.5:

$$\beta = \operatorname{cov}(\eta; \xi) / \operatorname{var}(\xi) \tag{3.404}$$
$$\boldsymbol{\tau} = \operatorname{cov}(\boldsymbol{x}; \xi) / \operatorname{var}(\xi) \tag{3.405}$$
$$\boldsymbol{\pi} = \operatorname{cov}(\boldsymbol{y}; \xi) / \operatorname{var}(\eta) \tag{3.406}$$

If the latent variables are not known, some additional specifications are required to make either the parameters or the variables estimable. For example, if the joint distribution of the generating variables $\xi, v, \boldsymbol{\delta}, \boldsymbol{\epsilon}$, is completely specified, maximum likelihood estimates for the parameters may become available.

The two-block component model. As estimates of the latent variables ξ and η, two composites are defined,

$$\xi = \boldsymbol{\gamma}'\boldsymbol{x}, \qquad \eta = \boldsymbol{\omega}'\boldsymbol{y}, \tag{3.407}$$

where $\boldsymbol{\gamma}$, $\boldsymbol{\omega}$ are the weight vectors. Because the rest of Sc.3.4 is concerned not with the theoretical LVs of Eqs.3.401,..., 3.403 but only with the composites, no notational distinction is made between them. The covariances of observed variables and estimated LVs are

$$\operatorname{var}\begin{bmatrix} \xi \\ \eta \\ \boldsymbol{x} \\ \boldsymbol{y} \end{bmatrix} = \begin{bmatrix} \boldsymbol{\gamma}'\boldsymbol{\Sigma}_{xx}\boldsymbol{\gamma} & \boldsymbol{\gamma}'\boldsymbol{\Sigma}_{xy}\boldsymbol{\omega} & \boldsymbol{\gamma}'\boldsymbol{\Sigma}_{xx} & \boldsymbol{\gamma}'\boldsymbol{\Sigma}_{xy} \\ \boldsymbol{\omega}'\boldsymbol{\Sigma}_{yx}\boldsymbol{\gamma} & \boldsymbol{\omega}'\boldsymbol{\Sigma}_{yy}\boldsymbol{\omega} & \boldsymbol{\omega}'\boldsymbol{\Sigma}_{yx} & \boldsymbol{\omega}'\boldsymbol{\Sigma}_{yy} \\ \boldsymbol{\Sigma}_{xx}\boldsymbol{\gamma} & \boldsymbol{\Sigma}_{xy}\boldsymbol{\omega} & \boldsymbol{\Sigma}_{xx} & \boldsymbol{\Sigma}_{xy} \\ \boldsymbol{\Sigma}_{yx}\boldsymbol{\gamma} & \boldsymbol{\Sigma}_{yy}\boldsymbol{\omega} & \boldsymbol{\Sigma}_{yx} & \boldsymbol{\Sigma}_{yy} \end{bmatrix} \tag{3.408}$$

Depending on further specifications, three different estimates for the latent variables are obtained.

3.4.2 The Canonical Correlation Model

Step 1, model relations. The canonical correlation model (see Tb.3.6b) disregards the fact that the one block is the predictor and the other the criteria block. Both blocks are treated in the same way. Both aggregates ξ, η are scaled to unit variance.

$$\rho_\xi^2 \equiv \operatorname{var}(\xi) = \boldsymbol{\gamma}'\boldsymbol{\Sigma}_{xx}\boldsymbol{\gamma} = 1 \tag{3.409}$$
$$\rho_\eta^2 \equiv \operatorname{var}(\eta) = \boldsymbol{\omega}'\boldsymbol{\Sigma}_{yy}\boldsymbol{\omega} = 1 \tag{3.410}$$
$$\begin{aligned} \rho_{\xi\eta} &\equiv \operatorname{cov}(\xi, \eta) = \boldsymbol{\gamma}'\boldsymbol{\Sigma}_{xy}\boldsymbol{\omega} \\ &= \frac{\operatorname{cov}(\xi, \eta)}{\sqrt{\operatorname{var}(\xi)\operatorname{var}(\eta)}} = \operatorname{cor}(\xi, \eta) = \alpha = \beta = \text{unknown} \end{aligned} \tag{3.411}$$
$$\eta = \beta\xi + v = \beta\boldsymbol{\gamma}'\boldsymbol{x} + v \tag{3.412}$$
$$\xi = \alpha\eta + \zeta = \beta\boldsymbol{\omega}'\boldsymbol{y} + \zeta \tag{3.413}$$

Step 2, predictor specification. In order to make the weights γ and ω estimable, ModeB predictor specifications are applied to the inner relations Eqs.3.412, 3.413, after the latent predictors have been substituted by the weighted aggregates from Eq.3.407.

$$\begin{aligned}\eta = \beta\gamma'\boldsymbol{x} + \upsilon, \qquad & \tilde{\eta} \equiv \mathrm{E}(\eta|\boldsymbol{x}) = \beta\gamma'\boldsymbol{x} \\ & \Longrightarrow \mathrm{E}(\upsilon) = 0 \\ & \Longrightarrow \mathrm{cov}(\upsilon;\boldsymbol{x}) = \mathrm{cov}(\upsilon;\tilde{\eta}) = \mathbf{0} \\ & \Longrightarrow \mathrm{cov}(\eta;\boldsymbol{x}) = \mathrm{cov}(\tilde{\eta};\boldsymbol{x}) = \beta\gamma'\,\mathrm{var}(\boldsymbol{x})\end{aligned} \tag{3.414}$$

$$\begin{aligned}\xi = \alpha\omega'\boldsymbol{y} + \zeta, \qquad & \tilde{\xi} \equiv \mathrm{E}(\xi|\boldsymbol{y}) = \alpha\omega'\boldsymbol{y} \\ & \Longrightarrow \mathrm{E}(\zeta) = 0 \\ & \Longrightarrow \mathrm{cov}(\zeta;\boldsymbol{y}) = \mathrm{cov}(\zeta;\tilde{\xi}) = \mathbf{0} \\ & \Longrightarrow \mathrm{cov}(\xi;\boldsymbol{y}) = \mathrm{cov}(\tilde{\xi};\boldsymbol{y}) = \alpha\omega'\,\mathrm{var}(\boldsymbol{y}).\end{aligned} \tag{3.415}$$

Step 3, predictand-predictor covariance. Two expression for $\mathrm{cov}(\boldsymbol{x};\eta)$, the inductive expression $\boldsymbol{\Sigma}_{xy}\omega$ from Eq.3.408 and the deductive expression $\beta\boldsymbol{\Sigma}_{xx}\gamma$ from Eq.3.414, are equated to give:

$$\boldsymbol{\Sigma}_{xy}\omega = \beta\boldsymbol{\Sigma}_{xx}\gamma$$

and two expressions for $\mathrm{cov}(\boldsymbol{y};\xi)$, the inductive expression $\boldsymbol{\Sigma}_{yx}\gamma$ from Eq.3.408 and the inductive expression $\beta\omega\boldsymbol{\Sigma}_{yy}$ from Eq.3.415, are equated to give:

$$\boldsymbol{\Sigma}_{yx}\gamma = \alpha\boldsymbol{\Sigma}_{yy}\omega.$$

Both LVs are standardized, so the regressions are equal to the covariance, $\rho \equiv \rho_{\xi\eta} = \alpha = \beta$, and we write the two equations jointly as:

$$\begin{bmatrix} \mathbf{0} & \boldsymbol{\Sigma}_{xy} \\ \boldsymbol{\Sigma}_{yx} & \mathbf{0} \end{bmatrix}\begin{bmatrix} \gamma \\ \omega \end{bmatrix} = \rho\begin{bmatrix} \boldsymbol{\Sigma}_{xx} & \mathbf{0} \\ \mathbf{0} & \boldsymbol{\Sigma}_{yy} \end{bmatrix}\begin{bmatrix} \gamma \\ \omega \end{bmatrix} \tag{3.416}$$

Step 4, stationary relation. When the two equations of Step 3 are premultiplied by the inverse of the symmetric covariance matrix on the right-hand side,

$$\begin{aligned}\boldsymbol{\Sigma}_{xx}^{-1}\boldsymbol{\Sigma}_{xy}\omega &= \rho\gamma \\ \boldsymbol{\Sigma}_{yy}^{-1}\boldsymbol{\Sigma}_{yx}\gamma &= \rho\omega,\end{aligned} \tag{3.417}$$

they can be written jointly as

$$\begin{bmatrix} \mathbf{0} & \boldsymbol{\Sigma}_{xx}^{-1}\boldsymbol{\Sigma}_{xy} \\ \boldsymbol{\Sigma}_{yy}^{-1}\boldsymbol{\Sigma}_{yx} & \mathbf{0} \end{bmatrix}\begin{bmatrix} \gamma \\ \omega \end{bmatrix} = \rho\begin{bmatrix} \gamma \\ \omega \end{bmatrix}, \tag{3.418}$$

which has the form of the asymmetric one-matrix eigenvalue problem, $\boldsymbol{A}\boldsymbol{w} = \lambda\boldsymbol{g}$, with left and right eigenvectors being equal.

On the other hand, the two equations Eq.3.417 can be inserted into each other,

$$\boldsymbol{\Sigma}_{yy}^{-1}\boldsymbol{\Sigma}_{yx}\boldsymbol{\Sigma}_{xx}^{-1}\boldsymbol{\Sigma}_{xy}\omega = \rho^2\omega, \tag{3.419}$$

and a symmetric eigenvalue problem emerges, with eigenvector ω and eigenvalue ρ^2. After ω has been found from Eq.3.419, the other vector γ can be obtained by Eq.3.417.

Table 3.7: PLS raw data algorithm for canonical correlation model

$\boldsymbol{x} = [x_\ell] = [x_{\ell n}]$	
$\boldsymbol{y} = [y_k] = [x_{kn}]$	Manifest variables are centered to zero mean.
${}_0\boldsymbol{w}$	Starting values are (nearly) arbitrary.
Repeat Steps a to d, beginning at Step d, for cycles $c = 0, 1, 2, \ldots$	
a	Compute weights $\boldsymbol{g}$
$\hat{\eta}_n = \sum_\ell \breve{g}_\ell x_{\ell n} + \text{residuum}_n$	gives $\breve{g}_\ell$ by multiple regression
$\breve{\boldsymbol{g}} := \boldsymbol{S}_{xx}^{-1} \operatorname{cov}(\boldsymbol{x}, \hat{\eta}) = \boldsymbol{S}_{xx}^{-1} \boldsymbol{S}_{xy} \boldsymbol{w}$	
b	Compute standardized $\hat{\xi}$
$\breve{\xi}_n := \sum_k \breve{g}_\ell x_{\ell n}$	
$\breve{r}_\xi^2 := \bar{\sum}_n \breve{\xi}_n^2 = \breve{\boldsymbol{g}}' \boldsymbol{S}_{xx} \breve{\boldsymbol{g}}$	variance of $\breve{\xi}$
$g_\ell := \breve{g}_\ell / \breve{r}_\xi$	rescaled weights
$\hat{\xi}_n := \breve{\xi}_n / \breve{r}_\xi = \sum_k g_\ell x_{\ell n}$	now $\operatorname{var}(\hat{\xi}) = \boldsymbol{g}' \boldsymbol{S}_{xx} \boldsymbol{g} = 1$
c	Compute weights $\boldsymbol{w}$
$\hat{\xi}_n = \breve{w}_k y_{kn} + \text{residuum}_n$	gives $\breve{w}_k$ by multiple regression
$\breve{\boldsymbol{w}} := \boldsymbol{S}_{yy}^{-1} \operatorname{cov}(\boldsymbol{y}, \hat{\xi}) = \boldsymbol{S}_{yy}^{-1} \boldsymbol{S}_{yx} \boldsymbol{g}$	
d	Compute standardized $\hat{\eta}$
$\breve{\eta}_n := \sum_k \breve{w}_k y_{kn}$	
$\breve{r}_\eta^2 := \bar{\sum}_n \breve{\eta}_n^2 = \breve{\boldsymbol{w}}' \boldsymbol{S}_{yy} \breve{\boldsymbol{w}}$	variance of $\breve{\eta}$
$w_k := \breve{w}_k / \breve{r}_\eta$	rescaled weights
$\hat{\eta}_n := \breve{\eta}_n / \breve{r}_\eta = \sum_k w_k y_{kn}$	now $\operatorname{var}(\hat{\eta}) = \boldsymbol{w}' \boldsymbol{\Sigma}_{yy} \boldsymbol{w} = 1$
until the weights stabilize	
	Note: Iteration index ${}_c(\cdot)$ omitted from $\hat{\xi}$, $\hat{\eta}$, g, w, $\breve{r}$.
	Note: Intermediate values with "wrong" scale are are marked with a brĕve accent.

Step 5a, PLS algorithm for raw data. The PLS algorithm in Table 3.7 for the canonical correlation model operates with raw data $x_{\ell n}$ and y_{kn}. In each iteration cycle (Steps b and d) the LV scores $\hat{\xi}_n$ and $\hat{\eta}_n$ are computed and properly scaled. The algorithm consists of two regressions (Steps a and c) corresponding to predictors specifications Eqs.3.415, 3.414. The computation of the regression coefficients is repeated iteratively. After the iterative estimation of the weight vectors the loadings are estimated by Eq.3.404*ss*.

Table 3.8: PLS covariance algorithm for canonical correlation model

	S_{xx}, S_{yy}, S_{xy}	Covariance (or correlation) matrices.
	${}_0\boldsymbol{w}$	Starting values are (nearly) arbitrary.
Repeat Steps a to d, beginning at Step d, for cycles $c = 0, 1, 2, \ldots$		
a	$\breve{\boldsymbol{g}} \equiv r\boldsymbol{g} := \boldsymbol{S}_{xx}^{-1}\boldsymbol{S}_{xy}\boldsymbol{w}$	
b	$\boldsymbol{g} := \frac{\breve{\boldsymbol{g}}}{\sqrt{\breve{\boldsymbol{g}}'\boldsymbol{S}_{xx}\breve{\boldsymbol{g}}}}$	from Eq.3.409
c	$\breve{\boldsymbol{w}} \equiv r\boldsymbol{w} := \boldsymbol{S}_{yy}^{-1}\boldsymbol{S}_{yx}\boldsymbol{g}$	
d	$\boldsymbol{w} := \frac{\breve{\boldsymbol{w}}}{\sqrt{\breve{\boldsymbol{w}}'\boldsymbol{S}_{yy}\breve{\boldsymbol{w}}}}$	from Eq.3.410
until the weights stabilize		
		Note: Iteration index ${}_c(\cdot)$ omitted from $\boldsymbol{g}$, $\boldsymbol{w}$.
		Note: Intermediate values with "wrong" scale are are marked with a brĕve accent.

Step 5b, PLS algorithm for covariance. The LV scores in Tb.3.7 are used to compute covariances and variances. These statistics, however, can be computed from other covariances, without using the raw data: $\text{cov}(\boldsymbol{x}; \hat{\eta}) = \boldsymbol{S}_{xy}\boldsymbol{w}$, $\text{var}(\breve{\xi}) = \breve{\boldsymbol{g}}\boldsymbol{S}_{xx}\breve{\boldsymbol{g}}$. The algorithm in Tb.3.8 omits the raw data and uses the covariances. Otherwise, the algorithms in Tb.3.7 and Tb.3.8 are completely equivalent and lead to the same $\boldsymbol{g}$ and $\boldsymbol{w}$.

Step 5c, Power algorithm. The four steps in Tb.3.8 can be combined into two steps:

$$\begin{aligned} \breve{\boldsymbol{w}} \equiv r^2\boldsymbol{w} &:= \boldsymbol{S}_{yy}^{-1}\boldsymbol{S}_{yx}\boldsymbol{S}_{xx}^{-1}\boldsymbol{S}_{xy}\boldsymbol{w} \\ \boldsymbol{w} &:= \frac{\breve{\boldsymbol{w}}}{\sqrt{\breve{\boldsymbol{w}}'\boldsymbol{S}_{yy}\breve{\boldsymbol{w}}}}. \end{aligned} \tag{3.420}$$

This is clearly the power algorithm corresponding to the symmetric eigenproblem Eq.3.419. The algorithm converges to the dominant eigenvalue r^2 and the corresponding eigenvector $\boldsymbol{w}$. The convergence property holds for the two-line algorithm Eq.3.420 as well as the equivalent four-line algorithm Tb.3.8 and the twelve-line algorithm Tb.3.7.

Remark 3★13 Eq.3.416 can be rearranged into the normal equations system,

$$\left(\begin{bmatrix} 0 & \Sigma_{xy} \\ \Sigma_{yx} & 0 \end{bmatrix} - \rho \begin{bmatrix} \Sigma_{xx} & 0 \\ 0 & \Sigma_{yy} \end{bmatrix} \right) \begin{bmatrix} \gamma \\ \omega \end{bmatrix} = 0. \tag{3.421}$$

This is the two-matrix eigenproblem, like $(\boldsymbol{A} - \lambda\boldsymbol{B})\boldsymbol{w} = \boldsymbol{0}$, cf. Eq.3.177. A vector $(\gamma', \omega')'$ satisfying Eq.3.421 can be found when a value ρ has been found that makes the determinant of the matrix in Eq.3.421 zero, i.e.

$$\begin{vmatrix} -\rho\boldsymbol{\Sigma}_{xx} & \boldsymbol{\Sigma}_{xy} \\ \boldsymbol{\Sigma}_{yx} & -\rho\boldsymbol{\Sigma}_{yy} \end{vmatrix} = 0 \tag{3.422}$$

These two equations are the traditional representation of the canonical correlation model. This proves that the model derived by predictor specification amounts in fact to the CC model. The equation system has as many solutions as there are distinct eigenvalues in this system. The power algorithm (and the PLS algorithm) determine only the dominant solution.

Remark 3⋆14 Instead of the passage from Eq.3.416 to Eq.3.418 one may develop Eq.3.421 along the lines of Eq.3.179 into a symmetric, one-matrix eigenvalue problem:

$$\begin{bmatrix} \mathbf{0} & \mathbf{\Sigma}^*_{xy} \\ \mathbf{\Sigma}^*_{yx} & \mathbf{0} \end{bmatrix} \begin{bmatrix} \gamma^* \\ \omega^* \end{bmatrix} = \rho \begin{bmatrix} \gamma^* \\ \omega^* \end{bmatrix} \tag{3.423}$$

with $\mathbf{\Sigma}^*_{xy} = \mathbf{\Sigma}^{-1/2}_{xx}\mathbf{\Sigma}_{xy}\mathbf{\Sigma}^{-1/2}_{yy}$, $\boldsymbol{\gamma}^* = \mathbf{\Sigma}^{1/2}_{xx}\boldsymbol{\gamma}$, and $\boldsymbol{\omega}^* = \mathbf{\Sigma}^{1/2}_{yy}\boldsymbol{\omega}$. The resulting vectors $\boldsymbol{\gamma}^*$ and $\boldsymbol{\omega}^*$ are estimated in "estimation metric" and have to be backtransformed into "interpretation metric" by $\boldsymbol{\gamma} := \mathbf{\Sigma}^{-1/2}_{xx}\boldsymbol{\gamma}^*$ and $\boldsymbol{\omega} := \mathbf{\Sigma}^{-1/2}_{yy}\boldsymbol{\omega}^*$. This is the approach that Horst (1965) extended to multiblock models.

Remark 3⋆15 This representation of the problem allows for another solution, after the equation system has been split up into $\mathbf{\Sigma}^*_{yx}\boldsymbol{\gamma}^* = \rho\boldsymbol{\omega}^*$ and $\mathbf{\Sigma}^*_{yx}\boldsymbol{\omega}^* = \rho\boldsymbol{\gamma}$. Now we have the rectangular singular value problem. To solve this instead of the eigenvalue problem presented before is recommended by numerical analysts (cf. Golub & Van Loan 1983), but it does not allow for a generalization for multi-block models.

Remark 3⋆16 There are different ways of defining the canonical correlation model. Progressing from different starting points, which all provide different meanings and interpretations, the algebraic elaborations of the models end in equivalent conditions, namely Eq.3.422. Two of these methods will now be presented.

The very name "Canonical Correlation" implies seeking a transformation that gives the correlation matrix $\mathbf{\Sigma}_{xy}$ the so-called canonical, normal, or Jordan-Weierstraß canonical form (see Zurmühl 1964:257, Bodewig 1956:79). To this end, the symmetric matrix $\boldsymbol{S}$ of the eigenproblem $(\boldsymbol{S} - \lambda\mathbf{I})\boldsymbol{x} = \mathbf{0}$ is transformed into the canonical form $\mathbf{\Lambda} = \boldsymbol{T}^{-1}\boldsymbol{S}\boldsymbol{T}$. In case of no multiple eigenvalues $\boldsymbol{T}$ is the matrix of eigenvectors and $\mathbf{\Lambda}$ is the diagonal matrix of eigenvalues.

In the two-matrix eigenproblem $(\boldsymbol{S} - \lambda\boldsymbol{D})\boldsymbol{x} = \mathbf{0}$ the canonical form is defined as $\mathbf{\Lambda} = \boldsymbol{T}^{-1}\boldsymbol{D}^{-1}\boldsymbol{S}\boldsymbol{T}$, where again, in case of there no being multiple eigenvalues, $\mathbf{\Lambda}$ has diagonal form. If we take $\boldsymbol{S} = \mathbf{\Sigma}_{xy}\mathbf{\Sigma}^{-1}_{yy}\mathbf{\Sigma}_{yx}$ and $\boldsymbol{D} = \mathbf{\Sigma}_{xx}$, the canonical form $\mathbf{\Lambda}$ is the diagonal matrix of canonical correlations. In this strict mathematical sense only principal components (one-matrix eigenproblem) and the canonical correlation model (two-matrix eigenproblem) are "canonical" (while neither the canonical redundancy model nor the generalized canonical correlation for more than two blocks of variables are canonical in this strict sense).

Remark 3⋆17 Another approach to CC starts from the maximal possible correlation between the two sets of variables. This means that a standardized composite $\hat{\xi}$ of the $\boldsymbol{x}$-variables and a standardized composite $\hat{\eta}$ of the $\boldsymbol{y}$-variables are sought such that they correlate maximally, $\text{cor}(\xi;\eta) = \boldsymbol{\gamma}'\mathbf{\Sigma}_{xy}\boldsymbol{\omega} \overset{!}{=} \max$. The quantity to be maximized is:

$$g = \boldsymbol{\gamma}'\boldsymbol{\Sigma}_{xy}\boldsymbol{\omega} - \frac{\beta}{2}(\boldsymbol{\gamma}'\boldsymbol{\Sigma}_{xx}\boldsymbol{\gamma} - 1) - \frac{\alpha}{2}(\boldsymbol{\omega}'\boldsymbol{\Sigma}_{yy}\boldsymbol{\omega} - 1) \stackrel{!}{=} \max, \tag{3.424}$$

where α and β are unknown Lagrangian multipliers. The maximum of g is obtained when the first derivatives with respect to the parameters are equal to zero:

$$\frac{\partial g}{\partial \boldsymbol{\gamma}}' = \boldsymbol{\Sigma}_{xy}\boldsymbol{\omega} - \beta\boldsymbol{\Sigma}_{yy}\boldsymbol{\omega} = 0 \tag{3.425}$$

$$\frac{\partial g}{\partial \boldsymbol{\omega}}' = \boldsymbol{\Sigma}_{yx}\boldsymbol{\gamma} - \alpha\boldsymbol{\Sigma}_{xx}\boldsymbol{\gamma} = 0, \tag{3.426}$$

This, then, results in the normal equation system Eq.3.421, after it has been established that $\alpha = \beta$.

Despite the title of his article, Hotelling's (1935) derivation for the estimation of "the most predictable criterion" did not follow the lines of the PLS procedure but the second approach which maximizes the correlation between the composites.

3.4.3 The Principal Predictor Model

Step 1, model relations. Using the notation for two-block models as defined in Sc.3.4.1 the structural relations of the Principal Predictor model are depicted in Tb.3.6c. It simplifies the following arguments when we choose the scales of the components ξ and η as follows:

$$\rho_\xi^2 \equiv \mathrm{var}(\xi) = \boldsymbol{\gamma}'\boldsymbol{\Sigma}_{xx}\boldsymbol{\gamma} = 1 \tag{3.427}$$

$$\rho_\eta^2 \equiv \mathrm{var}(\eta) = \boldsymbol{\omega}'\boldsymbol{\Sigma}_{yy}\boldsymbol{\omega} = \text{unknown, but } \boldsymbol{\omega}'\boldsymbol{\omega} = 1 \tag{3.428}$$

$$\rho_{\xi\eta} \equiv \mathrm{cov}(\xi;\eta) = \boldsymbol{\gamma}'\boldsymbol{\Sigma}_{xy}\boldsymbol{\omega} = \beta. \tag{3.429}$$

Step 2, predictor specification.

(i) The inner relation $\eta = \beta\xi + \upsilon$ is subject to predictor specification, after ξ has been replaced by $\boldsymbol{\gamma}'\boldsymbol{x}$:

$$\begin{aligned} \eta = \beta\boldsymbol{\gamma}'\boldsymbol{x} + \upsilon, \qquad & \tilde{\eta} \equiv \mathrm{E}(\eta|\boldsymbol{x}) = \beta\boldsymbol{\gamma}'\boldsymbol{x} \\ & \Longrightarrow \mathrm{E}(\upsilon) = 0 \\ & \Longrightarrow \mathrm{cov}(\upsilon;\boldsymbol{x}) = \mathrm{cov}(\upsilon;\tilde{\eta}) = \mathbf{0} \\ & \Longrightarrow \mathrm{cov}(\eta;\boldsymbol{x}) = \mathrm{cov}(\tilde{\eta};\boldsymbol{x}) = \beta\boldsymbol{\gamma}'\,\mathrm{var}(\boldsymbol{x}). \end{aligned} \tag{3.430}$$

(ii) The outer relation for $\boldsymbol{y}$ is subject to interlocking predictor specification (see Sc.3.1.6), after η has been replaced from the inner relation Eq.3.403:

$$\boldsymbol{y} = \tilde{\boldsymbol{\pi}}\eta + \tilde{\boldsymbol{\epsilon}} = \tilde{\boldsymbol{\pi}}(\beta\xi + \upsilon) + \tilde{\boldsymbol{\epsilon}} = \beta\tilde{\boldsymbol{\pi}}\xi + (\tilde{\boldsymbol{\pi}}\upsilon + \tilde{\boldsymbol{\epsilon}}) \tag{3.431}$$

$$\begin{aligned} \boldsymbol{y} = \tilde{\boldsymbol{\pi}}\beta\xi + \tilde{\boldsymbol{\pi}}\upsilon + \tilde{\boldsymbol{\epsilon}}, \qquad & \tilde{\boldsymbol{y}} \equiv \mathrm{E}(\boldsymbol{y}|\xi) = \tilde{\boldsymbol{\pi}}\beta\xi \\ & \Longrightarrow \mathrm{E}(\tilde{\boldsymbol{\pi}}\upsilon + \tilde{\boldsymbol{\epsilon}}) = \mathbf{0} \\ & \Longrightarrow \mathrm{cov}(\tilde{\boldsymbol{\pi}}\upsilon + \tilde{\boldsymbol{\epsilon}};\xi) = \mathrm{cov}(\tilde{\boldsymbol{\pi}}\upsilon + \tilde{\boldsymbol{\epsilon}};\tilde{\boldsymbol{y}}) = \mathbf{0} \\ & \Longrightarrow \mathrm{cov}(\boldsymbol{y};\xi) = \mathrm{cov}(\tilde{\boldsymbol{y}};\xi) = \tilde{\boldsymbol{\pi}}\beta\,\mathrm{var}(\xi). \end{aligned} \tag{3.432}$$

The additional covariances (by Eq.3.162) are:

$$\text{cov}(\xi; \tilde{\boldsymbol{\epsilon}}) = \mathbf{0} \tag{3.433}$$
$$\text{cov}(\eta; \tilde{\boldsymbol{\epsilon}}) = \text{cov}(\boldsymbol{\upsilon}; \tilde{\boldsymbol{\epsilon}}). \tag{3.434}$$

Notice that $\tilde{\boldsymbol{\pi}}$ is not the loading of the relation $\boldsymbol{y} = \boldsymbol{\pi}\eta + \boldsymbol{\epsilon}$, but the "cross-loading", or "redundancy", for the relation $\boldsymbol{y} = \tilde{\boldsymbol{\pi}}\beta\xi + \tilde{\boldsymbol{\epsilon}}$. It is the loading not on the LV of the own block, but of the other block. Under the rationale for ModeA weights (cf. Sc.3.3.4) we choose the weights to be proportional to the cross-loadings,

$$\boldsymbol{\omega} := \tilde{\boldsymbol{\pi}}(\tilde{\boldsymbol{\pi}}'\tilde{\boldsymbol{\pi}})^{-1/2}, \tag{3.435}$$

so that $\boldsymbol{\omega}'\boldsymbol{\omega} = 1$, $\tilde{\boldsymbol{\pi}} = \boldsymbol{\omega}(\tilde{\boldsymbol{\pi}}'\tilde{\boldsymbol{\pi}})^{1/2}$, and $\boldsymbol{\omega}'\tilde{\boldsymbol{\pi}} = (\tilde{\boldsymbol{\pi}}'\tilde{\boldsymbol{\pi}})^{1/2}$.

Step 3, predictand-predictor covariance.

(i) The two expression for $\text{cov}(\eta; \boldsymbol{x})$, the inductive expression $\boldsymbol{\Sigma}_{xy}\boldsymbol{\omega}$ from Eq.3.408 and the deductive expression $\beta\boldsymbol{\Sigma}_{xx}\boldsymbol{\gamma}$ from Eq.3.430 are equated to give

$$\boldsymbol{\Sigma}_{xy}\boldsymbol{\omega} = \beta\boldsymbol{\Sigma}_{xx}\boldsymbol{\gamma}. \tag{3.436}$$

Premultiplication by $\boldsymbol{\gamma}'$ gives for the left-hand side $\boldsymbol{\gamma}'\boldsymbol{\Sigma}_{xy}\boldsymbol{\omega} = \text{cov}(\xi; \eta) \equiv \rho$ and for the right-hand side $\boldsymbol{\gamma}'\beta\boldsymbol{\Sigma}_{xx}\boldsymbol{\gamma} = \beta\rho_\xi^2 = \beta$, which assures us that $\rho = \beta$, and we rewrite the equation:

$$\boldsymbol{\Sigma}_{xy}\boldsymbol{\omega} = \rho\boldsymbol{\Sigma}_{xx}\boldsymbol{\gamma}. \tag{3.437}$$

(ii) The two expressions for $\text{cov}(\boldsymbol{y}; \xi)$, the inductive expression $\boldsymbol{\Sigma}_{yx}\boldsymbol{\gamma}$ from Eq.3.408 and the deductive expression $\tilde{\boldsymbol{\pi}}\beta\rho_\xi^2$ from Eq.3.432 are equated to give

$$\boldsymbol{\Sigma}_{yx}\boldsymbol{\gamma} = \tilde{\boldsymbol{\pi}}\beta\rho_\xi^2. \tag{3.438}$$

Premultiplication by $\boldsymbol{\omega}'$ gives for the left-hand side $\boldsymbol{\omega}'\boldsymbol{\Sigma}_{yx}\boldsymbol{\gamma} = \text{cov}(\xi; \eta) \equiv \rho$ and for the right-hand side $\boldsymbol{\omega}'\tilde{\boldsymbol{\pi}}\beta\rho_\xi^2 = (\tilde{\boldsymbol{\pi}}'\tilde{\boldsymbol{\pi}})^{1/2}\beta\rho_\xi^2$, Using $\tilde{\boldsymbol{\pi}} = \boldsymbol{\omega}(\tilde{\boldsymbol{\pi}}'\tilde{\boldsymbol{\pi}})^{1/2}$ from Eq.3.435 we rewrite the equation this relation is written as:

$$\boldsymbol{\Sigma}_{yx}\boldsymbol{\gamma} = \boldsymbol{\omega}(\tilde{\boldsymbol{\pi}}'\tilde{\boldsymbol{\pi}})^{1/2}\beta\rho_\xi^2 = \rho\boldsymbol{\omega}. \tag{3.439}$$

The two equations Eqs.3.437, 3.439 are written jointly as:

$$\begin{bmatrix} \mathbf{0} & \boldsymbol{\Sigma}_{xy} \\ \boldsymbol{\Sigma}_{yx} & \mathbf{0} \end{bmatrix} \begin{bmatrix} \boldsymbol{\gamma} \\ \boldsymbol{\omega} \end{bmatrix} = \rho \begin{bmatrix} \boldsymbol{\Sigma}_{xx} & \mathbf{0} \\ \mathbf{0} & \mathbf{I} \end{bmatrix} \begin{bmatrix} \boldsymbol{\gamma} \\ \boldsymbol{\omega} \end{bmatrix} \tag{3.440}$$

Step 4, stationary relation. We premultiply Eq.3.437 by the inverse of the matrix on the right-hand side, and the two equations are written as:

$$\begin{aligned} \boldsymbol{\Sigma}_{xx}^{-1}\boldsymbol{\Sigma}_{xy}\boldsymbol{\omega} &= \rho\boldsymbol{\gamma} \\ \boldsymbol{\Sigma}_{yx}\boldsymbol{\gamma} &= \rho\boldsymbol{\omega}. \end{aligned} \tag{3.441}$$

If we write it as one equation system,

$$\begin{bmatrix} \mathbf{0} & \boldsymbol{\Sigma}_{xx}^{-1}\boldsymbol{\Sigma}_{xy} \\ \boldsymbol{\Sigma}_{yx} & \mathbf{0} \end{bmatrix} \begin{bmatrix} \boldsymbol{\gamma} \\ \boldsymbol{\omega} \end{bmatrix} = \rho \begin{bmatrix} \boldsymbol{\gamma} \\ \boldsymbol{\omega} \end{bmatrix}, \tag{3.442}$$

an asymmetric eigenvalue problem with equal right and left eigenvector emerges. If we insert the first line of Eq.3.441 into the second line, we get a symmetric eigenvalue problem:

$$\boldsymbol{\Sigma}_{yx}\boldsymbol{\Sigma}_{xx}^{-1}\boldsymbol{\Sigma}_{xy}\boldsymbol{\omega} = \rho^2\boldsymbol{\omega}. \tag{3.443}$$

Step 5, PLS algorithms. The PLS algorithm for the Principal Predictor model is nearly the same as for the Canonical Correlation model, the only difference being in Step c. The weights $\boldsymbol{w}$ are not regression coefficients, but covariances. Step c of the raw data algorithm (Tb.3.7) is replaced by:

$$\breve{w}_k := \bar{\sum_n} y_{kn}\hat{\xi}_n = \text{cov}(y_k; \hat{\xi})$$

Step c of the covariance algorithm (Tb.3.8) is replaced by:

$$\breve{\boldsymbol{w}} \equiv r\boldsymbol{w} := \boldsymbol{S}_{yx}\boldsymbol{g}$$

And the power algorithm Eq.3.420 is written as

$$\breve{\boldsymbol{w}} \equiv r^2\boldsymbol{w} := \boldsymbol{S}_{yx}\boldsymbol{S}_{xx}^{-1}\boldsymbol{S}_{xy}\boldsymbol{w},$$

which corresponds to the eigenproblem Eq.3.443. Hence the convergence of the three variants of the algorithm is assured under the conditions of the power algorithm.

Remark 3⋆18 Eq.3.440 can be reordered into the normal equation system

$$\left(\begin{bmatrix} \mathbf{0} & \boldsymbol{\Sigma}_{xy} \\ \boldsymbol{\Sigma}_{yx} & \mathbf{0} \end{bmatrix} - \rho \begin{bmatrix} \boldsymbol{\Sigma}_{xx} & \mathbf{0} \\ \mathbf{0} & \mathbf{I} \end{bmatrix} \right) \begin{bmatrix} \boldsymbol{\gamma} \\ \boldsymbol{\omega} \end{bmatrix} = \mathbf{0} \tag{3.444}$$

which clearly is a two-matrix eigenproblem. A vector $(\boldsymbol{\gamma}', \boldsymbol{\omega}')'$ satisfying Eq.3.444 can be found after a value ρ has been found that makes the determinant of the matrix in Eq.3.444 zero, i.e.

$$\begin{vmatrix} -\rho\boldsymbol{\Sigma}_{xx} & \boldsymbol{\Sigma}_{xy} \\ \boldsymbol{\Sigma}_{yx} & -\rho\mathbf{I} \end{vmatrix} = \mathbf{0}. \tag{3.445}$$

This determinant can be expressed as

$$\begin{aligned} |-\rho\mathbf{I}|\left|-\rho\boldsymbol{\Sigma}_{xx} - \boldsymbol{\Sigma}_{xy}(-\rho\mathbf{I})^{-1}\boldsymbol{\Sigma}_{yx}\right| &= -\rho\left|-\rho\boldsymbol{\Sigma}_{xx} - (-\rho)^{-1}\boldsymbol{\Sigma}_{xy}\boldsymbol{\Sigma}_{yx}\right| \\ &= \left|\rho^2\boldsymbol{\Sigma}_{xx} - \boldsymbol{\Sigma}_{xy}\boldsymbol{\Sigma}_{yx}\right|, \end{aligned}$$

and hence this condition is equivalent to the one presented by Wold (1975c:Eqs. 49a, 50).

Remark 3⋆19 Instead of the passage from Eq.3.440 to Eq.3.442 one may develop Eq.3.440 along the lines of Eq.3.179 into a symmetric, one-matrix eigenvalue problem:

$$\begin{bmatrix} \mathbf{0} & \mathbf{\Sigma}^*_{xy} \\ \mathbf{\Sigma}^*_{yx} & \mathbf{0} \end{bmatrix} \begin{bmatrix} \gamma^* \\ \omega^* \end{bmatrix} = \rho \begin{bmatrix} \gamma^* \\ \omega^* \end{bmatrix} \tag{3.446}$$

with $\mathbf{\Sigma}^*_{xy} = \mathbf{\Sigma}^{-1/2}_{xx}\mathbf{\Sigma}_{xy}$, $\gamma^* = \mathbf{\Sigma}^{1/2}_{xx}\gamma$, and $\omega^* = \omega$. This approach can be extended to multiblock models.

Remark 3⋆20 The unrestricted regression matrix $\mathbf{B}$ for all predictors and predictands $\boldsymbol{y} = \mathbf{B}\boldsymbol{x} + \epsilon$ is found by $\mathbf{B} = \mathbf{\Sigma}_{yx}\mathbf{\Sigma}^{-1}_{xx}$ and has the same rank as $\mathbf{\Sigma}_{yx}$, $A = \text{rank}(\mathbf{\Sigma}_{xy}) = \text{rank}(\mathbf{B})$. Like any matrix, $\mathbf{B}$ can be expressed as the product of two matrices, $\mathbf{B}_{[K\times L]} = \boldsymbol{\pi}_{[K\times A]}\gamma'_{[L\times A]}$. Often it may be possible to approximate $\mathbf{B}$ by two matrices $\boldsymbol{\gamma}^*$ and $\boldsymbol{\pi}^*$ with fewer columns, say $A^* < A$. The matrix $\mathbf{B}^* = \boldsymbol{\pi}^*_{[K\times A^*]}\gamma^{*\prime}_{[L\times A^*]}$ is called then a Rank Restricted Regression matrix, and the model Eqs.3.430,..., 3.435 may be called the Rank Restricted Regression model. The method of estimating $\mathbf{B}^*$ is shown above. The name "Principal Predictor model" used here is a short-cut notation for "principal component of a full- rank predictor" and will be explained in Sc.3.4.5

Remark 3⋆21 If only one dimension of the LVs ξ and η is estimated, we have, as a special case of the Rank Restricted Regression model, the Rank One Regression model, which is treated by Whittle and Adelman (1982) and Dijkstra (1982). An even more special case was discussed by Guttman (1941b), with $A^* = 1$ and the values of $\boldsymbol{\gamma}$ assumed as known and fixed. Guttman named his model the "most common regression", as being different and in most instances more appropriate than the "most predictable criterion" of canonical correlation analysis.

Remark 3⋆22 van den Wollenberg (1977) introduced the Principal Predictor model calling it the Redundancy model, because the criterion it maximizes is the Stewart and Love (1968) redundancy index. Two years earlier than van den Wollenberg the model has been treated by Wold (1975b) as one of three two-block models. However, he provided no special name for this model.

The cross-loadings $\beta\tilde{\boldsymbol{\pi}} = \text{cov}(\xi, \boldsymbol{y})$ are called redundancies. The very name redundancy means that the information is double, once in the $\boldsymbol{x}$- and ξ-variables and once in the $\boldsymbol{y}$- and η-variables. So it can be used for prediction. Without redundant information no prediction is possible. The redundancy index is the sum of squared cross-loadings $\beta^2\tilde{\boldsymbol{\pi}}'\tilde{\boldsymbol{\pi}}$ which happens to be the squared covariance $\rho_{\xi\eta}$ and to be maximized by the Principal Predictor model, cf. Eq.3.443. On the redundancy coefficient cf. Cohen & Cohen (1977) and Gleason (1976); on transformation of the LVs cf. Dawson-Saunders & Tatsuoka (1983), Krus, Reynolds & Krus (1976) and Tyler (1982).

Remark 3⋆23 For $\boldsymbol{x}$ we have scale-freeness. Any of the variables x_k may be multiplied by a nonzero constant c; the weight w_k will be different by a factor $1/c$, and the loading by a factor c. The LV score ξ_n, however, and the rest of the model will be the same. For the y-variables the property of scale-freeness does not hold.

3.4.4 The Interbattery Factor Model

Tucker's (1958) Interbattery Factor (IBF) Model (see Tb.3.6c) assumes a factor ζ which is measured by two batteries of tests, $\boldsymbol{x}$ and $\boldsymbol{y}$. Either of the batteries can be taken as measurement of ζ:

$$\xi = \gamma'\boldsymbol{x}, \qquad \gamma'\gamma = 1 \tag{3.447}$$

$$\eta = \omega'\boldsymbol{y}, \qquad \omega'\omega = 1; \tag{3.448}$$

or else both batteries can be put together into one measurement:

$$\zeta = \xi + \eta \tag{3.449}$$

Step 1, model relations.

$$\xi = \breve{\alpha}\zeta + \breve{\upsilon}_x = \breve{\alpha}(\xi + \eta) + \breve{\upsilon}_x = \breve{\alpha}\xi + \breve{\alpha}\eta + \breve{\upsilon}_x \equiv \alpha\eta + \upsilon_x \tag{3.450}$$

with $\alpha = \breve{\alpha}/(1 - \breve{\alpha})$ and $\upsilon_x = \breve{\upsilon}_x/(1 - \breve{\alpha})$. Similarly,

$$\eta = \breve{\beta}\zeta + \breve{\upsilon}_y = \breve{\beta}(\xi + \eta) + \breve{\upsilon}_y = \breve{\beta}\zeta + \breve{\beta}\eta + \breve{\upsilon}_y \equiv \beta\xi + \upsilon_y \tag{3.451}$$

$$\boldsymbol{x} = \tilde{\boldsymbol{\tau}}\xi + \tilde{\boldsymbol{\delta}} = \alpha\tilde{\boldsymbol{\tau}}\eta + \tilde{\boldsymbol{\tau}}\upsilon_x + \tilde{\boldsymbol{\delta}} \tag{3.452}$$

$$\boldsymbol{y} = \tilde{\boldsymbol{\pi}}\eta + \tilde{\boldsymbol{\epsilon}} = \beta\tilde{\boldsymbol{\pi}}\xi + \tilde{\boldsymbol{\pi}}\upsilon_y + \tilde{\boldsymbol{\epsilon}}. \tag{3.453}$$

For the meaning of υ_y, υ_x cf. Sc.3.3.4. The interbattery factor ζ as well as the loadings α and β are not estimated; we assume $\alpha = \beta$.

Step 2, predictor specification. We apply ModeA predictors specification to both blocks. Thereby we require the LVs to be predictors of the opposite block of MVs:

$$\begin{aligned} \boldsymbol{x} = \alpha\tilde{\boldsymbol{\tau}}\eta + \tilde{\boldsymbol{\tau}}\upsilon_x + \tilde{\boldsymbol{\delta}}, \qquad & \tilde{\boldsymbol{x}} \equiv \mathrm{E}(\boldsymbol{x}|\eta) = \alpha\tilde{\boldsymbol{\tau}}\eta \\ & \Longrightarrow \mathrm{E}(\tilde{\boldsymbol{\tau}}\upsilon_x + \tilde{\boldsymbol{\delta}}) = \mathbf{0} \\ & \Longrightarrow \mathrm{cov}(\tilde{\boldsymbol{\tau}}\upsilon_x + \tilde{\boldsymbol{\delta}}; \eta) = \mathrm{cov}(\tilde{\boldsymbol{\tau}}\upsilon_x + \tilde{\boldsymbol{\delta}}; \tilde{\boldsymbol{x}}) = \mathbf{0} \\ & \Longrightarrow \mathrm{cov}(\boldsymbol{x}; \eta) = \mathrm{cov}(\tilde{\boldsymbol{x}}; \eta) = \alpha\tilde{\boldsymbol{\tau}}\,\mathrm{var}(\eta) \end{aligned} \tag{3.454}$$

$$\begin{aligned} \boldsymbol{y} = \beta\tilde{\boldsymbol{\pi}}\xi + \tilde{\boldsymbol{\pi}}\upsilon_y + \tilde{\boldsymbol{\epsilon}}, \qquad & \tilde{\boldsymbol{y}} \equiv \mathrm{E}(\boldsymbol{y}|\xi) = \beta\tilde{\boldsymbol{\pi}}\xi \\ & \Longrightarrow \mathrm{E}(\tilde{\boldsymbol{\pi}}\upsilon_y + \tilde{\boldsymbol{\epsilon}}) = \mathbf{0} \\ & \Longrightarrow \mathrm{cov}(\tilde{\boldsymbol{\pi}}\upsilon_y + \tilde{\boldsymbol{\epsilon}}; \xi) = \mathrm{cov}(\tilde{\boldsymbol{\pi}}\upsilon_y + \tilde{\boldsymbol{\epsilon}}; \tilde{\boldsymbol{y}}) = \mathbf{0} \\ & \Longrightarrow \mathrm{cov}(\boldsymbol{y}; \xi) = \mathrm{cov}(\tilde{\boldsymbol{y}}; \xi) = \beta\tilde{\boldsymbol{\pi}}\,\mathrm{var}(\xi). \end{aligned} \tag{3.455}$$

The ensuing cross-loadings $\tilde{\boldsymbol{\tau}}$, $\tilde{\boldsymbol{\pi}}$ are used as weights, after a proper rescaling:

$$\omega := \tilde{\boldsymbol{\pi}}(\tilde{\boldsymbol{\pi}}'\tilde{\boldsymbol{\pi}})^{-1/2} \qquad \gamma := \tilde{\boldsymbol{\tau}}(\tilde{\boldsymbol{\tau}}'\tilde{\boldsymbol{\tau}})^{-1/2}$$

$$\begin{aligned} \omega'\omega &= 1 & \gamma'\gamma &= 1 \\ \tilde{\boldsymbol{\pi}} &= \omega(\tilde{\boldsymbol{\pi}}'\tilde{\boldsymbol{\pi}})^{1/2} & \tilde{\boldsymbol{\tau}} &= \gamma(\tilde{\boldsymbol{\tau}}'\tilde{\boldsymbol{\tau}})^{1/2} \\ \omega'\tilde{\boldsymbol{\pi}} &= (\tilde{\boldsymbol{\pi}}'\tilde{\boldsymbol{\pi}})^{1/2} & \gamma'\tilde{\boldsymbol{\tau}} &= (\tilde{\boldsymbol{\tau}}'\tilde{\boldsymbol{\tau}})^{1/2} \end{aligned} \tag{3.456}$$

Step 3, predictand-predictor covariance. Two expression for cov($\boldsymbol{x};\eta$), the inductive expression $\boldsymbol{\Sigma}_{xy}\boldsymbol{\omega}$ from Eq.3.408 and the deductive expression $\alpha\tilde{\boldsymbol{\tau}}\rho_\eta^2$ from Eq.3.454 are equated to give:

$$\boldsymbol{\Sigma}_{xy}\boldsymbol{\omega} = \alpha\tilde{\boldsymbol{\tau}}\rho_\eta^2. \tag{3.457}$$

Premultiplication by $\boldsymbol{\gamma}'$ gives $\rho = \boldsymbol{\gamma}'\boldsymbol{\Sigma}_{xy}\boldsymbol{\omega} = \boldsymbol{\gamma}'\tilde{\boldsymbol{\tau}}\alpha\rho_\eta^2 = (\tilde{\boldsymbol{\tau}}'\tilde{\boldsymbol{\tau}})^{1/2}\alpha\rho_\eta^2$. We replace $\tilde{\boldsymbol{\tau}}$ from Eq.3.456:

$$\boldsymbol{\Sigma}_{xy}\boldsymbol{\omega} = \boldsymbol{\gamma}(\tilde{\boldsymbol{\tau}}'\tilde{\boldsymbol{\tau}})^{1/2}\rho_\eta^2 = \rho\boldsymbol{\gamma} \tag{3.458}$$

Two expressions for cov($\boldsymbol{y};\xi$), the inductive expression $\boldsymbol{\Sigma}_{yx}\boldsymbol{\gamma}$ from Eq.3.408 and the inductive expression $\beta\tilde{\boldsymbol{\pi}}\rho_\xi^2$ from Eq.3.455 are equated to give:

$$\boldsymbol{\Sigma}_{yx}\boldsymbol{\gamma} = \beta\tilde{\boldsymbol{\pi}}\rho_\xi^2. \tag{3.459}$$

Premultiplication by $\boldsymbol{\omega}'$ gives $\rho = \boldsymbol{\omega}'\boldsymbol{\Sigma}_{yx}\boldsymbol{\gamma} = \boldsymbol{\omega}'\tilde{\boldsymbol{\pi}}\beta\rho_\xi^2 = (\tilde{\boldsymbol{\pi}}'\tilde{\boldsymbol{\pi}})^{1/2}\beta\rho_\xi^2$ We replace $\tilde{\boldsymbol{\pi}}$ from Eq.3.456:

$$\boldsymbol{\Sigma}_{yx}\boldsymbol{\gamma} = \boldsymbol{\omega}(\tilde{\boldsymbol{\pi}}'\tilde{\boldsymbol{\pi}})^{1/2}\rho_\xi^2 = \rho\boldsymbol{\omega} \tag{3.460}$$

Step 4, the stationary condition. The two equations Eqs.3.458, 3.460 are written jointly in the form of a one-matrix eigenvalue problem:

$$\begin{bmatrix} \mathbf{0} & \boldsymbol{\Sigma}_{xy} \\ \boldsymbol{\Sigma}_{yx} & \mathbf{0} \end{bmatrix} \begin{bmatrix} \boldsymbol{\gamma} \\ \boldsymbol{\omega} \end{bmatrix} = \rho \begin{bmatrix} \boldsymbol{\gamma} \\ \boldsymbol{\omega} \end{bmatrix}. \tag{3.461}$$

The normal equation system is:

$$\left(\begin{bmatrix} \mathbf{0} & \boldsymbol{\Sigma}_{xy} \\ \boldsymbol{\Sigma}_{yx} & \mathbf{0} \end{bmatrix} - \rho \begin{bmatrix} \mathbf{I} & \mathbf{0} \\ \mathbf{0} & \mathbf{I} \end{bmatrix} \right) \begin{bmatrix} \boldsymbol{\gamma} \\ \boldsymbol{\omega} \end{bmatrix} = \mathbf{0}. \tag{3.462}$$

Step 5, PLS algorithm. Starting with arbitrary starting values $\boldsymbol{\omega}$, Eqs.3.463,..., 3.466 are repeated until convergence:

$$\breve{\boldsymbol{\gamma}} = \boldsymbol{\gamma}\rho := \boldsymbol{\Sigma}_{xy}\boldsymbol{\omega} \qquad \text{by Eq.3.458} \tag{3.463}$$

$$\boldsymbol{\gamma} := \breve{\boldsymbol{\gamma}}(\breve{\boldsymbol{\gamma}}'\breve{\boldsymbol{\gamma}})^{-1/2} \tag{3.464}$$

$$\breve{\boldsymbol{\omega}} = \boldsymbol{\omega}\rho := \boldsymbol{\Sigma}_{yx}\boldsymbol{\gamma} \qquad \text{by Eq.3.460} \tag{3.465}$$

$$\boldsymbol{\omega} := \breve{\boldsymbol{\omega}}(\breve{\boldsymbol{\omega}}'\breve{\boldsymbol{\omega}})^{-1/2} \tag{3.466}$$

Then $\rho_\eta^2 = \boldsymbol{\omega}'\boldsymbol{\Sigma}_{yy}\boldsymbol{\omega}$, $\rho_\xi^2 = \boldsymbol{\gamma}'\boldsymbol{\Sigma}_{xx}\boldsymbol{\gamma}$, and $\rho = \boldsymbol{\gamma}'\boldsymbol{\Sigma}_{xy}\boldsymbol{\omega}$ are computed, and the equations above are used to find $\alpha, \beta, \xi, \eta, \zeta$.

Remark 3⋆24 Tucker (1958) derived his IBF analysis from two different interpretations, the first being a covariance structure approach, the second one a data structure approach. The PLS approach above is in accordance with the data-analytical derivation.

The data analytical approach assumes two composites with normed weights, $\xi = \boldsymbol{\gamma}'\boldsymbol{x}$ and $\eta = \boldsymbol{\omega}'\boldsymbol{y}$ with $\boldsymbol{\gamma}'\boldsymbol{\gamma} = \boldsymbol{\omega}'\boldsymbol{\omega} = 1$, which are determined so as to have maximum covariance $\rho = \gamma'\Sigma_{xy}\omega$. If one is using Lagrangian multipliers to account for the constraints, the expression to be maximized is:

$$f = \gamma'\Sigma_{xy}\omega - \frac{\alpha}{2}(\gamma'\gamma - 1) - \frac{\beta}{2}(\omega'\omega - 1) \tag{3.467}$$

When the partial derivatives

$$\frac{\partial f}{\partial \gamma'} = \Sigma_{xy}\omega - \alpha\gamma \tag{3.468}$$

$$\frac{\partial f}{\partial \omega'} = \Sigma_{yx}\gamma - \beta\omega \tag{3.469}$$

have been set to zero and after it is established that $\alpha = \beta$, Eqs.3.468, 3.469 can be put together to result in the normal equation system Eq.3.462.

Remark 3⋆25 In the covariance analytical approach it is assumed that a factor model holds for all variables together, where the factors are partitioned into three blocks of uncorrelated standardized factors,

(i) the inter-battery factor ζ,
(ii) the "factors" $\boldsymbol{\delta}$ related only to the $\boldsymbol{x}$-variables and
(iii) the "factors" $\boldsymbol{\epsilon}$ related only to the $\boldsymbol{y}$-variables.

$\boldsymbol{\delta}$ and $\boldsymbol{\epsilon}$ may be understood as containing unique factors, doublets, triplets and all sorts of battery-specific factors. The model is thus:

$$\begin{bmatrix} \boldsymbol{x} \\ \boldsymbol{y} \end{bmatrix} = \begin{bmatrix} \boldsymbol{T} & \boldsymbol{A} & 0 \\ \boldsymbol{P} & 0 & \boldsymbol{B} \end{bmatrix} \begin{bmatrix} \zeta \\ \boldsymbol{\delta} \\ \boldsymbol{\epsilon} \end{bmatrix} \tag{3.470}$$

Consequently the covariance matrix of the MVs has the form

$$\begin{bmatrix} \Sigma_{xx} & \Sigma_{yx} \\ \Sigma_{xy} & \Sigma_{yy} \end{bmatrix} \begin{bmatrix} \boldsymbol{TT}' + \boldsymbol{AA}' & \boldsymbol{TP}' \\ \boldsymbol{PT}' & \boldsymbol{PP}' + \boldsymbol{BB}' \end{bmatrix} \tag{3.471}$$

Only the matrix $\Sigma_{yx} = \boldsymbol{PT}'$ turns out to be of any interest for the estimation of the interbattery loading matrices; $\boldsymbol{P}$ and $\boldsymbol{T}$ are estimated so as to give a LS fit to Σ_{yx}.

Remark 3⋆26 All three approaches, the PLS approach, Tucker's data analytical and Tucker's covariance analytical approach end in the same or equivalent eigenproblem. A different approach has been developed by Browne (1979), which is a Maximum Likelihood (ML) solution. The estimation is very simple, because it implies only rescaling the Canonical Correlation results. The interbattery factor ζ (Eqs.3.450, 3.451) is introduced into the Canonical Correlation model, where $\rho_\xi^2 = \rho_\eta^2 = 1$. Consequently, the canonical correlation coefficient must be the product of the two loading coefficients $\breve{\alpha}$ and $\breve{\beta}$, $\rho = \breve{\alpha}\breve{\beta}$. In the absence of better assumptions, $\breve{\alpha} = \breve{\beta} = \sqrt{\rho}$ is assumed, and the loadings of $\boldsymbol{x} = \boldsymbol{\tau}\xi + \boldsymbol{\delta}$ and $\boldsymbol{y} = \boldsymbol{\pi}\eta + \boldsymbol{\epsilon}$ are rescaled, $\tilde{\boldsymbol{\tau}} = \sqrt{\rho}\boldsymbol{\tau}$ and $\tilde{\boldsymbol{\pi}} = \sqrt{\rho}\boldsymbol{\tau}$, to fit into the models $\boldsymbol{x} = \tilde{\boldsymbol{\tau}}\zeta + \tilde{\boldsymbol{\delta}}$ and $\boldsymbol{y} = \tilde{\boldsymbol{\pi}}\zeta + \tilde{\boldsymbol{\epsilon}}$.

Remark 3⋆27 Notice that the definition $\zeta = \xi + \eta$ does not belong to either Tucker's or Browne's form of the Interbattery Factor model. Thus the Interbattery factor is not assumed as measured, whereas ξ and η are measured. This is not always clear when the IBF model is applied. Harris and McArthur (1974) report an IBF analysis of the Marjoribanks (1972) correlation matrix (presented here in Sc.2.5). What they report as "Correlations of Variables with Interbattery Factor" in their Table 1, is, in fact, the correlation with ξ, the first-set LV.

Remark 3⋆28 An application and comparison of Canonical Correlation model and Browne's version of IBF model is reported by Huba, Newcomb and Bentler (1981). They find that when the number of dimensions is increased, a rotation of the LVs (cf. Bentler & Huba 1979; Lohmöller 1977; Kristof 1967) preserves the integrity of the LVs better in Canonical Correlation than it does in IBF analysis.

3.4.5 Fortier's Simultaneous Linear Prediction Model

Given a set $\boldsymbol{y}$ of criteria to be predicted from a set $\boldsymbol{x}$ of predictors, the matrix $\mathbf{B}$ of unrestricted regression weight coefficients

$$\boldsymbol{y} = \mathbf{B}\boldsymbol{x} + \boldsymbol{\epsilon} = \tilde{\boldsymbol{y}} + \boldsymbol{\epsilon} \tag{3.472}$$

is found by multiple regression

$$\mathbf{B} = \boldsymbol{\Sigma}_{yx}\boldsymbol{\Sigma}_{xx}^{-1} \tag{3.473}$$

which is a closed expression for all $\boldsymbol{y}$-variables and is identical to *separate* regression for each variable y_k. The predictions are

$$\tilde{\boldsymbol{y}} = \boldsymbol{\Sigma}_{yx}\boldsymbol{\Sigma}_{xx}^{-1}\boldsymbol{x} \tag{3.474}$$

with covariance

$$\operatorname{cov}(\tilde{\boldsymbol{y}}) = \mathbf{B}\boldsymbol{\Sigma}_{xx}\mathbf{B}' = \boldsymbol{\Sigma}_{yx}\boldsymbol{\Sigma}_{xx}^{-1}\boldsymbol{\Sigma}_{xx}\boldsymbol{\Sigma}_{xx}^{-1}\boldsymbol{\Sigma}_{xy} = \boldsymbol{\Sigma}_{yx}\boldsymbol{\Sigma}_{xx}^{-1}\boldsymbol{\Sigma}_{xy}. \tag{3.475}$$

Matrix $\mathbf{B}$ has the same size and rank as $\boldsymbol{\Sigma}_{yx}$.

Fortier (1966) proposed finding a linear combination $\xi = \boldsymbol{\gamma}'\boldsymbol{x}$ which give the best prediction of all MVs $\boldsymbol{y}$ *simultaneously* (see Tb.3.6d). The absolutely highest fit to the $\boldsymbol{y}$-variables is provided by the separate predictions $\tilde{\boldsymbol{y}}$ in Eq.3.474 which are linear combinations of the $\boldsymbol{x}$-variables. But $\tilde{\boldsymbol{y}}$ has $L = \operatorname{rank}(\boldsymbol{\Sigma}_{xy})$ dimensions instead of the requested unidimensional predictor ξ (full-rank regression). Fortier's solution is to take the first principal component of $\tilde{\boldsymbol{y}}$ as ξ (cf. Rao 1964). The weights $\boldsymbol{\gamma}$ are found by an eigenanalysis of $\operatorname{cov}(\tilde{\boldsymbol{y}})$. Fortier proves that the weights $\boldsymbol{\gamma}$ are found by maximizing the ratio of quadratic forms

$$\frac{\boldsymbol{\gamma}'\boldsymbol{\Sigma}_{xy}\boldsymbol{\Sigma}_{yx}\boldsymbol{\gamma}}{\boldsymbol{\gamma}'\boldsymbol{\Sigma}_{xx}\boldsymbol{\gamma}} = \max_{\boldsymbol{\gamma}} \tag{3.476}$$

which can be done by solving the eigenproblem

$$(\boldsymbol{\Sigma}_{xx}^{-1}\boldsymbol{\Sigma}_{xy}\boldsymbol{\Sigma}_{yx} - \beta\mathbf{I})\boldsymbol{\gamma} = 0 \tag{3.477}$$

or equivalently

$$(\boldsymbol{\Sigma}_{xy}\boldsymbol{\Sigma}_{xx}^{-1}\boldsymbol{\Sigma}_{yx} - \beta\mathbf{I})\boldsymbol{\gamma} = 0 \tag{3.478}$$

The solution to Fortier's model is identical to the Principal Predictor model, as presented in Sc.3.4.3. The derivation of PLS estimators would require introducing $\eta = \boldsymbol{\omega}'\boldsymbol{y}$ as target variable for $\eta = \boldsymbol{\gamma}'\boldsymbol{x} + \upsilon$. The solution given in Sc.3.4.3 is equivalent to Eq.3.478.

The LV of the criteria is omitted in Fortier's model. This would seem to be quite reasonable. As shown in the equation $\eta = \beta\xi + \upsilon$, η can only partly be predicted from the predictor LV ξ; the residual variable υ is uncorrelated to $\boldsymbol{x}$ and ξ and is a function of the criterion MVs. Thus $\boldsymbol{y}$, when predicted by η, is partly predicted by itself (tautological prediction). This odd feature is avoided when η is omitted and $\boldsymbol{y}$ is predicted directly by ξ (Johansson 1981:102). If η is omitted in the Principal Predictor model, the result is Fortier's prediction model.

3.4.6 The MIMIC Model

Hauser & Goldberger's (1971) article is one of the first to treat unobservable variables in path analysis, though they confine themselves to models with two and one LV. Looking back over fifteen years of rapid development in this field, their models can now be regarded as special cases of LV analysis.

The first of the three models Hauser and Goldberger present is an Interbattery Factor model, with the additional assumption of uncorrelated residuals. The estimation of the parameters is done by an ML-like procedure Hauser and Goldberger call "efficient estimation". Thus, the IBF model becomes a factor analytic model with block-diagonal loading matrix. Hauser and Goldberger fail to mention that only full recursive path models can be estimated like restricted factor models. Thus, contrary to what they claim in a footnote on p.99, they have not solved all the problems of LV path analysis. (Reanalyzing their respective example by LISREL, I have found estimates different from the ML-estimates reported by Hauser and Goldberger).

The second Hauser and Goldberger model is similar to Fortier's (1966) Simultaneous Linear Prediction model but with the additional constraint that the prediction residuals are uncorrelated. As for the estimation method, Hauser and Goldberger refer to Jöreskog's (1970) general Maximum Likelihood (ML) method for the analysis of covariance structures (cf. Jöreskog & Goldberger 1975).

The third Hauser and Goldberger model is the same as Fortier's (1966) Simultaneous Linear Prediction model,

$$\xi = \boldsymbol{\gamma}'\boldsymbol{x} \qquad \boldsymbol{\Sigma}_{xx} = \text{cov}(\boldsymbol{x}) \tag{3.479}$$

$$\boldsymbol{y} = \boldsymbol{\pi}\xi + \boldsymbol{\epsilon} \qquad \boldsymbol{\Theta} = \text{cov}(\boldsymbol{\epsilon}) \tag{3.480}$$

with no (explicit) restriction on the residual covariance matrix $\boldsymbol{\Theta}$. However, the model seems to imply that:

$$\text{cov}(\boldsymbol{x};\boldsymbol{\epsilon}) = \mathbf{0}. \tag{3.481}$$

The unrestricted regression matrix in $\boldsymbol{y} = \mathbf{B}\boldsymbol{x} + \tilde{\boldsymbol{\epsilon}}$ is found by:

$$\mathbf{B} = \boldsymbol{\Sigma}_{yx}\boldsymbol{\Sigma}_{xx}^{-1}. \tag{3.482}$$

The regression moment is

$$\tilde{\boldsymbol{\Sigma}}_{xx} \equiv \mathrm{cov}(\tilde{\boldsymbol{y}}) \equiv \mathrm{cov}(\mathbf{B}\boldsymbol{x}) = \mathbf{B}\boldsymbol{\Sigma}_{xx}\mathbf{B}' = \boldsymbol{\Sigma}_{yx}\boldsymbol{\Sigma}_{xx}^{-1}\boldsymbol{\Sigma}_{xy}, \tag{3.483}$$

the regression residual covariance:

$$\tilde{\boldsymbol{\Theta}} \equiv \mathrm{cov}(\tilde{\boldsymbol{\epsilon}}) = \boldsymbol{\Sigma}_{yy} - \tilde{\boldsymbol{\Sigma}}_{yy} = \boldsymbol{\Sigma}_{yy} - \mathbf{B}\boldsymbol{\Sigma}_{xx}\mathbf{B}' \tag{3.484}$$

Combining Eq.3.479 with Eq.3.480 gives the rank-one regression model

$$\boldsymbol{y} = \boldsymbol{\pi}\boldsymbol{\gamma}'\boldsymbol{x} + \boldsymbol{\epsilon}. \tag{3.485}$$

with regression residual

$$\boldsymbol{\Theta} = \boldsymbol{\Sigma}_{yy} - \boldsymbol{\pi}\boldsymbol{\gamma}'\boldsymbol{\Sigma}_{xx}\boldsymbol{\gamma}\boldsymbol{\pi}' \tag{3.486}$$

The difference of the regression residuals of the full-rank model with residual covariance $\boldsymbol{\Theta}$ and the restricted-rank model with residual $\tilde{\boldsymbol{\Theta}}$ is:

$$\boldsymbol{\Delta} \equiv \boldsymbol{\Theta} - \tilde{\boldsymbol{\Theta}} = (\mathbf{B} - \boldsymbol{\pi}\boldsymbol{\gamma}')\boldsymbol{\Sigma}_{xx}(\mathbf{B} - \boldsymbol{\pi}\boldsymbol{\gamma}')'. \tag{3.487}$$

Fortier's LS-approach estimates $\boldsymbol{\pi}$ and $\boldsymbol{\gamma}$ by minimizing

$$\mathrm{LS} = \mathrm{trace}(\boldsymbol{\Delta}) \overset{!}{=} \min. \tag{3.488}$$

Hauser and Goldberger's Generalized Least Squares (GLS) approach estimates the parameters by minimizing the weighted sum of the elements of $\boldsymbol{\Delta}$.

$$\mathrm{GLS} = \mathrm{trace}\left(\tilde{\boldsymbol{\Theta}}^{-1}\boldsymbol{\Delta}\right)^2 \overset{!}{=} \min. \tag{3.489}$$

As Hauser and Goldberger derive, the expression Eq.3.489 is equal to the sum of all possible pair products of the elements of $(\mathbf{B} - \boldsymbol{\pi}\boldsymbol{\gamma}')$ divided by their standard deviation.

The estimation turns out to imply the solution of an eigenproblem. As Hauser and Goldberger state in the appendix on p.112, the solution is a rescaled canonical correlation, the weight vector $\boldsymbol{\gamma}$ being identical to the Canonical Correlation solution and the loading vector $\boldsymbol{\pi}$ being the Canonical Correlation loading vector multiplied by the canonical correlation. (This was confirmed by reanalyzing the related Hauser and Goldberger example by PLS).

Hauser and Goldberger's model, dubbed MIMIC for "**M**ultiple effect **I**ndicators $\boldsymbol{y}$ for **M**ultiple **C**auses $\boldsymbol{x}$", is not a completely new model, for the results are identical to the Canonical Correlation analysis. What it offers is a new interpretation for an old model. Again, the Canonical Correlation model turns out to be at the same time a ML and a GLS model, as well as a PLS model also.

The estimation criterion Eq.3.482 involves prediction and the unrestricted regression matrix $\mathbf{B}$ and hence $\tilde{\boldsymbol{\Theta}}$ are found by the LS assumption $\mathrm{E}(\boldsymbol{y}|\boldsymbol{x}) = \mathbf{B}\boldsymbol{x}$. Prediction involves minimization of residual variances, which is typical for "data structure" models. On the other hand, Eq.3.489 involves the minimization of residual covariances, which is typical for "covariance structure" models. The Canonical Correlation model seems to have properties of both types of modeling.

3.4.7 Discussion

Fifty years after Hotelling's (1935) four page paper on the "most predictable criterion" the interest in two-block models is experiencing a boom (Areskoug 1975, 1982 Cohen 1982, Coxhead 1974, Cramer 1974, DeSarbo 1981, Darlington, Weinberg & Walberg 1973, Frederiksen 1974, Hakstian 1973, 1974, 1976, 1977, Johansson 1981, Lunneborg 1967, Meredith 1964, Muller 1981, Rao 1969, Thompson 1982, Thorndike 1977) and has hit the textbook level (Levine 1977, Gittins 1985, van de Geer 1986). The new interest in two-block models may have been initiated by van den Wollenberg's (1977) paper on redundancy analysis, which implies basically the directed relationship between two LVs. However, the literature lacks a unifying theory and approach that capable of covering the models treated in Sc.3.4. Years before this discussion started Wold (1975b) presented his PLS method providing both a unifying theory and a simple estimation theory. Because his interest was directed more towards LVs in multi-block models, he did not bother to link up with the traditional two-block models.

Fornell (1978) distinguishes three types of canonical analysis, each constituted by different operative use and different interpretation:

(a) Canonical correlation analysis: Weights and cross-correlations are used to examine the variable relationships.

(b) Canonical variate analysis: Loadings are used to identify "meaningful" constructs from the variate structure.

(c) Canonical regression analysis: Both weights and loadings are used, weights to estimate the regression for predictor variables and loadings for the comprehension of the variate structure in the criterion variables.

PLS offers three different models for the three different purposes distinguished by Fornell, (a) the Canonical Correlation model, (b) the Interbattery Factor model and (c) the Principal Predictor model. The three PLS two-block models have different estimation algorithms and give different parameter estimates, which serve their purpose better than reformulated values from the canonical correlation model (see also van de Geer 1971).

3.5 Split Principal Components

Section Sc.3.5 is intended to show that for some special cases of multi-block PLS models the partial criteria amount to a global criterion or, at least, are approaching asymptotically a global criterion. The main idea is that the first principal component (PC) of a total set of variables is split up into several parts, each part being measured by a separate block of MVs. The theorem given in Eq.3.539 states that the weighted sum of the correlations of the parts of the PC is maximal (Sc.3.5.2). Applications of this theorem to various traditional models are shown in Sc.3.5.7.

The notation is changed in this section to fit the three levels of variables. The MVs (first level) are denoted by $\boldsymbol{y} = [y_{k_j}]$, the lower-order LVs (second level) by $\boldsymbol{\eta} = [\eta_j]$ and the higher-order LVs (third level) by ξ. The Hierarchical Component model is set up in Sc.3.5.1, using the conditional expectation as modeling device and the standard PLS algorithm for estimation. The following subsection contain equivalent models which, however, are specified by "maximize something" devices, amount to well-defined eigenproblems and lead to convergent variations of the power algorithm.

3.5.1 Hierarchical Component Model

The special- and general-factor model. The Hierarchical Common Factor model has its roots in the modeling of mental abilities (intelligence).

$$\begin{array}{c}\text{General}\\\text{ability}\end{array}\left\{\begin{array}{l}\begin{array}{c}\text{Verbal}\\\text{ability}\end{array}\left\{\begin{array}{l}\text{Verbal1}\\\text{Verbal2}\\\text{Verbal3}\end{array}\right.\\[2ex]\begin{array}{c}\text{Numerical}\\\text{ability}\end{array}\left\{\begin{array}{l}\text{Numerical1}\\\text{Numerical2}\\\text{Numerical3}\end{array}\right.\\[2ex]\begin{array}{c}\text{Spatial}\\\text{ability}\end{array}\left\{\begin{array}{l}\text{Spatial1}\\\text{Spatial2}\\\text{Spatial3}\end{array}\right.\end{array}\right. \tag{3.501}$$

This sketch has three sorts of variables:

(i) The observed variables, here named Verbal1, Verbal2, etc., below denoted by y. The variables typically are scales, aggregated from a handful of simple passed-failed items.

(ii) The lower-order factors, here named Verbal, Numerical and Spatial abilities, later denoted by η. These variables are not directly observed. They represent special abilities which can be distinguished, but nevertheless are correlated to some extend.

(iii) The higher-order factor, here named General Ability, later denoted by ξ. This is what the special abilities have in common and what makes them correlated.

The multiple battery model. The ordering of the variables in Eq.3.501 is not the only one that makes sense. There is another type of a hierarchical model that serves a different purpose.

$$\text{General abilities} \left\{ \begin{array}{l} \text{Battery1} \left\{ \begin{array}{l} \text{Verbal1} \\ \text{Numerical1} \\ \text{Spatial1} \end{array} \right. \\ \\ \text{Battery2} \left\{ \begin{array}{l} \text{Verbal2} \\ \text{Numerical2} \\ \text{Spatial2} \end{array} \right. \\ \\ \text{Battery3} \left\{ \begin{array}{l} \text{Verbal3} \\ \text{Numerical3} \\ \text{Spatial3} \end{array} \right. \end{array} \right. \tag{3.502}$$

The 12 subtests are used to construct 3 equivalent test batteries which are designed to be used interchangeably as parallel tests of general ability. A statistical analysis of this setup should demonstrate that the three batteries are equivalent. The equivalence should be proven not to rely on a single ingredient of the battery, say on the base of well-measured numerical subtests while the other tests are flaw.

Table 3.9: A two-level hierarchical LV model

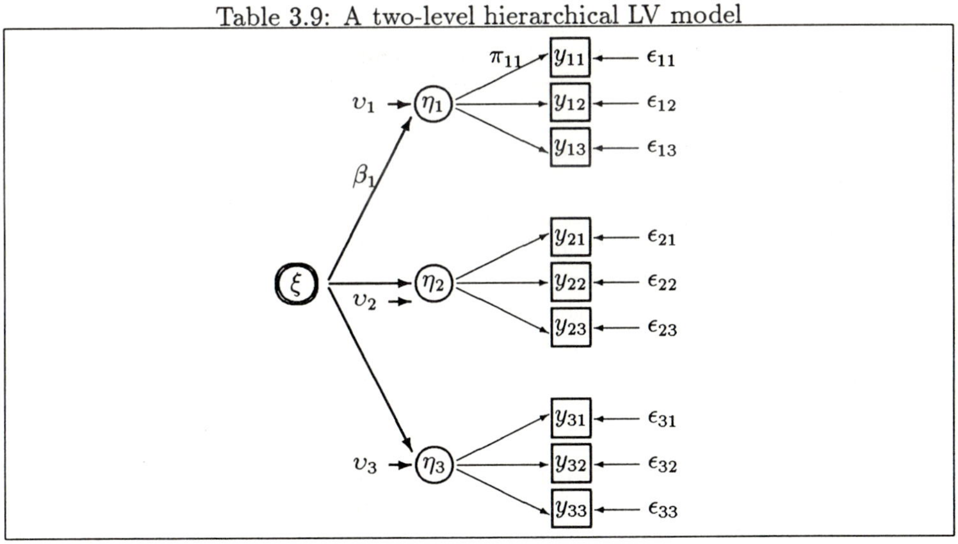

The model relations. Whether the subtests are grouped into equivalent batteries or into homogeneous abilities, the formal notation of the model relations looks alike (see Table 3.9) and is a combination of two factor models.

The second-level relation

$$\eta_j = \beta_j \xi + \upsilon_j, \qquad \forall j \tag{3.503}$$

represents the lower-order factors η_j as a linear function of the general factor ξ, with loading coefficient β_j, and a variable υ_j that stands for the unique and special contents of η_j. The lower-order factors are correlated due the influence of ξ that is shared between all lower-order factors, and they are not perfectly correlated due to the influence of their specific factor υ_j. The unique factors υ_j are what makes the lower-order factors different from the general factor and from each other.

The first-level relation

$$y_{k_j} = \pi_{k_j}\eta_j + \epsilon_{k_j}, \qquad \forall j, k_j \tag{3.504}$$

represents the observed variable (manifest variable, MV) y_{k_j} as a linear function of the lower-order factor η_j, with loading coefficient π_{k_j}, and a variable ϵ_{k_j} which stands for the peculiarities of y_{k_j} which are not shared by the other MVs. The variables ϵ_{k_j} often are interpreted as measurement errors.

Weighted aggregates. PLS estimation requires two additional specifications. The first one specifies the estimates of the LVs to be weighted aggregates:

$$\xi = \sum_j g_j \eta_j, \qquad \eta_j = \sum_k w_{k_j} y_{k_j} \tag{3.505}$$

The second specification scales the estimated LVs ξ and η_j to unit variance.

$$\mathrm{E}\,\xi^2 = 1, \qquad \mathrm{E}\,\eta_j^2 = 1, \quad \forall j. \tag{3.506}$$

This restriction for scale unambiguity (RSU) can be replaced by equivalent specifications — and will be in following subsections.

Repeated MVs. Herman Wold (1982c) proposed to estimate the Hierarchical Component model by repeated use of the MVs. This is demonstrated by the path diagram Table 3.10: The nine manifest variables are used twice, once as indicators for ξ, and once as indicators for η_j. The two sorts of outer residuals are noted by ϵ and ϵ^*. The higher-order factor ξ is directly measured by observed variables, so the model can be estimated by the usual PLS procedure.

Besides the second-level Eq.3.503 and the first-level Eq.3.504, this model has an additional relation that can be called the cross-level relation.

$$y_{k_j} = \alpha_{k_j}\xi + \epsilon^*_{k_j}, \qquad \forall k_j \tag{3.507}$$

If η_j in Eq.3.504 is substituted from Eq.3.503, we get the expression

$$\begin{aligned} y_{k_j} &= \pi_{k_j}\eta_j + \epsilon_{k_j} \\ &= \pi_{k_j}(\beta_j\xi + \upsilon_j) + \epsilon_{k_j} \\ &= (\pi_{k_j}\beta_j)\xi + (\pi_{k_j}\upsilon_j + \epsilon_{k_j}) = \alpha_{k_j}\xi + \epsilon^*_{k_j} \qquad \forall j, k_j \end{aligned} \tag{3.508}$$

We apply predictor specification to Eqs.3.503, 3.508, and, due to the interlocking conditional expectations (cf. Sc.3.1.7), we get three sets of zero correlations:

$$\mathrm{E}(\xi\epsilon_{k_j}) = 0, \qquad \mathrm{E}(\xi\epsilon^*_{k_j}) = 0, \qquad \mathrm{E}(\xi\upsilon_j) = 0. \tag{3.509}$$

Unfortunately, $\mathrm{E}(\upsilon_j\epsilon_{k_j}) = \mathrm{E}(\eta_j\epsilon_{k_j}) \neq 0$.

Table 3.10: Hierarchical LV model with repeated MVs

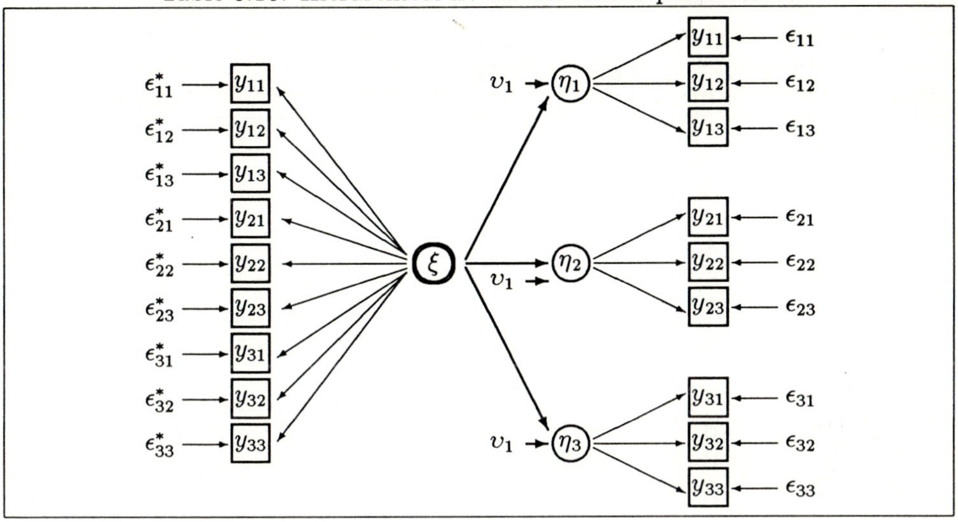

Table 3.11: PLS algorithm for Hierarchical Component models with repeated MVs

	$\boldsymbol{y} = [\boldsymbol{y}_j] = [y_{k_j}]$	The MVs are partitioned into blocks.
	${}_0w_{k_j} = {}_0g_{k_j} = ?$	Starting values are (nearly) arbitrary.
Repeat steps #1 to #4b, starting with #4a, for cycles $c = 0, 1, 2, \ldots$, and for blocks $j = 1 \ldots J$:		
		Iteration index ${}_c()$ omitted from $\tilde{\xi}$, $\tilde{\eta}$, g, w, v, f for readability.
#1	$\boldsymbol{v} := \begin{cases} \mathrm{cor}(\hat{\boldsymbol{\eta}}; \hat{\xi}) \\ \mathrm{reg}(\hat{\boldsymbol{\eta}} \mid \hat{\xi}) \\ \mathrm{reg}(\hat{\xi} \mid \hat{\boldsymbol{\eta}}) \end{cases}$	Inner weights by factor, or path, weighting scheme.
#2a	$\tilde{\xi} := \boldsymbol{v}'\hat{\boldsymbol{\eta}}$	Inside approximation for general factor.
#2b	$\tilde{\eta}_j := \hat{\xi}$	Inside approximation for special factors.
#3a	$\boldsymbol{g} := \begin{cases} \mathrm{reg}(\boldsymbol{y} \mid \tilde{\xi}) & \mathrm{A} \\ \mathrm{reg}(\tilde{\xi} \mid \boldsymbol{y}) & \mathrm{B} \end{cases}$	Outer weights for general factor by ModeA or ModeB
#3a	$\boldsymbol{w}_j := \begin{cases} \mathrm{reg}(\boldsymbol{y}_j \mid \tilde{\eta}_j) & \mathrm{A} \\ \mathrm{reg}(\tilde{\eta}_j \mid \boldsymbol{y}_j) & \mathrm{B} \end{cases}$	Outer weights for special factors by ModeA or ModeB
#4a	${}_{c+1}\hat{\xi} := f\boldsymbol{g}'\boldsymbol{y}$	f makes $\hat{\xi}$ standardized.
#4b	${}_{c+1}\hat{\eta}_j := f_j\boldsymbol{w}_j'\boldsymbol{y}_j$	f_j makes $\hat{\eta}_j$ standardized.
until the weights stabilize.		

Table 3.12: Hierarchical LV model and PLS solutions

Row	Weight Mode for ξ	Weight Mode for η_j	Inner model	Inner Weight Mode	Diagram	Solution	$R^2_{\xi\vert\eta}$
1	A	-	-	-	$y \leftarrow \xi$	#2 - PC	–
2	-	A	-	-	$\eta_j \rightarrow y_j$	#0	–
3	A	A	Spread	Factor	$y \leftarrow \xi \Leftrightarrow \eta_j \rightarrow y_j$	#1	$=1$
4	A	A	Spread	Path	$y \leftarrow \xi \Rightarrow \eta_j \rightarrow y_j$	#1	$=1$
5	A	A	Collect	Factor	$y \leftarrow \xi \Leftrightarrow \eta_j \rightarrow y_j$	#1	$=1$
6	A	A	Collect	Path	$y \leftarrow \xi \Leftarrow \eta_j \rightarrow y_j$	#2 - PC	$=1$
7	A	B	Spread	Factor	$y \leftarrow \xi \Leftrightarrow \eta_j \leftarrow y_j$	#3	<1
8	A	B	Spread	Path	$y \leftarrow \xi \Rightarrow \eta_j \leftarrow y_j$	#3	<1
9	A	B	Collect	Factor	$y \leftarrow \xi \Leftrightarrow \eta_j \leftarrow y_j$	#3	<1
10	A	B	Collect	Path	$y \leftarrow \xi \Leftarrow \eta_j \leftarrow y_j$	#4	<1
11	B	A	Spread	Factor	$y \leftarrow \xi \Leftrightarrow \eta_j \leftarrow y_j$	#5	$=1$
12	B	A	Spread	Path	$y \leftarrow \xi \Rightarrow \eta_j \leftarrow y_j$	#5	$=1$
13	B	A	Collect	Factor	$y \leftarrow \xi \Leftrightarrow \eta_j \leftarrow y_j$	#5	$=1$
14	B	A	Collect	Path	$y \leftarrow \xi \Leftarrow \eta_j \leftarrow y_j$	#6	$=1$
15	B	B	Spread	Factor	$y \rightarrow \xi \Leftrightarrow \eta_j \leftarrow y_j$	#7 - CC	$=1$
16	B	B	Spread	Path	$y \rightarrow \xi \Rightarrow \eta_j \leftarrow y_j$	#7 - CC	$=1$
17	B	B	Collect	Factor	$y \rightarrow \xi \Leftrightarrow \eta_j \leftarrow y_j$	#7 - CC	$=1$
18	B	B	Collect	Path	$y \rightarrow \xi \Leftarrow \eta_j \leftarrow y_j$	#8	$=1$
19	-	B	-	-	$\Rightarrow \eta_j \leftarrow y_j$	#7 - CC	$=1$

"Spread" is shown in Tb.3.10: ξ spreads out into the η_j. In "Collect" the inner arrows are turned around, so ξ collects the η_j.

The arrows $\rightarrow$ and $\leftarrow$ point outwards (to the MVs) which is ModeA, or inwards (to the LVs) which is ModeB.

The solutions #6 and #8 depend on the starting values and do not neccesarily converge to a same point.

PLS algorithm with repeated MVs. The PLS algorithm for the Hierarchical Component model is outlined in Table 3.11. The algorithm has several options regarding the inner and outer weights. Furthermore, one can understand the general factor as "spreading out" into the special factors, as shown in Tb.3.9, or as "collecting" the special factors. The sixteen different PLS specifications, which are listed in Table 3.12, rows 3 to 18, give not more than eight different solutions. Two of the solutions show "fake convergence": the algorithm converges, but the results depend on the starting values[1] and do not come to an optimum. One of the solutions (marked PC) is identical to the principal component of the total set of MVs. One solution (marked CC) is identical to Horst's Generalized Canonical Correlation model. This solution can be obtained by two other algorithms without repeated MVs, which will be developed in the following subsections.

3.5.2 Splitting Principal Components

Optimal weighting. McDonald (1968) has presented a general method of determining optimal weights for different problems and models, including principal component and canonical correlation models. The general algebraic expression for the weights $\boldsymbol{w}$ is given by the Rayleigh quotient

$$\lambda^2 = \frac{\boldsymbol{w}'\boldsymbol{S}\boldsymbol{w}}{\boldsymbol{w}'\boldsymbol{D}\boldsymbol{w}} = \max_{\boldsymbol{w}}, \tag{3.510}$$

where $\boldsymbol{S} = \mathrm{E}(\boldsymbol{y}\boldsymbol{y}')$ is the $K \times K$ product matrix. If λ^2 is maximal over any choice of a real vector $\boldsymbol{w}$, then λ^2 is the first eigenvalue of $\boldsymbol{S}$ with respect to $\boldsymbol{D}$, and $\boldsymbol{w}$ is proportional to the first eigenvector of $\boldsymbol{S}$. Some special cases are noted:

(i) If $\boldsymbol{D} = \mathbf{I}$, then $\boldsymbol{w}$ will be the weight vector for the first principal component of the covariance matrix $\boldsymbol{S}$.

(ii) If $\boldsymbol{D} = \mathbf{I} * \boldsymbol{S}$ is the diagonal of a covariance matrix $\boldsymbol{S}$, $\boldsymbol{w}$ will be the weight vector for the first principal component of the correlation matrix $\boldsymbol{S}^* = \boldsymbol{D}^{-1/2}\boldsymbol{S}\boldsymbol{D}^{-1/2}$.

(iii) If $\boldsymbol{y}$ is split up into two subsets $\boldsymbol{y}_1$ and $\boldsymbol{y}_2$, and $\boldsymbol{D}$ is the block-diagonal matrix $\boldsymbol{D} = \boldsymbol{S}_{11} \oplus \boldsymbol{S}_{22}$, we have the Canonical Correlation model.

(iv) If $\boldsymbol{y}$ is split up into more subsets $\boldsymbol{y}_j$, and $\boldsymbol{D} = [\cdots \oplus \boldsymbol{S}_{jj} \oplus \cdots]$, Horst's (1965) Maximum Variance algorithm for a Multi-Set Factor model is obtained.

(v) If $\boldsymbol{S} = \boldsymbol{S}_C + \boldsymbol{D}$, where $\boldsymbol{S}_C$ is the common covariance and the diagonal matrix $\boldsymbol{D}$ is the residual variance, Eq.3.510 is the Common Factor model.

The special cases (ii) and (iv) will be treated below. To determine the weights $\boldsymbol{w}$ so as to maximize λ^2, the partial derivatives of λ^2 with respect to the elements of $\boldsymbol{w}$ are equated to zero, which after a few algebraic operations results in the stationary expression

$$(\boldsymbol{S} - \lambda^2\boldsymbol{D})\boldsymbol{w} = \boldsymbol{0}. \tag{3.511}$$

The principal component model. Consider a column vector of random variables $\boldsymbol{y}_{[K\times1]}$ with covariance matrix $\boldsymbol{S}_{[K\times K]} = \mathrm{E}(\boldsymbol{y}\boldsymbol{y}')$, and the first principal component ξ with variance q^2,

$$\xi = \boldsymbol{w}^{*\prime}\boldsymbol{y} \qquad \text{weight relation} \tag{3.512}$$

$$\boldsymbol{y} = \boldsymbol{p}\xi + \boldsymbol{\epsilon} \qquad \text{loading relation} \tag{3.513}$$

$$\boldsymbol{w}^{*\prime}\boldsymbol{w}^* = 1. \qquad \text{scale condition} \tag{3.514}$$

[1]The solutions for different starting values are often identical on the first two, sometimes even three decimals. It took me three days to verify that they are essentially different.

The first principal component is characterized by

$$\lambda^2 = \frac{\boldsymbol{w}^{*\prime}\boldsymbol{S}\boldsymbol{w}^*}{\boldsymbol{w}^{*\prime}\boldsymbol{w}^*} = \max_{\boldsymbol{w}^*}, \tag{3.515}$$

i.e. if λ^2 is maximal over any choice of a real vector $\boldsymbol{w}^*$, then λ^2 is the first eigenvalue, and $\boldsymbol{w}^*$ is proportional to the first eigenvector of $\boldsymbol{S}$.

Splitting the principal component. Let the vector $\boldsymbol{y}$ be partitioned into non-overlapping and exhaustive sub-vectors $\boldsymbol{y}_{j[K_j \times 1]}, j = 1 \dots J$, and the weight vector $\boldsymbol{w}^*$ be split up in parts $\boldsymbol{w}^*_{j[K_j \times 1]}$ corresponding to the $\boldsymbol{y}_j$. These subvectors are used to form a diagonal super-matrix of weights $\boldsymbol{W}^*_{[K \times J]}$:

$$\boldsymbol{W}^* = [\boldsymbol{w}_1^* \oplus \boldsymbol{w}_2^* \oplus \cdots \oplus \boldsymbol{w}_J^*] = \begin{pmatrix} \boldsymbol{w}_1^* & \mathbf{0} & \cdots & \mathbf{0} \\ \mathbf{0} & \boldsymbol{w}_2^* & \cdots & \mathbf{0} \\ \vdots & \vdots & \ddots & \vdots \\ \mathbf{0} & \mathbf{0} & \cdots & \boldsymbol{w}_J^* \end{pmatrix} \tag{3.516}$$

with the property $\boldsymbol{w}^* = \boldsymbol{W}^*\mathbf{1}$, where $\mathbf{1}_{[J \times 1]}$ is a unit vector. Applying the weight matrix $\boldsymbol{W}^*$ to $\boldsymbol{y}$ gives a set of variables $\boldsymbol{\eta}^*_{[J \times 1]} = [\eta_j^*]$,

$$\boldsymbol{\eta}^* = \boldsymbol{W}^{*\prime}\boldsymbol{y}, \tag{3.517}$$

with covariance matrix

$$\boldsymbol{R}^*_{[K \times K]} = \mathrm{E}(\boldsymbol{\eta}^*\boldsymbol{\eta}^{*\prime}) = \mathrm{E}(\boldsymbol{W}^{*\prime}\boldsymbol{y}^*\boldsymbol{y}^{*\prime}\boldsymbol{W}) = \boldsymbol{W}^{*\prime}\boldsymbol{S}\boldsymbol{W}^*. \tag{3.518}$$

Each one of the variables η_j^* is a fraction (or a slice, part, fragment, splinter) of the total principal component and is measured by just one block of the MVs $\boldsymbol{y}_j$. The unweighted sum of all fractions gives the total principal component,

$$\xi = \boldsymbol{w}^{*\prime}\boldsymbol{y} = \mathbf{1}'\boldsymbol{W}^{*\prime}\boldsymbol{y} = \mathbf{1}'\boldsymbol{\eta}^* = \sum_j \eta_j^*, \tag{3.519}$$

and the sum of all elements of the covariance matrix $\boldsymbol{R}^*$ gives the variance of the total principal component,

$$q^2 = \boldsymbol{w}^{*\prime}\boldsymbol{S}\boldsymbol{w}^* = \mathbf{1}'\boldsymbol{W}^{*\prime}\boldsymbol{S}\boldsymbol{W}^*\mathbf{1} = \mathbf{1}'\boldsymbol{R}^*\mathbf{1} = \sum_i \sum_j r_{ij}. \tag{3.520}$$

Standardized fragments. The standard deviation of each fragment η_j^* is used to attain the corresponding standardized variable η_j. We define a diagonal matrix $\boldsymbol{V}_{[J \times J]}$ with the standard deviations $v_{jj} = +\sqrt{r_{jj}^*}$ as diagonal elements,

$$\boldsymbol{V} = (\mathbf{I} * \boldsymbol{R}^*)^{1/2}, \tag{3.521}$$

where the inner star denotes the Hadamard product operator and $\mathbf{I} * \boldsymbol{R}^*$ contains just the diagonal elements of $\boldsymbol{R}^*$. For convenience, we define a vector of standard deviations by

$$\boldsymbol{v} = \boldsymbol{V}\mathbf{1}. \tag{3.522}$$

The standardized principal component fragment, then, are

$$\boldsymbol{\eta} = \boldsymbol{V}^{-1}\boldsymbol{\eta}^* = \boldsymbol{V}^{-1}\boldsymbol{W}^{*\prime}\boldsymbol{y} = \boldsymbol{W}'\boldsymbol{y}, \tag{3.523}$$

where $\boldsymbol{W} = \boldsymbol{W}^*\boldsymbol{V}^{-1}$ is the weight matrix for acquiring standardized principal component fragments directly from the MVs. The correlation, and covariance, matrix of $\boldsymbol{\eta}$ is:

$$\boldsymbol{R} = \mathrm{E}(\boldsymbol{\eta}\boldsymbol{\eta}') = \boldsymbol{V}^{-1\prime}\boldsymbol{R}^*\boldsymbol{V}^{-1} = \boldsymbol{V}^{-1\prime}\boldsymbol{W}^{*\prime}\boldsymbol{S}\boldsymbol{W}^*\boldsymbol{V}^{-1} = \boldsymbol{W}'\boldsymbol{S}\boldsymbol{W}. \tag{3.524}$$

In order to form the total principal component (as introduced in Eq.3.512) from the standardized fragments $\boldsymbol{\eta}$ we must not use the simple unweighted summation as in Eq.3.519. We have to use $\boldsymbol{V}$ or $\boldsymbol{v}$ as weights,

$$\xi = \boldsymbol{v}'\boldsymbol{\eta} = \mathbf{1}'\boldsymbol{V}'\boldsymbol{\eta} = \mathbf{1}'\boldsymbol{V}'\boldsymbol{V}^{-1\prime}\boldsymbol{\eta}^* = \mathbf{1}'\boldsymbol{\eta}^*. \tag{3.525}$$

The expression for the variance of the total principal component (Eq.3.520) becomes:

$$q^2 = \boldsymbol{v}'\boldsymbol{R}\boldsymbol{v} = \boldsymbol{v}'\boldsymbol{W}'\boldsymbol{S}\boldsymbol{W}\boldsymbol{v}. \tag{3.526}$$

Block-orthonormal MVs. If the MVs $\boldsymbol{y}$ are uncorrelated within each block,

$$\boldsymbol{S}_{jj} = \mathrm{E}(\boldsymbol{y}_j\boldsymbol{y}_j') = \mathbf{I}, \qquad \forall j, \tag{3.527}$$

then the variance r_{jj} of η_j is easily computed

$$r_{jj} = \boldsymbol{w}_j^{*\prime}\boldsymbol{S}_{jj}\boldsymbol{w}_j^* = \boldsymbol{w}_j^{*\prime}\mathbf{I}\boldsymbol{w}_j^* = \boldsymbol{w}_j^{*\prime}\boldsymbol{w}_j^* = v_{jj}^2, \tag{3.528}$$

or equivalently:

$$\mathbf{I} * \boldsymbol{R}^* = \boldsymbol{V}^2 = \boldsymbol{W}^{*\prime}\boldsymbol{W}^*. \tag{3.529}$$

Furthermore, the weight matrix $\boldsymbol{W}$ is orthonormal, due to the orthonormality of $\boldsymbol{y}$:

$$\boldsymbol{W}'\boldsymbol{W} = \boldsymbol{V}^{-1\prime}\boldsymbol{W}^{*\prime}\boldsymbol{W}^*\boldsymbol{V}^{-1} = \boldsymbol{V}^{-1\prime}\boldsymbol{V}^2\boldsymbol{V}^{-1} = \mathbf{I}. \tag{3.530}$$

For the case of correlated MVs we define

$$\boldsymbol{\Delta}^2 = \boldsymbol{W}'\boldsymbol{W} = [\cdots \oplus r_{jj}^*\boldsymbol{w}_j^{*\prime}\boldsymbol{w}_j^* \oplus \cdots]. \tag{3.531}$$

Blockwise orthonormalization. In most applications the variables are correlated within the blocks. The correlated variables, here denoted by $\breve{\boldsymbol{y}}_j$, with covariance matrix $\breve{\boldsymbol{S}}_{jj}$, may be transformed to give uncorrelated substitutes

$$\boldsymbol{y}_j = \breve{\boldsymbol{S}}_{jj}^{-1/2}\breve{\boldsymbol{y}}_j. \tag{3.532}$$

The weights $\boldsymbol{w}_j$ relating the orthonormal substitutes $\boldsymbol{y}_j$ to the higher order variables (Eq.3.523) can be transformed to relate the original correlated variables $\breve{\boldsymbol{y}}_j$ to the higher order variables by

$$\breve{\boldsymbol{w}}_j = \breve{\boldsymbol{S}}_{jj}^{1/2}\boldsymbol{w}_j. \tag{3.533}$$

In fact,

$$\eta_j = \boldsymbol{w}_j'\boldsymbol{y}_j = \breve{\boldsymbol{w}}_j'\breve{\boldsymbol{S}}_{jj}^{1/2}\breve{\boldsymbol{S}}_{jj}^{-1/2}\breve{\boldsymbol{y}}_j = \breve{\boldsymbol{w}}_j'\breve{\boldsymbol{y}}_j. \tag{3.534}$$

Hierarchical principal components. Eqs.3.523, 3.525 are repeated to introduce a Hierarchical Principal Component model with weight relations:

$$\boldsymbol{\eta} = \boldsymbol{W}'\boldsymbol{y}; \qquad \boldsymbol{R} = \mathrm{E}(\boldsymbol{\eta}\boldsymbol{\eta}') = \boldsymbol{W}'\boldsymbol{S}\boldsymbol{W} \tag{3.535}$$

$$\xi = \boldsymbol{v}'\boldsymbol{\eta}; \qquad q^2 = \mathrm{E}(\xi\xi') = \boldsymbol{v}'\boldsymbol{R}\boldsymbol{v} \tag{3.536}$$

and loading relations:

$$\boldsymbol{y} = \boldsymbol{P}\boldsymbol{\eta} + \boldsymbol{e}; \qquad \mathrm{E}(\boldsymbol{y}|\boldsymbol{\eta}) = \boldsymbol{P}\boldsymbol{\eta} \tag{3.537}$$

$$\boldsymbol{\eta} = \boldsymbol{b}\xi + \boldsymbol{u}; \qquad \mathrm{E}(\boldsymbol{\eta}|\xi) = \boldsymbol{b}\xi, \tag{3.538}$$

where $\boldsymbol{P}_{[K\times J]}$ is the block-diagonal first level loading matrix and $\boldsymbol{b}_{[K\times 1]}$ the second level loading matrix.

Theorem: Split Principal Components: If $\boldsymbol{y} = [\boldsymbol{y}_j]$ is block-orthonormal, and if the principal component ξ of $\boldsymbol{y}$ is split into the parts $\boldsymbol{\eta} = [\eta_j]$, then the first eigenvalue $\lambda^2(\boldsymbol{S})$ of $\boldsymbol{S} = \mathrm{E}(\boldsymbol{y}\boldsymbol{y}')$ is identical to the first eigenvalue $\lambda^2(\boldsymbol{R}) = \boldsymbol{v}'\boldsymbol{R}\boldsymbol{v}$ of $\boldsymbol{R} = \mathrm{E}(\boldsymbol{\eta}\boldsymbol{\eta}')$.

Proof: Let $\boldsymbol{w}^*$ denote the first eigenvector of $\boldsymbol{S}$, so that the first eigenvalue is

$$\begin{aligned}
\lambda^2(\boldsymbol{S}) &= \frac{\boldsymbol{w}^{*\prime}\boldsymbol{S}\boldsymbol{w}^*}{\boldsymbol{w}^{*\prime}\boldsymbol{w}^*} && \text{by Eq.3.515}\\
&= \frac{\mathbf{1}'\boldsymbol{W}^{*\prime}\boldsymbol{S}\boldsymbol{W}^*\mathbf{1}}{\mathbf{1}'\boldsymbol{W}^{*\prime}\boldsymbol{W}^*\mathbf{1}} && \text{by Eq.3.516}\\
&= \frac{\mathbf{1}'\boldsymbol{V}'\boldsymbol{W}'\boldsymbol{S}\boldsymbol{W}\boldsymbol{V}\mathbf{1}}{\mathbf{1}'\boldsymbol{V}'\boldsymbol{W}'\boldsymbol{W}\boldsymbol{V}\mathbf{1}} && \text{by Eq.3.523}\\
&= \frac{\boldsymbol{v}'\boldsymbol{W}'\boldsymbol{S}\boldsymbol{W}\boldsymbol{v}}{\boldsymbol{v}'\boldsymbol{W}'\boldsymbol{W}\boldsymbol{v}} && \text{by Eq.3.522}\\
&= \frac{\boldsymbol{v}'\boldsymbol{R}\boldsymbol{v}}{\boldsymbol{v}'\mathbf{I}\boldsymbol{v}} && \text{by Eqs.3.530, 3.536}\\
&= \lambda^2(\boldsymbol{R}) && \text{q.e.d.}
\end{aligned} \tag{3.539}$$

For any choice of weight vectors $\boldsymbol{v} = \boldsymbol{v}_0$ the ratio $\boldsymbol{v}'\boldsymbol{R}\boldsymbol{v}/\boldsymbol{v}'\boldsymbol{v}$ will be smaller than for the eigenvectors $\boldsymbol{v} = \boldsymbol{v}_{PC}$, and consequently the ratio of the quadratic forms in Eq.3.539 will be smaller than the first eigenvalue of $\boldsymbol{R}$.

To put it in geometrical terms: The variables $\boldsymbol{\eta}$ represent a J-dimensional subspace of the K-dimensional space spanned by the MVs $\boldsymbol{y}$. ξ is the dominant axis of $\boldsymbol{y}$, with length (eigenvalue, variance) $\lambda^2(\boldsymbol{S})$. The dominant axis of $\boldsymbol{\eta}$, as a subspace of $\boldsymbol{y}$, can not have greater length $\lambda^2(\boldsymbol{R})$ than the dominant axis of $\boldsymbol{y}$, but could have smaller length. However, due to the choice of the subspace, ξ is part of this subspace, because it can be determined as a linear aggregate of the variables $\boldsymbol{\eta}$.

The theorem can be extended to the case of variables which are not block-orthonormal. In this case the term $\boldsymbol{W}'\boldsymbol{W}$ does not vanish in Eq.3.539, and we get the Rayleigh quotient $\lambda^2(\boldsymbol{R}) = \boldsymbol{v}'\boldsymbol{R}\boldsymbol{v}/\boldsymbol{v}'\boldsymbol{\Delta}^2\boldsymbol{v}$, i.e., we get the principal component of $\boldsymbol{R}$ with respect to a scaling matrix $\boldsymbol{\Delta}^2$ — which depends by Eq.3.531 on the weights $\boldsymbol{W}$.

$\boldsymbol{R}$ is a correlation matrix and, the larger the off-diagonal elements, the larger $\lambda^2(\boldsymbol{R})$ will become. Conversely, if $\lambda^2(\boldsymbol{R}) = \lambda^2(\boldsymbol{S})$ is maximal, the correlations between the η_j-variables are as high as possible. The highest possible value of $\lambda^2(\boldsymbol{R})$ is J, i.e. the number of blocks. In this case $\boldsymbol{R}$ is a unit matrix, the correlation matrix of perfectly correlated variables η_j, and $\boldsymbol{R}$ has rank one.

3.5.3 Horst's Maximum Variance Algorithm

The preceding considerations of Hierarchical PC models are implemented in the following non-iterative algorithm which is Horst's (1965) Maximum Variance algorithm for Generalized Canonical Correlation models:

1. Make the variables $\breve{\boldsymbol{y}}$ blockwise orthonormal by Eq.3.532.
2. Compute the principal component of the block-orthonormal variables, i.e. get $\boldsymbol{w}^*$ by Eq.3.515.
3. Split the principal component ξ into pieces η_j^* by Eq.3.516.
4. Standardize the pieces by Eq.3.523.
5. Backtransform the weights so they relate the standardized LVs η_j to the original variables $\breve{\boldsymbol{y}}$ by Eq.3.533.

The definition of principal components via the Rayleigh quotient leaves the scale unit of the weights and the component ambiguous. If we restrict the scale of ξ by $\boldsymbol{w}^{*\prime}\boldsymbol{w}^* = 1$ as indicated in Eq.3.514, then the variance of ξ is identical to the eigenvalue λ^2, and the principal component has maximum variance.

Blockwise orthonormal variables are purified from all factors which operate only within blocks, there remaining only factors which go across blocks (Inter-battery Factor model, cf. Eq.3.470). If the principal component of blockwise orthonormal variables has $\lambda^2 > 1$, it has captured only between-block, but not within-block factors, and it has captured as much correlation as possible. If the variables $\boldsymbol{y}$ are uncorrelated, the results is $\lambda^2 = 1$; if $\lambda^2 > 1$ this due to the fact that the principal component uses correlations between the MVs.

We split the principal component ξ of blockwise orthonormal variables into the unstandardized pieces η_j^*. By Eq.3.520 the variance q^2 is equal to the sum of the variances and covariances of pieces η_j. And by Eq.3.526 the variance q^2 is equal to the $\boldsymbol{v}$-weighted sum of the correlations of the η_js. So we know that the lower-order LVs have maximum covariance (maximum weighted correlation) if the higher-order LV has maximum variance. This comes close to the concept of canonical correlation of several sets of variables, and as a matter of fact, this method amounts to the canonical correlation in a strict sense if there are just two blocks.

3.5.4 The Principle of Constant Proportionality

We link up again with Eq.3.510 and treat two of the special cases mentioned above. It will lead us to a general principle that facilitates the estimation of Hierarchical Component models without explicit computation of the Higher-order principal component. The ensuing algorithms are special cases of the general PLS algorithm, and due to the reference to the eigenvalue problem, a proof of convergence is available.

Special case (ii): Diagonal $\boldsymbol{D}$. If $\boldsymbol{D}$ is a diagonal matrix of order 3, $\boldsymbol{D} = \mathbf{I} * \boldsymbol{S}$ and $J = 3$, we have these three equivalent expressions for Eq.3.511:

$$\boldsymbol{S}\boldsymbol{w} = \lambda^2 \boldsymbol{D}\boldsymbol{w} \tag{3.540}$$

$$\sum_{\ell} s_{k\ell} w_\ell = \lambda^2 d_{kk} w_k \qquad \forall k. \tag{3.541}$$

$$\begin{aligned} s_{11}w_1 + s_{12}w_2 + s_{13}w_3 &= \lambda^2 d_{11} w_1 \\ s_{21}w_1 + s_{22}w_2 + s_{23}w_3 &= \lambda^2 d_{22} w_2 \\ s_{31}w_1 + s_{32}w_2 + s_{33}w_3 &= \lambda^2 d_{33} w_3 \end{aligned} \tag{3.542}$$

Subtraction of the diagonal elements $d_{kk}w_k = s_{kk}w_k$ on both sides of Eq.3.542 gives

$$\begin{aligned} 0 + s_{12}w_2 + s_{13}w_3 &= (\lambda^2 - 1) d_{11} w_1 \\ s_{21}w_1 + 0 + s_{23}w_3 &= (\lambda^2 - 1) d_{22} w_2 \\ s_{31}w_1 + s_{32}w_2 + 0 &= (\lambda^2 - 1) d_{33} w_3, \end{aligned} \tag{3.543}$$

or in general form as

$$\sum_{\ell \neq k} s_{k\ell} w_\ell = (\lambda^2 - 1) d_{kk} w_k, \qquad \forall k, \tag{3.544}$$

$$(\boldsymbol{S} - \boldsymbol{D})\boldsymbol{w} = (\lambda^2 - 1)\boldsymbol{D}\boldsymbol{w}, \tag{3.545}$$

$$\frac{\sum_{\ell \neq k} s_{k\ell} w_\ell}{s_{kk}} = (\lambda^2 - 1) w_k, \qquad \forall k. \tag{3.546}$$

This relation will be used in Sc.3.5.5 to develop a different algorithm for principal components.

Special case (iv): Block-diagonal $\boldsymbol{D}$. Let the variables $\boldsymbol{y} = [y_k, k = 1 \ldots K]$ be partitioned into subsets, $\boldsymbol{y} = [\boldsymbol{y}_j], \boldsymbol{y}_j = [\boldsymbol{y}_{k_j}, k_j = 1 \ldots K_j]$, and $\boldsymbol{D}$ defined as the diagonal supermatrix with elements

$$\boldsymbol{D}_{jj} = \boldsymbol{S}_{jj} = \mathrm{E}(\boldsymbol{y}_j \boldsymbol{y}_j'). \tag{3.547}$$

The stationary expression Eq.3.542 can now be written for three blocks:

$$\begin{aligned} \boldsymbol{S}_{11}\boldsymbol{w}_1 + \boldsymbol{S}_{12}\boldsymbol{w}_2 + \boldsymbol{S}_{13}\boldsymbol{w}_3 &= \lambda^2 \boldsymbol{D}_{11} \boldsymbol{w}_1 \\ \boldsymbol{S}_{21}\boldsymbol{w}_1 + \boldsymbol{S}_{22}\boldsymbol{w}_2 + \boldsymbol{S}_{23}\boldsymbol{w}_3 &= \lambda^2 \boldsymbol{D}_{22} \boldsymbol{w}_2 \\ \boldsymbol{S}_{31}\boldsymbol{w}_1 + \boldsymbol{S}_{32}\boldsymbol{w}_2 + \boldsymbol{S}_{33}\boldsymbol{w}_3 &= \lambda^2 \boldsymbol{D}_{33} \boldsymbol{w}_3 \end{aligned} \tag{3.548}$$

$$\sum_i \boldsymbol{S}_{ji} \boldsymbol{w}_i = \lambda^2 \boldsymbol{D}_{jj} \boldsymbol{w}_j, \qquad \forall j. \tag{3.549}$$

Subtraction of $\boldsymbol{S}_{jj}\boldsymbol{w}_j = \boldsymbol{D}_{jj}\boldsymbol{w}_j$ gives, in the same way in the passage from Eq.3.542 to Eq.3.544,

$$\sum_{i \neq j} \boldsymbol{S}_{ji} \boldsymbol{w}_i = (\lambda^2 - 1) \boldsymbol{D}_{jj} \boldsymbol{w}_j, \qquad \forall j. \tag{3.550}$$

Premultiplication by $\boldsymbol{w}_j'$ and division by the scalar $\boldsymbol{w}_j'\boldsymbol{D}_{jj}\boldsymbol{w}_j = \boldsymbol{w}_j'\boldsymbol{S}_{jj}\boldsymbol{w}_j$ gives

$$\lambda^2 - 1 = \frac{\sum_{i\neq j} \boldsymbol{w}_j'\boldsymbol{S}_{ji}\boldsymbol{w}_i}{\boldsymbol{w}_j'\boldsymbol{S}_{jj}\boldsymbol{w}_j}, \qquad \forall j. \tag{3.551}$$

This relation is named the principle of constant proportionality by Nishisato and Sheu (1980), because the constant $\lambda^2 - 1$ is the same for all blocks. We will use this principle in Sc.3.5.6 to develop a different algorithm for generalized canonical correlation.

The principle of constant proportionality is presented by its authors on moment matrices $\boldsymbol{S}$. In the following subsection we will reinterpret the principle in terms of variables so it can be applied for prediction, not only covariance reconstruction.

3.5.5 Principal Component, One Variable Omitted.

The setup Eqs.3.540,..., 3.546 describes the principal component of a correlation matrix. We will interpret the model in terms of variables $\boldsymbol{y}$ with correlation matrix $\boldsymbol{S} = \mathrm{E}(\boldsymbol{y}'\boldsymbol{y})$ and the first principal component $\xi = \boldsymbol{w}'\boldsymbol{y}$. We adopt the scaling $\boldsymbol{w}'\boldsymbol{w} = 1$ so that $\mathrm{var}(\xi) = \boldsymbol{w}'\boldsymbol{S}\boldsymbol{w} = \lambda^2$. Let ξ_{-k} denote the without-one composite when y_k is omitted from the summation,

$$\xi_{-k} = \sum_{\ell\neq k} w_\ell y_\ell, \tag{3.552}$$

with variance

$$\begin{aligned}\mathrm{var}(\xi_{-k}) &= \mathrm{var}(\xi - w_k y_k) = \mathrm{var}(\xi) - 2w_k\,\mathrm{cov}(\xi y_k) + w_k^2\,\mathrm{var}(y_k)\\ &= \lambda^2 - 2\lambda^2 w_k^2 + w_k^2.\end{aligned} \tag{3.553}$$

The covariance of the without-one principal component with the omitted MV is

$$\begin{aligned}\mathrm{E}(y_k\xi_{-k}) &= \mathrm{E}(y_k \sum_{\ell\neq k} w_\ell y_\ell) = \sum_{\ell\neq k} w_\ell s_{k\ell}\\ &= (\lambda^2 - 1)w_k \qquad \text{due to Eq.3.546}\end{aligned} \tag{3.554}$$

and the regression coefficient in

$$\xi_{-k} = \breve{w}_k y_k + e_k^* \tag{3.555}$$

is

$$\begin{aligned}\breve{w}_k &= \frac{\mathrm{E}(y_k\xi_{-k})}{\mathrm{E}(y_k^2)} = \frac{\sum_{\ell\neq k} w_\ell s_{k\ell}}{s_{kk}}\\ &= (\lambda^2 - 1)w_k.\end{aligned} \tag{3.556}$$

That is to say, if we know the eigenvalue λ^2 and all weights $\boldsymbol{w}$ with the exception of w_k, we can immediately solve Eq.3.556 for the one missing weight. We can omit any one of the MVs, and the regression of the omitted variable y_k on the properly weighted composite ξ_{-k} of the other variables will give the proper weight coefficient w_k, distorted by a bias factor $\lambda^2 - 1$. The bias factor is the same for all variables. (The dominant eigenvalue of a $K\times K$ correlation matrix is $1 \leq \lambda^2 \leq K$.)

PLS algorithm for principal components. The power algorithm, as described in Sc.3.1.7, can be modified by using Eq.3.556 to determine the weights for the next iteration cycle. The algorithm shown in Table 3.13 is a Partial Least Squares algorithm: While the parameter of the one "block" y_k are computed, the rest of the parameters are taken as fixed and known, even if they only current and improvable approximations.

Table 3.13: PLS algorithm for principal component, one MV omitted

	${}_0\boldsymbol{w}$	Starting values are (nearly) arbitrary.
Repeat for cycles $c = 0, 1, 2, \dots$ and for variables $k = 1 \dots K$:		
#2	${}_c\tilde{\xi}_{-k} := {}_cf \sum_{\ell \neq k} {}_cw_\ell y_\ell$	The without-one principal component is the inside approximation, which is scaled by a constant ${}_cf$, preferably so as to make $\mathrm{E}(\hat{\xi}^2) = 1$.
#3	${}_{c+1}w_k := \mathrm{E}(y_k \, {}_c\tilde{\xi}_{-k})$	The weights are computed by ModeA.
until the weights stabilize.		

Convergence of PLS algorithm. We consider the starting weights ${}_0\boldsymbol{w}$ as consisting of the first eigenvector $\boldsymbol{w}$ and of a weighted sum of all other eigenvectors. To treat the case of slowest convergence we let the sum of all others consist of only the second eigenvector $\breve{\boldsymbol{w}}$; the second eigenvalue is $\breve{\lambda}^2$, the second PC is $\breve{\xi}$. The two eigenvectors are mixed in an unknown proportion $a : \breve{a}$:

$$
{}_0\boldsymbol{w} = a\boldsymbol{w} + \breve{a}\breve{\boldsymbol{w}} \tag{3.557}
$$

$$
{}_0\xi = a\xi + \breve{a}\breve{\xi}, \qquad \mathrm{E}(\xi\breve{\xi}) = 0 \tag{3.558}
$$

$$
\begin{aligned}
\mathrm{E}(y_k \, {}_0\xi_{-k}) &= \mathrm{E}[y_k({}_0\xi - y_k \, {}_0w_k)] \\
&= \mathrm{E}[y_k(a\xi + \breve{a}\breve{\xi} - y_k(aw_k + \breve{a}\breve{w}_k)] \\
&= a\,\mathrm{E}(y_k\xi) - aw_k\,\mathrm{E}(y_k^2) + \breve{a}\,\mathrm{E}(y_k\breve{\xi}) - \breve{a}\breve{w}_k\,\mathrm{E}(y_k^2) \\
&= a\lambda^2 w_k - aw_k + \breve{a}\breve{\lambda}^2\breve{w}_k - \breve{a}\breve{w}_k \\
&= (\lambda^2 - 1)aw_k + (\breve{\lambda}^2 - 1)\breve{a}\breve{w}_k
\end{aligned} \tag{3.559}
$$

$$
{}_1\boldsymbol{w} = {}_0f\left((\lambda^2 - 1)a\boldsymbol{w} + (\breve{\lambda}^2 - 1)\breve{a}\breve{\boldsymbol{w}}\right) \tag{3.560}
$$

If we choose the scalar ${}_0f$ to be equal to $(\lambda^2 - 1)^{-1}$, the last equation becomes:

$$
{}_1\boldsymbol{w} = a\boldsymbol{w} + \left(\frac{\breve{\lambda}^2 - 1}{\lambda^2 - 1}\right)\breve{a}\breve{\boldsymbol{w}}. \tag{3.561}
$$

The approximation to the first eigenvalue changes from one iteration cycle to the next by the factor $(\breve{\lambda}^2 - 1)/(\lambda^2 - 1)$. If both eigenvalues are greater than one, this is an improvement, and the algorithm converges to the first principal component. This convergence is better than that of the usual power algorithm which is $\breve{\lambda}^2/\lambda^2$. However, this without-one power algorithm needs more computations. For more details on the side conditions cf. Sc.3.1.8.

3.5.6 Hierarchical Component Model, One Block Omitted

Constant regression. Let ξ denote the composite measured by the total set of variables $\boldsymbol{y} = [\boldsymbol{y}_j]$. The weights are determined by Eqs.3.510, 3.551, and the ambiguity of the scale is resolved by adopting the scaling $\boldsymbol{w}'\boldsymbol{D}\boldsymbol{w} = \sum_j \boldsymbol{w}_j'\boldsymbol{S}_{jj}\boldsymbol{w}_j = \sum_j r_{jj} = 1$:

$$\xi = \boldsymbol{w}'\boldsymbol{y}, \qquad \mathrm{E}(\xi^2) \equiv q^2 = \boldsymbol{w}'\boldsymbol{S}\boldsymbol{w} = \lambda^2. \tag{3.562}$$

η_j is the fraction of ξ measured by the subset $\boldsymbol{y}_j$,

$$\eta_j = \boldsymbol{w}_j'\boldsymbol{y}_j, \qquad \mathrm{E}(\eta_j^2) \equiv r_{jj} = \boldsymbol{w}_j'\boldsymbol{S}_{jj}\boldsymbol{w}_j, \tag{3.563}$$

and ξ_{-j} is the remaining part of ξ measured by the other variables

$$\xi_{-j} = \xi - \eta_j = \sum_{i\neq j} \boldsymbol{w}_i'\boldsymbol{y}_i, \qquad \mathrm{E}(\xi_{-j}^2) \equiv q_{-j}^2 = \sum_{i\neq j}\sum_{h\neq j} \boldsymbol{w}_i'\boldsymbol{S}_{ih}\boldsymbol{w}_h = \sum_{i\neq j}\sum_{h\neq j} r_{ih}. \tag{3.564}$$

The covariance of the piece (the fraction) η_j with the remaining part ξ_{-j} is

$$\begin{aligned}\mathrm{E}(\eta_j\xi_{-j}) &= \mathrm{E}[(\boldsymbol{w}'\boldsymbol{y}_j)(\sum_{i\neq j} \boldsymbol{w}_i'\boldsymbol{y}_i)] = \sum_{i\neq j} \mathrm{E}(\boldsymbol{w}_j'\boldsymbol{y}_j\boldsymbol{y}_i'\boldsymbol{w}_i) = \sum_{i\neq j} \boldsymbol{w}_j'\boldsymbol{S}_{ji}\boldsymbol{w}_i = \sum_{i\neq j} r_{ji} \\ &= (\lambda^2 - 1)r_{jj} \qquad \text{by Eq.3.551}\end{aligned} \tag{3.565}$$

by the principle of constant proportionality. The regression of the rest ξ_{-j} on the piece η_j, $\xi_{-j} = v_j\eta_j + \text{residuum}_j$, has as regression coefficient, by Eq.3.551

$$v_j = \frac{\mathrm{E}(\eta_j\xi_{-j})}{E(\eta_j^2)} = \frac{\sum_{i\neq j} \boldsymbol{w}_j'\boldsymbol{S}_{ji}\boldsymbol{w}_i}{\boldsymbol{w}_j'\boldsymbol{S}_{jj}\boldsymbol{w}_j} = \frac{\sum_{i\neq j} r_{ji}}{r_{jj}} = \lambda^2 - 1, \qquad \forall j. \tag{3.566}$$

The theorem Eq.3.551 together with Eq.3.566 states that the regression coefficient is $\lambda^2 - 1$ and is constant for any block $\boldsymbol{y}_j$ of the variables $\boldsymbol{y}$. A special case where all $\boldsymbol{D}_j = \mathbf{I}$ will be treated in Sc.4.1.3.

A PLS algorithm. The Hierarchical Principal Component model in Sc.3.5.1 was estimated by repeated use of MVs. We will show in this paragraph that a convergent PLS algorithm can be constructed without explicit computation of the higher-level component, and hence without repeated MVs. The algorithm uses the principle of constant proportionality. First we will present the algorithm in a restricted version for blockwise orthonormal variables, because this simplifies the proof of convergence. Then follows a more general algorithm that is equivalent to the simplified one and enjoys convergence, too.

Table 3.14 shows a modified Hierarchical Principal Component model that may be called a "crypto-hierarchical" model. The higher-order component ξ is deleted from the model, and connections are introduced between all first-level components η_j. That is to say, all LVs are adjacent to each other, and the inside approximation $\tilde{\eta}_j$ is the sum of all LVs, with the exception of η_j, i.e. $\tilde{\eta}_j = \xi_{-j}$. The algorithm is shown in Table 3.15. In one iteration cycle the computations #2 to #4 are performed for all blocks j. The sequence of the $3 \times J$ computations can be done in total steps (i.e. #2 for all blocks, then #3 for all blocks, then #4 for all blocks), or in single steps (i.e. the three computations first for block 1, then for block 2, etc.).

Table 3.14: A crypto-hierarchical LV model

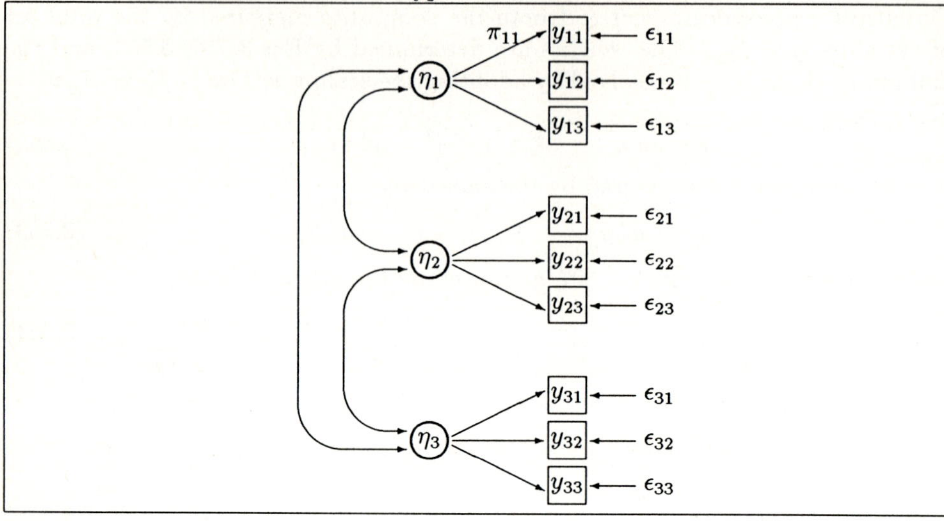

Table 3.15: PLS algorithm for hierarchical principal component, no repeated MVs, but blockwise orthonormal MVs

	$\mathrm{E}(\boldsymbol{y}_j\boldsymbol{y}_j') \equiv \boldsymbol{S}_{jj} = \mathbf{I}$	The MVs are blockwise orthonormal.
	${}_0\boldsymbol{w}_j$	Starting values are (nearly) arbitrary.
Repeat for cycles $c = 1, 2, \ldots$ and for blocks $j = 1 \ldots J$:		
#2	${}_c\tilde{\eta}_j := {}_cf \sum_{i \neq j} {}_c\hat{\eta}_i$	The inside approximation is the *unweighted* sum of all adjacent variables. It is scaled by a constant ${}_cf$, preferably so as to make $\mathrm{E}(\sum_j \tilde{\eta}_j^2) = 1$.
#3	${}_{c+1}\boldsymbol{w}_j := \mathrm{E}(\boldsymbol{y}_j\ {}_c\tilde{\eta}_j)$	The weights are computed by ModeA.
#4	${}_c\hat{\eta}_j := {}_c\boldsymbol{w}_j'\boldsymbol{y}_j$	The LVs are *not* standardized.
until the weights stabilize.		

Convergence of PLS algorithm. We consider the starting values ${}_0\boldsymbol{w}$ as consisting of the first eigenvector $\boldsymbol{w}$ and of a weighted sum of all other eigenvectors. To treat the case of worst convergence we let the sum of all others consist of only the second eigenvector $\breve{\boldsymbol{w}}$; the second eigenvalue is $\breve{\lambda}^2$, the second PC is $\breve{\xi}$. The two eigenvectors are mixed in an unknown proportion $a : \breve{a}$.

$$ {}_0\boldsymbol{w} = a\boldsymbol{w} + \breve{a}\breve{\boldsymbol{w}} \tag{3.567} $$

$$ {}_0\xi = a\boldsymbol{w}'\boldsymbol{y} + \breve{a}\breve{\boldsymbol{w}}'\boldsymbol{y} = a\xi + \breve{a}\breve{\xi}, \qquad \mathrm{E}(\xi\breve{\xi}) = 0 \tag{3.568} $$

$$ {}_0\xi_{-j} = {}_0\xi - \boldsymbol{y}_j'\,{}_0\boldsymbol{w}_j = a\xi - a\boldsymbol{y}_j'\boldsymbol{w}_j + \breve{a}\breve{\xi} - \breve{a}\boldsymbol{y}_j'\breve{\boldsymbol{w}}_j \tag{3.569} $$

$$ \begin{aligned} \mathrm{E}(\boldsymbol{y}_j\,{}_0\xi_{-j}) &= a\,\mathrm{E}(\boldsymbol{y}_j\xi) - a\,\mathrm{E}(\boldsymbol{y}_j\boldsymbol{y}_j')\boldsymbol{w}_j + \breve{a}\,\mathrm{E}(\boldsymbol{y}_j\breve{\xi}) - \breve{a}\,\mathrm{E}(\boldsymbol{y}_j\boldsymbol{y}_j')\breve{\boldsymbol{w}}_j \\ &= a\lambda^2\boldsymbol{w}_j - a\boldsymbol{S}_{jj}\boldsymbol{w}_j + \breve{a}\breve{\lambda}^2\breve{\boldsymbol{w}}_j - \breve{a}\boldsymbol{S}_{jj}\breve{\boldsymbol{w}}_j \end{aligned} $$

$$= (\lambda^2 - 1)a\boldsymbol{w}_j + (\check{\lambda}^2 - 1)\check{a}\check{\boldsymbol{w}}_j \tag{3.570}$$

$$_1\boldsymbol{w}_j = {}_0f\,\mathrm{E}(\boldsymbol{y}_j\,{}_0\xi_{-j}) = {}_0f\left[(\lambda^2 - 1)a\boldsymbol{w}_j + (\check{\lambda}^2 - 1)\check{a}\check{\boldsymbol{w}}_j\right]$$

$$_1\boldsymbol{w} = {}_0f\left[(\lambda^2 - 1)a\boldsymbol{w} + (\check{\lambda}^2 - 1)\check{a}\check{\boldsymbol{w}}\right] \tag{3.571}$$

If we choose the normalization constant $_0f$ equal to $(\lambda^2 - 1)^{-1}$, the last equation becomes

$$_1\boldsymbol{w} = a\boldsymbol{w} + \left(\frac{\check{\lambda}^2 - 1}{\lambda^2 - 1}\right)\check{a}\check{\boldsymbol{w}}. \tag{3.572}$$

We repeat the computation Eqs.3.567,..., 3.572, but now with $\check{a} := \check{a}(\check{\lambda}^2 - 1)/(\lambda^2 - 1)$ in lieu of $\check{a}$, and get the second approximation to the weights:

$$_2\boldsymbol{w} = a\boldsymbol{w} + \left(\frac{\check{\lambda}^2 - 1}{\lambda^2 - 1}\right)\check{\check{a}}\check{\boldsymbol{w}} = a\boldsymbol{w} + \left(\frac{\check{\lambda}^2 - 1}{\lambda^2 - 1}\right)^2\check{a}\check{\boldsymbol{w}}. \tag{3.573}$$

In general, in the cth iteration cycle the approximation becomes

$$_c\boldsymbol{w} = a\boldsymbol{w} + \left(\frac{\check{\lambda}^2 - 1}{\lambda^2 - 1}\right)^c\check{a}\check{\boldsymbol{w}}. \tag{3.574}$$

The ratio $(\check{\lambda}^2 - 1)/(\lambda^2 - 1)$ is smaller than one, hence it vanishes with increasing iteration cycle c:

$$\left(\frac{\check{\lambda}^2 - 1}{\lambda^2 - 1}\right)^c \rightarrow 0 \qquad \text{as} \qquad c \rightarrow \infty$$

and

$$_c\boldsymbol{w} \rightarrow a\boldsymbol{w} \qquad \text{as} \qquad c \rightarrow \infty,$$

which completes the proof of the convergence of the Partial Least Squares algorithm for Hierarchical Component model.

Generalization of the PLS algorithm. The proof of convergence simplifies if the MVs are blockwise orthonormal and $\boldsymbol{S}_{jj}$ vanishes in Eq.3.570. However, the algorithm works without this prerequisite.

Let $\check{\boldsymbol{y}}_j$ denote the correlated counterpart of $\boldsymbol{y}_j$, linked to it by a transformation $\boldsymbol{T}_j$ so that $\check{\boldsymbol{y}}_j = \boldsymbol{T}_j\boldsymbol{y}_j$ and $\mathrm{E}(\check{\boldsymbol{y}}_j\check{\boldsymbol{y}}_j') = \check{\boldsymbol{S}}_{jj} = \boldsymbol{T}_j\boldsymbol{T}_j'$. The weights are replaced by $\check{\boldsymbol{w}}_j = \boldsymbol{T}_j^{-1\prime}\boldsymbol{w}_j$ so that the LV is immune to the change of the MVs:

$$\hat{\eta}_j = \check{\boldsymbol{w}}_j'\check{\boldsymbol{y}}_j = (\boldsymbol{T}_j^{-1\prime}\boldsymbol{w}_j)'(\boldsymbol{T}_j\boldsymbol{y}_j) = \boldsymbol{w}_j'\boldsymbol{T}_j^{-1}\boldsymbol{T}_j\boldsymbol{y}_j = \boldsymbol{w}_j'\boldsymbol{y}_j.$$

The weights $\check{\boldsymbol{w}}_j$ are obtained by:

$$\mathrm{E}(\check{\boldsymbol{y}}_j\tilde{\eta}_j) = \mathrm{E}(\boldsymbol{T}_j\boldsymbol{y}_j\tilde{\eta}_j) = \boldsymbol{T}_j\,\mathrm{E}(\boldsymbol{y}_j\tilde{\eta}_j)$$

$$= \boldsymbol{T}_j\boldsymbol{w}_j \tag{3.575}$$

$$\boldsymbol{w}_j = \boldsymbol{T}_j^{-1}\,\mathrm{E}(\check{\boldsymbol{y}}_j\tilde{\eta}_j) \tag{3.576}$$

$$\check{\boldsymbol{w}}_j = \boldsymbol{T}_j^{-1\prime}\boldsymbol{T}_j^{-1}\,\mathrm{E}(\check{\boldsymbol{y}}_j\tilde{\eta}_j) = (\boldsymbol{T}_j\boldsymbol{T}_j')^{-1}\,\mathrm{E}(\check{\boldsymbol{y}}_j\tilde{\eta}_j)$$

$$= \check{\boldsymbol{S}}_{jj}^{-1}\,\mathrm{E}(\check{\boldsymbol{y}}_j\tilde{\eta}_j) \tag{3.577}$$

The last line ensures that the weights of the model can be determined from uncorrelated as well as correlated MVs. The weights are multiple regression coefficients, i.e. ModeB weights in all blocks. The algorithm for correlated MVs is shown in Table 3.16. The results from this algorithm are listed in Tb.3.13, row 19, and are identical to Wold's algorithm with repeated MVs and to Horst's Maximum Variance algorithm.

Table 3.16: PLS algorithm for hierarchical principal component

$\mathrm{E}(\breve{\boldsymbol{y}}_j \breve{\boldsymbol{y}}_j{}') \equiv \breve{\boldsymbol{S}}_{jj}$	The MVs are *not* required to be blockwise orthonormal.
${}_0\breve{\boldsymbol{w}}_j$	Starting values are (nearly) arbitrary.
Repeat for cycles $c = 1, 2, \ldots$ and for blocks $j = 1 \ldots J$:	
#2 ${}_c\tilde{\eta}_j := {}_cf \sum_{i \neq j} {}_c\hat{\eta}_i$	The inside approximation is the *unweighted* sum of all adjacent variables. It is scaled by a constant ${}_cf$, preferably so as to make $\mathrm{E}(\sum_j \tilde{\eta}_j^2) = 1$.
#3 ${}_{c+1}\breve{\boldsymbol{w}}_j := \breve{\boldsymbol{S}}_{jj}^{-1} \mathrm{E}(\breve{\boldsymbol{y}}_j \, {}_c\tilde{\eta}_j)$	The weights are computed by ModeB.
#4 ${}_c\hat{\eta}_j := {}_c\breve{\boldsymbol{w}}_j{}' \breve{\boldsymbol{y}}_j$	The LVs are *not* standardized.
until the weights stabilize.	

3.5.7 Applications of the Split PC Theorem

Categorical scaling. Let each one of several categorical variables y_j be replaced by a set of Boolean variables y_{k_j}. The product moment matrix of these Boolean variables is a super contingency table (see Sc.4.1.1) with all submatrices in the diagonal being themselves diagonal matrices. The principal component of the super contingency table is split up into standardized parts η_j. The weights serve as scale values for the categorical variables. If the originally categorical variables are rescaled in this way, they correlate maximally with each other, this being assured by the Split Principal Component theorem, Eq.3.539. The categorical scaling method will be discussed in more detail in Sc.4.1.

Generalized canonical correlation. The problem of generalizing Hotelling's (1935) Canonical Correlation model to more than two sets of variables has attracted many scholars, starting with Vinograde (1950) and Steel (1951); the first viable methods were developed by Horst (1961, 1965) and Carroll (1963); extensive discussions can be found with Browne (1980), Conger (1971), Golding and Seidman (1974), Jackson (1969, 1975, 1977), Kettenring (1971) Kristof and Wingersky (1971), McDonald (1968, 1970), McKeon (1966), Pruzek, Lehrer and Kim (1979), Rindskopf and Rose (1985), Rozeboom (1965), Saporta (1978), and van de Geer (1984). For reviews, see Skinner (1977, 1978) and Lohmöller (1977). Wilks (1935) developed a test of the hypothesis that all blocks are independent from each other. ten Berge (1984) and ten Berge and Knol (1984) treat the related problem of simultaneous congruence rotation of several factor loading matrices.

The problem of canonical correlation analysis, i.e. the maximization of the correlation of linear aggregates of two blocks of variables, becomes ambiguous when more than two blocks of MVs are involved. Some function of the correlation coefficients of all pairs of blocks must first be formulated which is then maximized. This can be the sum of all correlation coefficients ($\varphi = \mathbf{1}'\boldsymbol{R}\mathbf{1}$), or the weighted sum ($\varphi = \boldsymbol{v}'\boldsymbol{R}\boldsymbol{v}$), or the first eigenvalue $\varphi = \lambda^2(\boldsymbol{R})$, or some other function.

The first eigenvalue $\varphi = \lambda^2(\boldsymbol{R})$ is a bilinear function of $\boldsymbol{R}$ by $\varphi = \boldsymbol{v}'\boldsymbol{R}\boldsymbol{v}$ where $\boldsymbol{v}$ is the corresponding eigenvector. As shown above, the Split-PC method maximizes the first eigenvalue. Horst (1965), who first put forward this method, called it the RANKONE method, because the higher the first eigenvalue of a correlation matrix, the closer this matrix will be to a rank-one matrix.

Horst's (1965) SUMCOR method maximizes the unweighted sum of the correlations, $\varphi = \mathbf{1}'\boldsymbol{R}\mathbf{1}$. It appears as somewhat tedious to show that this, too, amounts to a Split PC method where the variables $\boldsymbol{\eta}$ have to be rescaled iteratively. Weight vectors consisting only of unit values (or -1 values to provide a reversion of the scale) produce linear combination $\xi = \mathbf{1}'\boldsymbol{\eta}$ which are the sums (or averages, centroids) of the $\boldsymbol{\eta}$-variables. Thus Horst's SUMCOR method maximizes the variance of the second-order centroid factor.

The algorithms based on the Split PC Theorem and proposed here for the first time give estimates for the Generalized Canonical Correlation model and for the Hierarchical Component model.

- The model is defined in terms of conditional expectations (Sc.3.5.1) and maximizes is the sum of variances and covariances of the LVs, $q^2 = \sum_i \sum_j r_{ij}$, under a side condition that limits the variance of the LVs, $\boldsymbol{w}_j'\boldsymbol{w}_j = 1$, say.
- q^2 is the eigenvalue of a blockwise orthogonal matrix $\boldsymbol{S}$, which can be obtained from blockwise uncorrelated MVs; the orthonormalization corresponds to ModeB weight estimation.
- Two ad-hoc devices of Wold's PLS algorithm, the inner weighting scheme and the standardization of the LVs, are given up in favour of a unified optimization criterion.
- The results are identical to the estimates by Wold's algorithm with repeated MVs and to Horst's Maximum Variance algorithm.

PLS factor weighting scheme. The weight vector $\boldsymbol{v}$ and the loading vector $\boldsymbol{b}$ of the PC model (Eqs.3.535, 3.537) are proportional to one other, $\boldsymbol{b} = \lambda^2\boldsymbol{v}$, which fact is used to estimate the PC iteratively: Rescaled loadings are used as weights. However, if the covariance matrix $\boldsymbol{R}$ really is a rank-one matrix, then each row and each column of $\boldsymbol{R}$ can be used as well. Consider the following vector $\boldsymbol{b}$ and rank-one matrix $\boldsymbol{R}$,

$$\boldsymbol{b} = \begin{bmatrix} .80 \\ .70 \\ .60 \\ .50 \end{bmatrix}, \qquad \boldsymbol{R} = \boldsymbol{b}\boldsymbol{b}' = \begin{bmatrix} .64 & .56 & .48 & .40 \\ .56 & .49 & .42 & .35 \\ .48 & .42 & .36 & .30 \\ .40 & .35 & .30 & .25 \end{bmatrix} \tag{3.578}$$

and observe that each row and each column of $\boldsymbol{R}$ is proportional to $\boldsymbol{b}$ and consequently proportional to $\boldsymbol{v}$.

$$\boldsymbol{b} \propto \boldsymbol{v} \propto \boldsymbol{R}_{.j} \propto \boldsymbol{R}'_{j.}. \tag{3.579}$$

Therefore the PC of $\boldsymbol{\eta}$ with $\text{cov}(\boldsymbol{\eta}) = \boldsymbol{R}$ is proportional to the correlation-weighted sum:

$$\xi = \boldsymbol{R}_{j.}\boldsymbol{\eta} \qquad \forall j. \tag{3.580}$$

By weighting the adjacent LVs with the correlations in order to form the "inside approximation" (see Sc.2.3), implicitly the variance of the second order principal component, ξ, and the first eigenvalue of the first order LVs, $\boldsymbol{\eta}$, is maximized. The usage of LV correlations as weights for the inside approximation is technically introduced by Wold (cf. Noonan & Wold 1977, 1982). The interpretation in terms of the second order PC was introduced by Lohmöller (1979b, 1981a:2-09) as the "factor weighting scheme."

PLS centroid weighting scheme. In Wold's basic method of PLS soft modeling, the MVs in block j are regressed on the inside approximation which is the sign-weighted sum $\tilde{\eta}_j$ of the LVs adjacent to the LV of block j. In the very special case of all LVs being adjacent to each other $\tilde{\eta}_j$ is the sum of all LVs with the exception of η_j itself, $\tilde{\eta}_j = \sum_{i \neq j} \eta_i$. Were η_j included in the summation, we would have Horst's (1965) SUMCOR method of multi-set factor analysis (cf. Kettenring 1971). The difference between a split principal component method and SUMCOR lies in the fact that Split PC assumes a second-order PC, whereas SUMCOR assumes a second-order centroid factor. Thus, the first-order LVs of SUMCOR can be dubbed split centroid factors.

Wold's method is different on two scores. The minor difference is the omission of η_j in the summation Eq.3.580, this difference becoming less important as the number of blocks is increased. This argument is used by Lohmöller (1981a:2-09) to justify the name "centroid" weighting scheme. The greater difference is in the usage of sign weights instead of unit weights. By introducing the sign weights negative correlations between LVs are preserved, whereas in Horst's method negative correlations are turned to become positive.

Noonan's higher order LVs. Noonan's program PLSGEN (cf. Noonan & Wold 1983:Ch.4.7) allows for the estimation of hierarchically ordered LVs in LVP models. "In the iterative estimation procedure ..., weights are computed at each iteration for the LVs at the highest level of hierarchy.... The weights of the lower level LVs ... are derived by multiplying the weights for the respective MVs in the highest level LV by a scaling factor so as to yield unit variance for the lower level LV estimates." This is the extension of Split Principal Components to Split Any Sort of Components.

3.6 Split Multiple Regression

In the preceding section, splitting the principal component into several parts was considered to be fruitful. The question arises as to whether the same holds for the Canonical Correlation model. A Hierarchic Canonical Correlation model is easily imaginable, where the MVs x_ℓ and y_k form two lower-order LVs ξ_i and η_j, which in turn form two higher-order LVs $\ddot{\xi}$ and $\ddot{\eta}$ having maximum correlation.

The Hierarchical Canonical Correlation model turns out to be a very useful superstructure for "structure vs. performance" models in systems analysis (Wold 1982d). It can be incorporated into models where the structure $\ddot{\xi}$ of the system is described by several LVs ξ_i and sets of MVs x_ℓ, where the performance $\ddot{\eta}$ of the system is described by several LVs η_j and attached MVs y_k, and where the goal of the analysis is to explain as well as possible the performance of the system as a function of its structure. The question is whether the results are the same for a PLS estimation of the three-level model at the one hand, and a first-and-third-level model with the second level introduced after the estimation by splitting the third-level LVs into the second-level LVs on the other hand.

To tell the result right away, the answer is negative. Split canonical correlation estimates are different from those of a three-level PLS approach. The results are different even for the case when there is only one variable on the y-side of the model, which is, in fact, the multiple regression model as a special case of the canonical correlation model. The Split Multiple Regression will, therefore, be examined in this section. The result will be a more thorough understanding of what it means to maximize the R^2 of an endogenous variable and to split the exogenous variables into several blocks.

3.6.1 Split Multiple Regression and PLS Approach

Total set of predictors. Let y be the (univariate) predictand, and $\boldsymbol{z}$ and $\boldsymbol{x}$ two multivariate sets of predictor variables with zero mean and covariance matrix

$$\operatorname{cov}\begin{bmatrix} \boldsymbol{z} \\ \boldsymbol{x} \\ y \end{bmatrix} = \begin{bmatrix} \boldsymbol{A} & \boldsymbol{B} & \boldsymbol{f} \\ \boldsymbol{B}' & \boldsymbol{C} & \boldsymbol{g} \\ \boldsymbol{f}' & \boldsymbol{g}' & s^2 \end{bmatrix} = \begin{bmatrix} \boldsymbol{S} & & \boldsymbol{h} \\ & & \\ \boldsymbol{h}' & & s^2 \end{bmatrix}. \tag{3.601}$$

The multiple regression equation

$$y = \boldsymbol{b}'\begin{bmatrix} \boldsymbol{z} \\ \boldsymbol{x} \end{bmatrix} + e \tag{3.602}$$

has the well known solution

$$\boldsymbol{b} = \boldsymbol{S}^{-1}\boldsymbol{h}. \tag{3.603}$$

Considering the predictor covariance matrix $\boldsymbol{S}$ as partitioned into four parts as shown in Eq.3.601, the inverse of $\boldsymbol{S}$ can be expressed in terms of the parts $\boldsymbol{A}, \boldsymbol{B}, \boldsymbol{C}$, using Frobenius' relation (cf. Bodewig 1956:188), and Eq.3.603 becomes:

$$\boldsymbol{b} = \begin{bmatrix} (\boldsymbol{A}-\boldsymbol{B}\boldsymbol{C}^{-1}\boldsymbol{B}')^{-1} & -\boldsymbol{A}^{-1}\boldsymbol{B}(\boldsymbol{C}-\boldsymbol{B}'\boldsymbol{A}^{-1}\boldsymbol{B})^{-1} \\ -\boldsymbol{C}^{-1}\boldsymbol{B}'(\boldsymbol{A}-\boldsymbol{B}\boldsymbol{C}^{-1}\boldsymbol{B}')^{-1} & (\boldsymbol{C}-\boldsymbol{B}'\boldsymbol{A}^{-1}\boldsymbol{B})^{-1} \end{bmatrix}\begin{bmatrix} \boldsymbol{f} \\ \boldsymbol{g} \end{bmatrix}. \tag{3.604}$$

PLS model. Instead of a multiple regression on the total set of predictors, one may attempt to use PLS with two separate blocks of predictor MVs $\boldsymbol{z}$ and $\boldsymbol{x}$, and two predictor LVs, ζ and ξ, which are formed by the weight relations

$$\zeta = \boldsymbol{v}'\boldsymbol{z} \tag{3.605}$$
$$\xi = \boldsymbol{w}'\boldsymbol{x}. \tag{3.606}$$

Regression Eq.3.602 is now replaced by

$$y = a\zeta + c\xi + d = a\boldsymbol{v}'\boldsymbol{z} + c\boldsymbol{w}'\boldsymbol{x} + d, \tag{3.607}$$

where a and c are regression coefficients and d is the residual variable. Let the weights $\boldsymbol{v}$ (and $\boldsymbol{w}$) be attained by ModeB, i.e. by multiple regressions of y on $\boldsymbol{z}$ (and on $\boldsymbol{x}$),

$$\boldsymbol{v} = \boldsymbol{A}^{-1}\boldsymbol{f} \tag{3.608}$$
$$\boldsymbol{w} = \boldsymbol{C}^{-1}\boldsymbol{g}, \tag{3.609}$$

which can be written jointly as

$$\boldsymbol{u} = \begin{bmatrix} \boldsymbol{v} & 0 \\ 0 & \boldsymbol{w} \end{bmatrix} = \begin{bmatrix} \boldsymbol{A}^{-1}\boldsymbol{f} & 0 \\ 0 & \boldsymbol{C}^{-1}\boldsymbol{g} \end{bmatrix}. \tag{3.610}$$

The covariance matrix of the predictor LVs is then:

$$\begin{aligned} \boldsymbol{R} \equiv \operatorname{cov}(\zeta, \xi) &= \boldsymbol{u}'\boldsymbol{S}\boldsymbol{u} \\ &= \begin{bmatrix} \boldsymbol{v} & 0 \\ 0 & \boldsymbol{w} \end{bmatrix}' \begin{bmatrix} \boldsymbol{A} & \boldsymbol{B} \\ \boldsymbol{B}' & \boldsymbol{C} \end{bmatrix} \begin{bmatrix} \boldsymbol{v} & 0 \\ 0 & \boldsymbol{w} \end{bmatrix} \\ &= \begin{bmatrix} \boldsymbol{f}'\boldsymbol{A}^{-1}\boldsymbol{f} & \boldsymbol{f}'\boldsymbol{A}^{-1}\boldsymbol{B}\boldsymbol{C}^{-1}\boldsymbol{g} \\ \boldsymbol{g}'\boldsymbol{C}^{-1}\boldsymbol{B}'\boldsymbol{A}^{-1}\boldsymbol{f} & \boldsymbol{g}'\boldsymbol{C}^{-1}\boldsymbol{g} \end{bmatrix}, \end{aligned} \tag{3.611}$$

with determinant and inverse

$$\begin{aligned} |\boldsymbol{R}| &= r_{\zeta\zeta}r_{\xi\xi} - r_{\zeta\xi}r_{\xi\zeta} \\ &= \boldsymbol{f}'\boldsymbol{A}^{-1}\boldsymbol{f}\boldsymbol{g}'\boldsymbol{C}^{-1}\boldsymbol{g} - (\boldsymbol{f}'\boldsymbol{A}^{-1}\boldsymbol{B}\boldsymbol{C}^{-1}\boldsymbol{g})^2, \end{aligned} \tag{3.612}$$

$$\boldsymbol{R}^{-1} = \frac{1}{|\boldsymbol{R}|} \begin{bmatrix} r_{\xi\xi} & -r_{\zeta\xi} \\ -r_{\xi\zeta} & r_{\zeta\zeta} \end{bmatrix}. \tag{3.613}$$

The covariances between predictor LVs and predictand y are

$$\operatorname{cov}(\zeta; y) = \boldsymbol{f}'\boldsymbol{v} = \boldsymbol{f}'\boldsymbol{A}^{-1}\boldsymbol{f} \tag{3.614}$$
$$\operatorname{cov}(\xi; y) = \boldsymbol{g}'\boldsymbol{w} = \boldsymbol{g}'\boldsymbol{C}^{-1}\boldsymbol{g}. \tag{3.615}$$

The solution for Eq.3.607 is

$$\begin{aligned} \begin{bmatrix} a \\ c \end{bmatrix} &= \boldsymbol{R}^{-1} \begin{bmatrix} \operatorname{cov}(\zeta; y) \\ \operatorname{cov}(\xi; y) \end{bmatrix} \\ &= \frac{1}{|\boldsymbol{R}|} \begin{bmatrix} (\boldsymbol{g}'\boldsymbol{C}^{-1}\boldsymbol{g})(\boldsymbol{f}'\boldsymbol{A}^{-1}\boldsymbol{f}) + (\boldsymbol{g}'\boldsymbol{C}^{-1}\boldsymbol{B}'\boldsymbol{A}^{-1}\boldsymbol{f})(\boldsymbol{g}'\boldsymbol{C}^{-1}\boldsymbol{g}) \\ (\boldsymbol{f}'\boldsymbol{A}^{-1}\boldsymbol{B}\boldsymbol{C}^{-1}\boldsymbol{g})(\boldsymbol{f}'\boldsymbol{A}^{-1}\boldsymbol{f}) + (\boldsymbol{f}'\boldsymbol{A}^{-1}\boldsymbol{f})(\boldsymbol{g}'\boldsymbol{C}^{-1}\boldsymbol{g}) \end{bmatrix}, \end{aligned} \tag{3.616}$$

where the products in parentheses are scalar quantities. The PLS estimate $\boldsymbol{b}^*$ for the direct influence of the predictor MVs $\boldsymbol{z}$ and $\boldsymbol{x}$ on the predictand y is found by multiplying the weight vectors $\boldsymbol{v}$ and $\boldsymbol{w}$ by the path coefficients a and c.

$$\boldsymbol{b}^* \equiv \begin{bmatrix} a\boldsymbol{v} \\ c\boldsymbol{w} \end{bmatrix}$$
$$= \frac{1}{|\boldsymbol{R}|} \begin{bmatrix} (\boldsymbol{g}'\boldsymbol{C}^{-1}\boldsymbol{g})\boldsymbol{f}\boldsymbol{A}^{-1} & (\boldsymbol{f}'\boldsymbol{A}^{-1}\boldsymbol{B}\boldsymbol{C}^{-1}\boldsymbol{g})\boldsymbol{f}\boldsymbol{A}^{-1} \\ (\boldsymbol{g}'\boldsymbol{C}^{-1}\boldsymbol{B}'\boldsymbol{A}^{-1}\boldsymbol{f})\boldsymbol{g}'\boldsymbol{C}^{-1} & (\boldsymbol{f}'\boldsymbol{A}^{-1}\boldsymbol{f})\boldsymbol{g}'\boldsymbol{C}^{-1} \end{bmatrix} \begin{bmatrix} \boldsymbol{f} \\ \boldsymbol{g} \end{bmatrix} \quad (3.617)$$

Eq.3.617 is arranged so as to show maximal similarity to Eq.3.604. However, the differences are striking and only in very special cases will the multiple regression Eq.3.604 and the PLS estimation by two separate predictor blocks Eq.3.617 yield the same results. One such special case is $\boldsymbol{B} = \boldsymbol{0}$, i.e. when the two predictor blocks are completely uncorrelated. Other — equally infrequent — special cases are imaginable.

With real world data, the results of a multiple regression analysis with one set of predictors (Eq.3.604) and several sets of predictors (Eq.3.617) will almost always give different results. Moreover, because Eq.3.604 always gives the highest possible multiple correlation, a PLS estimation with several separate blocks will give lower multiple correlations.

PLS modeling starts with a careful definition of the LVs to enhance the interpretability of the results. The Split Multiple Regression approach leaves this aside in favour of a purely mathematical criterion, the minimization of residual variance. The PLS estimates with different blocks of predictors give lower R^2. However, the question remains as to what the drop in R^2 is in practical applications when the predictors are split into several sets. And it shall to be discussed from a substantive point of view what is gained by the introduction of LVs for different predictor blocks.

3.6.2 How Great is the Loss in R²?

From the purely mathematical argumentation in the preceding section it is clear that the squared multiple correlation R^2 is smaller when the set of predictors is split up into several subsets and the influence of the predictors is "channelled" by LVs. The question now is how big the loss in R^2 is and whether it is acceptable in relation to other sources of variation in the results. Two examples will serve to compare the influence of different model specification on the explanatory power R^2, the first example using artificial data, the second real world data.

Artificial data. The artificial correlation matrix shown in Table 3.17 will be used as a first example. The correlation matrix contains twelve predictor variables $x_{01}, \ldots, x_{12}$ and three predictands y_{40}, y_{50}, y_{60}. In total, $6 \times 4 \times 2$ different models for these data will be analysed. Firstly, the total set of predictors is split up into subsets in six different ways, namely into 1, 2, 3, 4, 6, 12 subsets with 12, 6, 4, 3, 2, 1 MVs, respectively. The correlation of the predictors is a simplex matrix; thus it is assured that for each split the correlations within subsets are higher than

Table 3.17: Artificial correlation matrix

x_{01}	100														
x_{02}	77	100													
x_{03}	70	77	100												
x_{04}	63	70	77	100											
x_{05}	54	63	70	77	100										
x_{06}	49	54	63	70	77	100									
x_{07}	42	49	54	63	70	77	100								
x_{08}	35	42	49	54	63	70	77	100							
x_{09}	28	35	42	49	54	63	70	77	100						
x_{10}	21	28	35	42	49	54	63	70	77	100					
x_{11}	14	21	28	35	42	49	54	63	70	77	100				
x_{12}	7	14	21	28	35	42	49	54	63	70	77	100			
y_{40}	40	40	40	40	40	40	40	40	40	40	40	40	100		
y_{50}	50	50	50	50	50	50	50	50	50	50	50	50	70	100	
y_{60}	60	60	60	60	60	60	60	60	60	60	60	60	70	70	100

between subsets. Secondly, four different choices for the predictands are made, one model with all three MVs forming a LV named η, and three models with only one of the predictands y_{40}, y_{50}, y_{60}. Thirdly, all models are estimated in two different ways, namely in ModeA (all outwards) and ModeB (all inwards). The results are shown in Table 3.18.

With ModeB the results for the first model (one block with twelve MVs) and the last model (twelve blocks, each one with only one MV) are essentially the same and identical with respect to the multiple regression. Thus R^2 acquires its maximum. The R^2 for all other models is lower, the second model (2 block with 6 MVs) giving the lowest R^2. The difference, however, is not dramatic; it is only about 10% for all choices of predictands.

With ModeA (all blocks outwards directed) the differences are much larger. The last model, again, with 12 predictor blocks is identical to the multiple regression model: Each predictor LV is identical to one MV and the multiple regression in the inner model is identical to a regression on the MVs. The results for the first model, however, are dominated by the single regressions in the outer model and the inner model is also estimated by single regressions; the drop in R^2 is noticeable and expected.

A second set of twelve models, which is not reported here, is estimated, with orthogonality constraints between the predictor LVs (cf. Sc.2.3.4). The effect of these constraints is not that the LVs come out as uncorrelated but that the correlations are lowered and the LVs are adjusted to some extent to become independent predictors. The R^2 increases under this procedure.

Noonan & Wold data. The second example is based on the correlation matrix published by Noonan and Wold (1977). Their model assumes two predictor blocks, Home and School, and one predictand block, Achievement. Again, we compare the original model (two predictor blocks) to a model where both blocks are joined,

Table 3.18: Split multiple regression for artificial correlations: R^2 for different numbers of blocks of predictors.

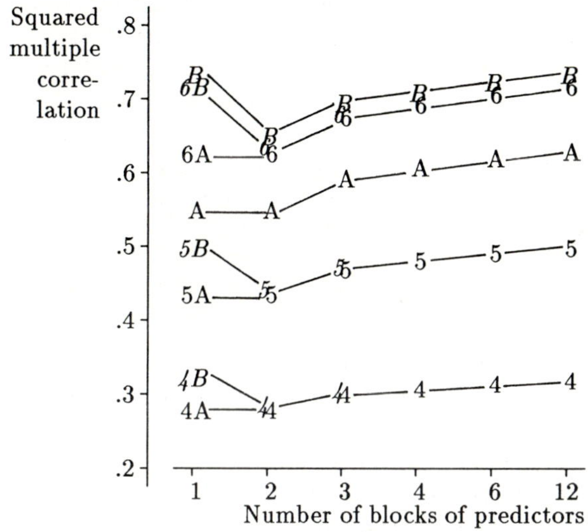

Predictor		Predictand							
		y_{40}		y_{50}		y_{60}		η	
blocks	vars	ModeB	ModeA	ModeB	ModeA	ModeB	ModeA	ModeB	ModeA
		4B	4A	*5B*	5A	*6B*	6A	*B*	A
1	12	.3183	.2781	.4974	.4346	.7163	.6258	.7324	.5494
2	6	.2824	.2781	.4412	.4346	.6353	.6258	.6496	.5494
3	4	.3018	.3000	.4716	.4686	.6791	.6747	.6944	.5923
4	3	.3080	.3074	.4812	.4803	.6929	.6917	.7085	.6072
6	2	.3139	.3139	.4905	.4905	.7063	.7063	.7222	.6200
12	1	.3183	.3183	.4974	.4974	.7162	.7162	.7324	.6288

Artificial correlations from Tb.3.17. Table entries are squared multiple correlations. Table contains results of $6 \times 4 \times 2$ one-relation models:

6 different predictors: 1 block with 12 MVs, 2 blocks with 6 MVs,...

4 different predictands: 3 with single MVs, 1 with one LV.

2 estimation Modes: B = all blocks inwards directed, or A = all outwards.

both models being estimated as ModeA and ModeB models. The squared multiple correlations R^2 and the averaged communalities H^2 are:

One predictor block	ModeB:	$R^2 = 0.615696$	$H^2 = 0.482171$
Two predictor blocks	ModeB:	$R^2 = 0.615414$	$H^2 = 0.594769$
One predictor block	ModeA:	$R^2 = 0.441948$	$H^2 = 0.599081$
Two predictor blocks	ModeA:	$R^2 = 0.446130$	$H^2 = 0.723301.$

If in ModeB estimation both predictor blocks are joined into one block, the model becomes a canonical correlation model, the R^2 being increased about 0.000282. The price paid for this unimportant gain is that the predictor LV has no clear interpretation; it is a mixture of Home and School conditions.

The differences behave differently in a ModeA model. The R^2 decreases from $R^2 = 0.446130$ with two predictor LVs to $R^2 = 0.441918$ with one predictor LV. The differences in the communalities are quite understandable: the less MVs in a block, the higher the communality; and if a block contains only one MV, the communality is $H^2 = 1$.

3.6.3 Conclusions and Recommendations

From the statistical argumentation in Sc.3.6.1 it follows that there is a drop in R^2 if a block of predictors is split into two or more blocks. The examples in Sc.3.6.2 show that the drop in R^2 can be very small and completely negligible in ModeB models. The Stone-Geisser test (cf. Lohmöller 1981a:Sc.5.1) might be helpful in deciding on the best tradeoff between predictive relevance on the one hand and increase of number of parameters and loss of degrees of freedom on the other hand.

These results indicate that the number of residual variables in a path diagram does not tell the whole story about what is minimized and maximized. The partition of predictors into subsets and the introduction of an LV for each block influence the solution and the results are not identical to a multiple regression on the joint subset of predictor MVs. The R^2 dropping is the price to be paid for increased interpretability of the results achieved by a careful introduction of LVs.

Unfortunately, this section on split canonical correlation and split multiple regression does not result in a proof of an optimality property, as it was in the case of the split principal component in Sc.3.5. It may be possible to establish some asymptotic optimality properties; for instance, inasmuch as the off-diagonal elements in Eq.3.604 or in Eq.3.617 become identical (or approach zero), the difference between one-set and multiple-set regression vanishes. But this will not be pursued here.

3.7 Uncorrelated Dimensions in Generalized Canonical Correlation Analysis

When the canonical correlation model is generalized to $J > 2$ blocks of manifest variables (MVs), then two problems are known to arise. Firstly, a new criterion to be maximized must be defined instead of the one correlation coefficient of the two canonical variates. This problem has been discussed in Sc.3.5.7.

The second problem is related to the orthogonality properties of the consecutive dimensions of the LVs. In the canonical correlation model, the residuals of the one block are uncorrelated with the LV of the own block as well as with the LV of the other block, say

$$\mathrm{cor}(\eta; \boldsymbol{e}) = \mathrm{cor}(\xi; \boldsymbol{e}) = \mathbf{0}, \qquad \mathrm{cor}(\xi; \boldsymbol{d}) = \mathrm{cor}(\eta; \boldsymbol{d}) = \mathbf{0}. \tag{3.701}$$

Therefore any linear combination of the residuals, say $\eta_2 = \boldsymbol{w}_2'\boldsymbol{e}$, is uncorrelated with the LVs of the own *and* of the other block. If $\boldsymbol{w}_2$ is the weight vector of the second dimension of the canonical variate in the $\boldsymbol{y}$-block, then the second dimension of η is uncorrelated with the first dimension of ξ, and similar holds for $\xi_2 = \boldsymbol{g}_2'\boldsymbol{d}$ and η_1.

Orthogonality of different dimensions in more than two blocks is not possible. (The orthogonality of different dimensions within the same block follows immediately from the deflation technique.) Orthogonality across dimensions and blocks uniquely defines the canonical correlation model, see Remark 3⋆16. The canonical variates $\xi_{\boldsymbol{y}}$ of a first set $\boldsymbol{x}$ with respect to a second set $\boldsymbol{y}$ are different from the canonical variates $\xi_{\boldsymbol{z}}$ of the first set $\boldsymbol{x}$ with respect to a third set $\boldsymbol{z}$, and the $\xi_{\boldsymbol{z}}$ variates are the only ones to allow for a diagonal correlation matrix with variates from the third set.

If the orthogonality of different dimensions across blocks is the predominant criterion and the maximization of the LV correlations the second one, the following procedure may be applied: Let $\boldsymbol{x} = [\boldsymbol{x}_j, j = 1 \dots J] = [\boldsymbol{x}_{k_j}, k_j = 1 \dots K_j, j = 1 \dots J]$ a vector of MVs, partitioned into J exhaustive and disjoint sets. Pick one set $\boldsymbol{x}_1$ and call the remaining variables $\boldsymbol{y}_1$, for a moment, and perform a canonical correlation analysis with $\boldsymbol{x}_1$ and the rest $\boldsymbol{y}_1$, to get the canonical variates ξ_1 and η_1. The residuals in block $\boldsymbol{x}_1$, called $\boldsymbol{d}_1$, are uncorrelated with η_1 and with any linear combination of $\boldsymbol{y}_1$. Then proceed to another set $\boldsymbol{x}_2$ and call the remaining variables $\boldsymbol{y}_2$, form the canonical variate ξ_2 and η_2, and it follows that the residuals $\boldsymbol{d}_2$ and any linear combinations thereof are uncorrelated with the remaining variables $\boldsymbol{y}_2$ and any linear combinations thereof. Then proceed to a third pair $\boldsymbol{x}_3$ and $\boldsymbol{y}_3$ and form the canonical variates ξ_3 and η_3, and so on. Finally, we have obtained the LVs $\xi_j, j = 1 \dots J$ such that the residuals in each block are uncorrelated to the variables of all other block. Consequently, the second dimension of the LV, which is formed as a linear aggregate of these residuals, is uncorrelated with the first dimension LV of all other blocks. This is easily generalized to the theorem that all dimensions of the one block are uncorrelated to all preceding dimension of all other blocks, if they are formed as canonical variates of the one block with the union of all others.

However, the uncorrelated variables are not necessarily those which correlate maximally. This can demonstrated by an artificial data set. The correlation matrix for three sets, each one consisting of two variables is

$$S = \frac{1}{100}\begin{bmatrix} 100 & . & . & . & . & \text{sym.} \\ 0 & 100 & . & . & . & . \\ -50 & 0 & 100 & . & . & . \\ 0 & 50 & 0 & 100 & . & . \\ 50 & 0 & -50 & 0 & 100 & . \\ 0 & -50 & 0 & 50 & 0 & 100 \end{bmatrix}. \tag{3.702}$$

The weight vectors for each pair of the three blocks are

$$\begin{bmatrix} \boldsymbol{w}_1 & \boldsymbol{g}_1 \end{bmatrix} = \begin{bmatrix} 1 & 0 \\ 1 & 0 \\ 0 & -1 \\ 0 & 1 \\ 0 & 1 \\ 0 & -1 \end{bmatrix}, \quad \begin{bmatrix} \boldsymbol{w}_2 & \boldsymbol{g}_2 \end{bmatrix} = \begin{bmatrix} 0 & -1 \\ 0 & 1 \\ 1 & 0 \\ 1 & 0 \\ 0 & -1 \\ 0 & 1 \end{bmatrix}, \quad \begin{bmatrix} \boldsymbol{w}_3 & \boldsymbol{g}_3 \end{bmatrix} = \begin{bmatrix} 0 & 1 \\ 0 & -1 \\ 0 & -1 \\ 0 & 1 \\ 1 & 0 \\ 1 & 0 \end{bmatrix}.$$

Collecting the weight vectors of the $\boldsymbol{x}$-blocks into a weight matrix $\boldsymbol{W} = (\boldsymbol{w}_1, \boldsymbol{w}_2, \boldsymbol{w}_3)$, we get

$$\boldsymbol{W} = \begin{bmatrix} \boldsymbol{w}_1 & \boldsymbol{w}_2 & \boldsymbol{w}_3 \end{bmatrix} = \begin{bmatrix} 1 & 0 & 0 \\ 1 & 0 & 0 \\ 0 & 1 & 0 \\ 0 & 1 & 0 \\ 0 & 0 & 1 \\ 0 & 0 & 1 \end{bmatrix}, \qquad \boldsymbol{R} = (\boldsymbol{W}'\boldsymbol{S}\boldsymbol{W})/2 = \begin{bmatrix} 100 & 0 & 0 \\ 0 & 100 & 0 \\ 0 & 0 & 100 \end{bmatrix} \tag{3.703}$$

with **zero** correlation between the LVs, whereas a solution like

$$\boldsymbol{w} = \begin{bmatrix} \boldsymbol{w}_1 & \boldsymbol{w}_2 & \boldsymbol{w}_3 \end{bmatrix} = \begin{bmatrix} 1 & 0 & 0 \\ 1 & 0 & 0 \\ 0 & -1 & 0 \\ 0 & 1 & 0 \\ 0 & 0 & 1 \\ 0 & 0 & -1 \end{bmatrix}, \qquad \boldsymbol{R} = (\boldsymbol{W}'\boldsymbol{S}\boldsymbol{W})/2 = \begin{bmatrix} 100 & 50 & 50 \\ 50 & 100 & 0 \\ 50 & 0 & 100 \end{bmatrix} \tag{3.704}$$

would give the highest possible correlation between the LVs. It is very unlikely that in real-world data the difference between the maximally correlated LVs and the LVs leaving the own-block residuals uncorrelated to the other-block LVs and MVs and residuals will be as eye-catching as in the artificial data example. However, the example proves that there is a difference, and that the choice between maximal correlated LVs and orthogonal higher dimensions must be made. If, by some device, LVs of higher dimensions are uncorrelated with LVs of lower dimensions of other blocks, then these LVs are the canonical variates in the sense described above.

Chapter 4

Mixed Measurement Level Multivariate Data

Linear equations are applicable only to interval-scaled variables which are the only ones to allow for linear transformations (Stevens 1951). Thus the standard procedures cannot be used for the categorical and the ordinal-scaled variables that are often encountered in the behavioral sciences. This chapter considers the extensions of LVP analysis so as to include categorical variables.

Two types of models for categorical variables have been applied in the last decades, (a) additive, or linear, and (b) multiplicative, or loglinear models. The origin of this differentiation can be traced back to 1900 (cf. Fienberg 1977:4ss, Gilula, Krieger & Ritov 1988), when Pearson and Yule developed their association measures for 2×2 tables. Pearson insisted that it always makes sense to assume an underlying continuum for a dichotomy or a polychotomy and assumed multinormality, e.g. for his tetrachoric coefficient. Yule chose to view the categories of a crossclassification as fixed; his cross-product ratio for a 2×2 table is not a substitute for an estimate of the true correlation coefficient of an underlying continuum.

The loglinear models, widely used in multivariate discrete analysis (cf. Bishop, Fienberg & Holland 1975; Grizzle, Starmer & Koch 1969; Falter & Ulbricht 1981; Flaig 1983; Fienberg 1977; Goodman 1978; Mooijaart 1982; McCullagh & Nelder 1983) is based on Yule's work and takes polychotomies to be genuinely polychotomous (by nature), not as polychotomized (by researchers). Linear models like regression analysis (Sockloff 1976), Guttman's (1941c) categorical scaling, the analysis of correspondence (Benzécri 1976) and the methods presented in this Chapter 4 are based on the assumption of an underlying continuum and stand as such in the Pearsonian tradition, without distributional assumptions, however.

4.1 Categorical Variables and LS Methods

Let us suppose a set of trichotomous categorical variables is to be analysed. If there are only two variables, all the information about the relationships between the two variables is collected into a $3 \times 3 = 3^2$ contingency table. If there are three variables, a contingency cube with $3 \times 3 \times 3 = 3^3 = 27$ cells contains all the information. In general, if there are J trichotomous variables, the contingency array contains 3^J cells. With $J = 5$ we have more cells ($3^5 = 245$) than cases in most experiments, with $J = 13$ we have more cells than most computers can store, i.e. $3^{13} = 1,594,323$.

Abandoning the idea of treating the higher order, let alone the 13th order interaction, the methods to be delineated in this section take into account only the first and second moment of the data matrix. Within this confinement only a square symmetric matrix of order $3J \times 3J$ need be laid out in order to store the relevant information about the relationships between J trichotomous variables. This matrix, called the super contingency table, will be introduced in Sc.4.1.1. Higher order interactions can easily be introduced into the model by forming product variables (Sc.4.4). Sc.4.1.2 presents the traditional LS technique for the analysis of super contingency tables (Guttman 1941c, Burt 1950, Benzécri 1976) and Sc.4.1.3 the PLS technique for path models with latent variables (Wold & Bertholet 1982a, 1982b, 1982c, Bertholet and Wold 1984, 1985, Bertholet 1987).

4.1.1 Super Contingency Tables

Boolean variables. A dichotomous variable can assume any two values; a Boolean variable as a special case of a dichotomous variable can take only the two values 0 and 1. A Boolean variable $\ddot{y}$ is formally defined by

$$\Pr(\ddot{y} = 1) = 1 - \Pr(\ddot{y} = 0) = \mu. \tag{4.101}$$

From this definition follows

$$\begin{aligned} E(\ddot{y}) &= \sum_k \Pr(\ddot{y} = k) \cdot k = \mu \cdot 1 + (1 - \mu) \cdot 0 \\ &= \mu \end{aligned} \tag{4.102}$$

$$\begin{aligned} E(\ddot{y}^2) &= \sum_k \Pr(\ddot{y} = k) \cdot k^2 = \mu \cdot 1^2 + (1 - \mu) \cdot 0^2 \\ &= \mu \end{aligned} \tag{4.103}$$

and in general for all $r > 0$

$$\begin{aligned} E(\ddot{y}^r) &= \sum_k \Pr(\ddot{y} = k) \cdot k^r = \mu \cdot 1^r + (1 - \mu) \cdot 0^r \\ &= \mu. \end{aligned} \tag{4.104}$$

The variance of $\ddot{y}$ is

$$\sigma^2 = \mu(1 - \mu) = \mu - \mu^2. \tag{4.105}$$

Categorical variables. A categorical variable $\dot{y}$ can assume several distinct values k with probabilities

$$\Pr(\dot{y} = k) = \mu_k, \qquad \sum_k \mu_k = 1. \tag{4.106}$$

The categorical variable $\dot{y}$ can be replaced by a set of Boolean variables $[\ddot{y}_k, k = 1 \ldots K]$ by the definition

$$\ddot{y}_k = \begin{cases} 1 & \text{if } \dot{y} = k \\ 0 & \text{otherwise} \end{cases} \tag{4.107}$$

and it follows that the expectation (mean) of the Boolean variable can be interpreted as a probability.

$$\begin{aligned} \mathrm{E}(\ddot{y}_k) = \mathrm{E}(\ddot{y}_k^2) &= \mathrm{E}(\ddot{y}_k^r) = \Pr(\dot{y} = k) \\ &= \mu_k. \end{aligned} \tag{4.108}$$

The expectation of any pair, triple etc. of these Boolean variables is

$$\mathrm{E}(\ddot{y}_k \cdot \ddot{y}_{\underline{k}} \cdot y_{\underline{\underline{k}}} \cdots) \begin{cases} \mu_k & \text{if } (\cdot)_k = (\cdot)_{\underline{k}} = (\cdot)_{\underline{\underline{k}}} \\ 0 & \text{otherwise.} \end{cases} \tag{4.109}$$

If $\ddot{\boldsymbol{y}}_{[K \times 1]} = [\ddot{y}_k, k = 1 \ldots K]$ denotes the column vector of Boolean variables replacing the categorical variable $\dot{y}$, then the expectations defined in Eq.4.108 form a vector $\boldsymbol{\mu}_{[K \times 1]} = \mathrm{E}(\ddot{\boldsymbol{y}}) = [\mu_k, k = 1 \ldots K]$ of "univariate probabilities".

Contingency table. Let $\ddot{\boldsymbol{x}}_{[L \times 1]}$ and $\ddot{\boldsymbol{y}}_{[K \times 1]}$ denote two vectors of Boolean variables replacing two categorical variables $\dot{x}$ and $\dot{y}$, with probability distribution

$$\Pr(\ddot{y}_k = 1, \ddot{x}_\ell = 1) = \mu_{k\ell}, \qquad k = 1 \ldots K, \ell = 1 \ldots L. \tag{4.110}$$

Hence

$$\mathrm{E}(\ddot{y}_k \ddot{x}_\ell) = \mu_{k\ell} \qquad \mathrm{E}(\ddot{\boldsymbol{y}} \ddot{\boldsymbol{x}}') = \boldsymbol{M}_{yx}. \tag{4.111}$$

The $K \times L$ matrix $\boldsymbol{M}_{yx}$ contains the "bivariate probabilities" and is called the contingency table.

Super contingency table. If $\ddot{\boldsymbol{x}}$ and $\ddot{\boldsymbol{y}}$ are stacked together to form a super-vector variable $\boldsymbol{u} = (\ddot{\boldsymbol{x}}', \ddot{\boldsymbol{y}}')' = (\ldots, \ddot{x}_\ell, \ldots, \ddot{y}_k, \ldots)'$, then the first moment is identical to the univariate probabilities,

$$\mathrm{E}(\boldsymbol{u}) = \boldsymbol{\mu}_u = \begin{bmatrix} \boldsymbol{\mu}_x \\ \boldsymbol{\mu}_y \end{bmatrix} = \begin{bmatrix} \vdots \\ \mu_\ell \\ \vdots \\ \mu_k \\ \vdots \end{bmatrix} \tag{4.112}$$

and the product matrix is the super contingency table

$$\mathrm{E}(\boldsymbol{u}\boldsymbol{u}') = \mathrm{E} \begin{bmatrix} \ddot{\boldsymbol{x}}\ddot{\boldsymbol{x}}' & \ddot{\boldsymbol{x}}\ddot{\boldsymbol{y}}' \\ \ddot{\boldsymbol{y}}\ddot{\boldsymbol{x}}' & \ddot{\boldsymbol{y}}\ddot{\boldsymbol{y}}' \end{bmatrix} = \begin{bmatrix} \boldsymbol{M}_{xx} & \boldsymbol{M}_{xy} \\ \boldsymbol{M}_{yx} & \boldsymbol{M}_{yy} \end{bmatrix} = \boldsymbol{M}_{\boldsymbol{u}} \tag{4.113}$$

which contains the ordinary contingency table $\boldsymbol{M}_{yx}$, its transpose $\boldsymbol{M}_{xy}$ and the matrices $\boldsymbol{M}_{xx}$ and $\boldsymbol{M}_{yy}$. $\boldsymbol{M}_{yy}$ is a diagonal matrix containing as diagonal elements the univariate probabilities $\boldsymbol{\mu}_y$, because $\mu_{kk} = \mu_k$ (due to Eq.4.109), and

Table 4.1: Super contingency table and conditional probabilities

Boolean variables	Contingency table $100 \times \boldsymbol{M}$						Conditional probabilities $100 \times \boldsymbol{B}$					
	$\ddot{x}_1$	$\ddot{x}_2$	$\ddot{x}_3$	$\ddot{y}_4$	$\ddot{y}_5$	$\ddot{y}_6$	$\ddot{x}_1$	$\ddot{x}_2$	$\ddot{x}_3$	$\ddot{y}_4$	$\ddot{y}_5$	$\ddot{y}_6$
$\ddot{x}_1$	25					sym.	100	.	.	80	0	20
$\ddot{x}_2$	.	50					.	100	.	60	40	0
$\ddot{x}_3$	.	.	25				.	.	100	0	10	90
$\ddot{y}_4$	20	30	0	50			80	60	0	100	.	.
$\ddot{y}_5$	0	20	5	.	25		0	40	20	.	100	.
$\ddot{y}_6$	5	0	20	.	.	25	20	0	80	.	.	100

Structural zeros are written as dots.

as nondiagonal elements zeros, because $\mathrm{E}(\ddot{y}_k\ddot{y}_{\underline{k}}) = 0$ if $k = \underline{k}$ (due to Eq.4.109). Analogously, $\boldsymbol{M}_{xx}$ is a diagonal matrix with elements of $\boldsymbol{\mu}_x$. As an illustration, Table 4.1 shows the super contingency table of two trichotomous variables.

The super contingency table for two variables contains $2^2 = 4$ submatrices. The procedure of forming the super contingency table can easily be extended to more variables. Suppose a set of J categorical variables $\dot{y}_j, j = 1 \ldots J$, each one with K_j categories, is replaced by a super vector $\ddot{\boldsymbol{y}} = [\ddot{\boldsymbol{y}}_j, j = 1 \ldots J] = [\ddot{y}_{k_j}, k_j = 1 \ldots K_j, j = 1 \ldots J]$ of Boolean variables. Then $\boldsymbol{\mu}_y = \mathrm{E}(\ddot{\boldsymbol{y}})$ contains J subvectors $\boldsymbol{\mu}_j$, and $\boldsymbol{M}_{yy}$ contains J^2 submatrices $\boldsymbol{M}_{ij}, i, j = 1 \ldots J$.

The marginals for the sub-tables of the supertable are univariate probabilities:

$$\mu_{k_j} = \mu_{k_j k_j} = \sum_{\ell_i} \mu_{k_j \ell_i}. \tag{4.114}$$

The conditional probabilities, conditional expectations, and regression coefficients are identical.

$$\beta_{k_j|\ell_i} = \Pr(\ddot{y}_{k_j} = 1 | \ddot{y}_{\ell_i} = 1) = \mathrm{E}(\ddot{y}_{k_j} | \ddot{y}_{\ell_i}) = \frac{\mu_{k_j \ell_i}}{\mu_{\ell_i}} \tag{4.115}$$

$$\boldsymbol{B}_{j|i} = \boldsymbol{M}_{ji}\boldsymbol{M}_{ii}^{-1}, \qquad i, j = 1 \ldots J \tag{4.116}$$

Data. Observations on the categorical variables $\dot{x}$ and $\dot{y}$ will be denoted by $\dot{x}_n$ and $\dot{y}_n, n = 1 \ldots N$ where N is the number of observational units (cases). Scores on the replacing Boolean variables are denoted by $\ddot{x}_{\ell n}$ and $\ddot{y}_{kn}$. The raw sum for the Boolean variables is:

$$n_\ell = \sum_n \ddot{x}_{\ell n}, \qquad n_k = \sum_n \ddot{y}_{kn}. \tag{4.117}$$

The average is:

$$m_\ell = \bar{\sum_n} \ddot{x}_{\ell n}, \qquad m_k = \bar{\sum_n} \ddot{y}_{kn}. \tag{4.118}$$

Table 4.2: Artificial data matrix with Boolean, categorical, and interval-scaled variables, $N = 20$ cases

Cases	$\dot{x}$	$\dot{y}$	z	$\ddot{x}_1$	$\ddot{x}_2$	$\ddot{x}_3$	$\ddot{y}_4$	$\ddot{y}_5$	$\ddot{y}_6$
01	1	4	2	1	.	.	1	.	.
02	1	4	1	1	.	.	1	.	.
03	1	4	1	1	.	.	1	.	.
04	1	4	1	1	.	.	1	.	.
05	1	6	1	1	.	.	.	.	1
06	2	4	1	.	1	.	1	.	.
07	2	4	1	.	1	.	1	.	.
08	2	4	0	.	1	.	1	.	.
09	2	4	0	.	1	.	1	.	.
10	2	4	0	.	1	.	1	.	.
11	2	4	0	.	1	.	1	.	.
12	2	5	0	.	1	.	.	1	.
13	2	5	0	.	1	.	.	1	.
14	2	5	−1	.	1	.	.	1	.
15	2	5	−1	.	1	.	.	1	.
16	3	5	−1	.	.	1	.	1	.
17	3	6	−1	.	.	1	.	.	1
18	3	6	−1	.	.	1	.	.	1
19	3	6	−1	.	.	1	.	.	1
20	3	6	−2	.	.	1	.	.	1
Sum			0	5	10	5	10	5	5
Square sum			20	5	10	5	10	5	5

Table 4.3: Contingency analysis for $\dot{x}$ and $\dot{y}$ from Tb.4.2

```
            X->    Count   |                          |  Row
                   Row Pct |      1 |      2 |      3 |  Total
       Y           --------+--------+--------+--------+
                         4 |   20   |   30   |        |   50
                           |  40.0  |  60.0  |        |   50.0
                           +--------+--------+--------+
                         5 |        |   20   |    5   |   25
                           |        |  80.0  |  20.0  |   25.0
                           +--------+--------+--------+
                         6 |    5   |        |   20   |   25
                           |  20.0  |        |  80.0  |   25.0
                           +--------+--------+--------+
                    Column      25       50       25      100
                     Total    25.0     50.0     25.0    100.0

       Chi-Square     D.F.        Significance
         72.00000       4            .0000

      Cramer's V                         .60000
      Contingency Coefficient            .64700
```

Table 4.4: Five product matrices of dichotomous variables

	$\bar{\sum_n} \ddot{y}_{kn}$ (a)	$\bar{\sum_n} \ddot{y}_{kn}^2$ (b)	$\operatorname{var} \ddot{y}_k$ (c)	$\bar{\sum_n} \ddot{y}_{kn}\ddot{x}_{\ell n}$ (d)	$\bar{\sum_n} \ddot{y}_{kn} z_n$ (e)
(i) $\ddot{y}_k \in \{0,1\}$	m_k	m_k	$m_k - m_k^2$	$m_{k\ell}$	$m_{kz} = m_k \bar{z}_k$
(ii) $\ddot{y}_k \in \left\{0, \frac{1}{\sqrt{m_k}}\right\}$	$\sqrt{m_k}$	1	$1 - m_k$	$q_{k\ell} = \frac{m_{k\ell}}{\sqrt{m_k ml}}$	$q_{kz} = \frac{m_{kz}}{\sqrt{m_k}}$
(iii) $\ddot{y}_k \in \{-m_k, 1-m_k\}$	0	$m_k - m_k^2$	$m_k - m_k^2$	$s_{k\ell} = m_{k\ell} - m_k ml$	$s_{kz} = m_{kz}$
(iv) $\ddot{y}_k \in \left\{\frac{-m_k}{\sqrt{m_k}}, \frac{1-m_k}{\sqrt{m_k}}\right\}$	0	$1 - m_k$	$1 - m_k$	$d_{k\ell} = \frac{m_{k\ell} - m_k ml}{\sqrt{m_k ml}}$	$d_{kz} = \frac{m_{kz}}{\sqrt{m_k}}$
(v) $\ddot{y}_k \in \left\{\frac{-m_k}{\sqrt{m_k - m_k^2}}, \frac{1-m_k}{\sqrt{m_k - m_k^2}}\right\}$	0	1	1	$r_{k\ell} = \frac{m_{k\ell} - m_k ml}{\sqrt{(m_k - m_k^2)(ml - ml^2)}}$	$r_{kz} = \frac{m_{kz}}{\sqrt{m_k - m_k^2}}$

Five dichotomous variables are defined in the five rows:
(i) a Boolean variable
(ii) a normed variable, square sum=1
(iii) a centered variable, mean=0
(iv) a normed-and-then-centered variable
(v) a standardized variable.
Five moments of these variables are given in the columns:
(a) the mean (first moment)
(b) the sum of squares (second raw moment)
(c) the variance (second central moment)
(d) the product moment of two dichotomous variables $\ddot{y}_k$ and $\ddot{x}l$
(e) the product moment of $\ddot{y}_k$ with a standardized variable z.

The raw product moment matrix $\boldsymbol{N}$ of the data matrix $\boldsymbol{u} = (\ddot{\boldsymbol{x}}', \ddot{\boldsymbol{y}}')'$ is

$$\boldsymbol{N}_u = \boldsymbol{u}\boldsymbol{u}' = \begin{bmatrix} \ddot{\boldsymbol{x}}\ddot{\boldsymbol{x}}' & \ddot{\boldsymbol{x}}\ddot{\boldsymbol{y}}' \\ \ddot{\boldsymbol{y}}\ddot{\boldsymbol{x}}' & \ddot{\boldsymbol{y}}\ddot{\boldsymbol{y}}' \end{bmatrix} = \begin{bmatrix} \boldsymbol{N}_{xx} & \boldsymbol{N}_{xy} \\ \boldsymbol{N}_{yx} & \boldsymbol{N}_{yy} \end{bmatrix}, \tag{4.119}$$

and the relative product moment matrix is

$$\boldsymbol{M}_u = \frac{1}{N}\boldsymbol{N}_u, \tag{4.120}$$

which is the data counterpart to the super contingency table $\boldsymbol{M}$ defined in Eq.4.113. The extension of Eq.4.119 to multivariate super contingency tables is straightforward and the properties stated in the previous paragraph (Eq.4.114, Eq.4.115) hold good for the data-generated $\boldsymbol{M}$ in analogy. For the rest of Chapter 4 the difference between population parameters (μ, σ) and sample statistics (m, s) is disregarded; hence we skip Greek symbols and use italics throughout.

Table 4.2 contains a data matrix with 20 cases, the categorical variables $\dot{x}$ and $\dot{y}$ and the corresponding Boolean sets $\ddot{x}_1, \ddot{x}_2, \ddot{x}_3$ and $\ddot{y}_4, \ddot{y}_5, \ddot{y}_6$. The super contingency table of $\ddot{\boldsymbol{x}}$ and $\ddot{\boldsymbol{y}}$ is the same as the one shown in Tb.4.1. The ordinary contingency table is presented in Table 4.3, together with a conventional contingency table analysis.

Table 4.5: The five product matrices for the data from Tb.4.2

	$\ddot{x}_1$	$\ddot{x}_2$	$\ddot{x}_3$	$\ddot{y}_4$	$\ddot{y}_5$	$\ddot{y}_6$	$\ddot{x}_1$	$\ddot{x}_2$	$\ddot{x}_3$	$\ddot{y}_4$	$\ddot{y}_5$	$\ddot{y}_6$
	M = Frequencies						***Q*** = Normed Product					
$\ddot{x}_1$	250						1000					
$\ddot{x}_2$	.	500					.	1000				
$\ddot{x}_3$	.	.	250				.	.	1000			
$\ddot{y}_4$	200	300	0	500			566	600	0	1000		
$\ddot{y}_5$	0	200	50	.	250		0	566	200	.	1000	
$\ddot{y}_6$	50	0	200	.	.	250	200	0	800	.	.	1000
	S = Covariance						***R*** = Correlation					
$\ddot{x}_1$	188						1000					
$\ddot{x}_2$	–125	250					–577	1000				
$\ddot{x}_3$	–63	–125	188				–333	–577	1000			
$\ddot{y}_4$	75	50	–125	250			346	200	–577	1000		
$\ddot{y}_5$	–63	75	–13	–125	188		–333	346	–67	–577	1000	
$\ddot{y}_6$	–13	–125	138	–125	–63	188	–67	–577	733	–577	–333	1000
	D = Deviation Product						***B*** = Conditional Probability					
$\ddot{x}_1$	750						1000	.	.	800	0	200
$\ddot{x}_2$	–354	500					.	1000	.	600	400	0
$\ddot{x}_3$	–250	–354	750				.	.	1000	0	100	900
$\ddot{y}_4$	212	100	–354	500			800	600	0	1000	.	.
$\ddot{y}_5$	–250	212	–50	–354	750		0	400	200	.	1000	.
$\ddot{y}_6$	–50	–354	550	–354	–250	750	200	0	800	.	.	1000

Structural zeros are written as dots. All table entries are multiplied by 1000.
Upper half omitted from symmetric matrices.

Other product matrices. The relative product moment matrix $\boldsymbol{M}$ can be used to compute the covariance matrix $\boldsymbol{S} = [s_{k\ell}]$, the correlation $\boldsymbol{R} = [r_{k\ell}]$, the conditional frequencies $\boldsymbol{B} = [b_{k|\ell}]$, as well as the normed product matrix $\boldsymbol{Q} = [q_{k\ell}]$ and the normed-and-then-centered product matrix $\boldsymbol{D} = [d_{k\ell}]$.[1] The elements of these matrices are defined in Table 4.4. Alternatively, $\boldsymbol{Q}, \boldsymbol{D}, \boldsymbol{S}, \boldsymbol{R}$ can be understood as products of dichotomous variables which are not Boolean but normed, centered, normed-and-then-centered, and centered-and-then-normed (i.e. standardized) respectively. Table 4.5 shows the different product matrices of the data from Tb.4.2. The elements of the normed deviation matrix $\boldsymbol{D}$ can be used to compute the mean squared contingency

$$\varphi^2 = \frac{\chi^2}{N} = \sum_{k\ell} d_{k\ell}^2 \tag{4.121}$$

[1]In Chapter 4 here, the symbols $\boldsymbol{Q}$ and $\boldsymbol{D}$ have a different meaning from in other sections.

and the well-known contingency coefficients

$$C_{\text{Cramér}} = \sqrt{\frac{\varphi^2}{\min(K,L)-1}} \tag{4.122}$$

$$C_{\text{Pearson}} = \sqrt{\frac{\varphi^2}{1+\varphi^2}} = \sqrt{\frac{\chi^2}{N+\chi^2}} \tag{4.123}$$

The diagonals of the matrices $\boldsymbol{M}, \boldsymbol{S}, \boldsymbol{D}$ are strongly dependent on the univariate frequencies because $m_{kk} = m_k$, $s_{kk} = m_k - m_k^2$, $d_{kk} = 1 - m_k$. The diagonals of $\boldsymbol{Q}$, $\boldsymbol{R}$, $\boldsymbol{B}$ equal to unity and thus are independent of the univariate frequencies. Matrices $\boldsymbol{S}, \boldsymbol{D}, \boldsymbol{R}$ are computed using the deviations from the mean $m_{k\ell} - m_k m_\ell$, so that the off-diagonal elements of these matrices ought to be independent of the means. However, a factor analysis of $\boldsymbol{R}$ is known to produce factors which are correlated with the means (called "difficulty factors" if the variables are achievement items).

4.1.2 Canonical Analysis of Contingency Tables

Although it is old standing, it is not widely known that the contingency coefficient Eq.4.122 can be understood against the background of a canonical analysis of any of the products $\boldsymbol{M}, \boldsymbol{Q}, \boldsymbol{S}, \boldsymbol{R}, \boldsymbol{D}$. This will be shown in Sc.4.1.2 here. The subsequent Sc.4.1.3 will demonstrate that the results of a canonical analysis are, in essence, identical to a principal component analysis (PCA) of the same matrix. The PCA can be generalized to multivariate super contingency tables (Sc.4.1.4), leading then to a rescaling of the original categorical variables.

The canonical correlation (CC) model for interval-scaled variables was introduced in Sc.3.4.2. The model will now be enlarged so as to include the case of non-centered manifest variables, thereby leading to the canonical analysis of contingency tables.

Let $\ddot{\boldsymbol{x}}$ and $\ddot{\boldsymbol{y}}$ denote two vectors of variables with first and second order moments:

$$\begin{aligned} m_x &= \mathrm{E}(\ddot{\boldsymbol{x}}) \\ m_y &= \mathrm{E}(\ddot{\boldsymbol{y}}) \\ \boldsymbol{M}_{xx} &= \mathrm{E}(\ddot{\boldsymbol{x}}\ddot{\boldsymbol{x}}') \\ \boldsymbol{M}_{xy} &= \mathrm{E}(\ddot{\boldsymbol{y}}\ddot{\boldsymbol{x}}') \\ \boldsymbol{M}_{yy} &= \mathrm{E}(\ddot{\boldsymbol{y}}\ddot{\boldsymbol{y}}') \end{aligned} \tag{4.124}$$

Two weight vectors $\boldsymbol{g}_{[L\times1]}$ and $\boldsymbol{w}_{[K\times1]}$ are sought to form the two LVs X and Y:

$$\begin{aligned} X &= \boldsymbol{g}'\ddot{\boldsymbol{x}}, & \boldsymbol{g}'\boldsymbol{M}_{xx}\boldsymbol{g} &= 1 \\ Y &= \boldsymbol{w}'\ddot{\boldsymbol{y}}, & \boldsymbol{w}'\boldsymbol{M}_{yy}\boldsymbol{w} &= 1. \end{aligned} \tag{4.125}$$

so that

$$r = \boldsymbol{g}'\boldsymbol{M}_{xy}\boldsymbol{w} = \max. \tag{4.126}$$

The model is augmented by the loading relations which define the loading vectors $\boldsymbol{t}_{[L\times1]}$ and $\boldsymbol{p}_{[K\times1]}$ and the residual variables $\boldsymbol{d}_{[L\times1]}$ and $\boldsymbol{e}_{[K\times1]}$:

$$\begin{aligned}\ddot{\boldsymbol{x}} &= \boldsymbol{t}X + \boldsymbol{d}\\ \ddot{\boldsymbol{y}} &= \boldsymbol{p}Y + \boldsymbol{e}.\end{aligned} \tag{4.127}$$

If $\ddot{\boldsymbol{x}}$ and $\ddot{\boldsymbol{y}}$ are centered around zero mean, then the LVs X and Y are centered, $\boldsymbol{M}_{xx}, \boldsymbol{M}_{yx}, \boldsymbol{M}_{yy}$ being covariance matrices and r a correlation coefficient. Expressed in symbols, this is:

$$r = \mathrm{cor}(X;Y) \quad \Longleftarrow \quad \mathrm{E}(\ddot{\boldsymbol{x}}) = 0 \;\&\; \mathrm{E}(\ddot{\boldsymbol{y}}) = 0. \tag{4.128}$$

If $\ddot{\boldsymbol{x}}$ and $\ddot{\boldsymbol{y}}$ are not centered, r is simply a bilinear form but no longer a correlation coefficient. It will instead be called the canonical association coefficient, the total model being called the Canonical Association (CA) model or, simply, the canonical model.

It turns out that the canonical analyses of all the different sorts of dichotomous variables defined in Tb.4.4 are related in a very simple way. The following theorem states the relationship between the results for the uncentered variables (with product $\boldsymbol{M}$) and centered variables (with product $\boldsymbol{S}$).

Theorem: First canonical variates of contingency table: If $\ddot{\boldsymbol{x}}$ and $\ddot{\boldsymbol{y}}$ are Boolean set variables, then the first dimensions of the canonical variates $X = \boldsymbol{g}'\ddot{\boldsymbol{x}}$ and $Y = \boldsymbol{w}'\ddot{\boldsymbol{y}}$ with $\mathrm{E}(X^2) = (Y^2) = 1$ and $r = \mathrm{E}(XY') = \max$ have the properties:

$$w_\ell = g_k = 1, \quad \forall \ell, k \qquad \text{unit weights} \tag{4.129}$$

$$X_n = Y_n = 1, \quad \forall n \qquad \text{constant scores} \tag{4.130}$$

$$\boldsymbol{t} = m_x, \quad \boldsymbol{p} = m_y \qquad \text{loadings} = \text{means} \tag{4.131}$$

$$\boldsymbol{S}_{yx} = \mathrm{E}(\boldsymbol{e}\boldsymbol{d}') \qquad \text{residual product.} \tag{4.132}$$

$$r = 1 \tag{4.133}$$

Proof: The elements of the contingency tables $\boldsymbol{M}_{xx}, \boldsymbol{M}_{yx}, \boldsymbol{M}_{yy}$ sum up to unity:

$$\mathbf{1}'\boldsymbol{M}_{xx}\mathbf{1} = \mathbf{1}'\boldsymbol{M}_{yx}\mathbf{1} = \mathbf{1}'\boldsymbol{M}_{yy}\mathbf{1} = 1. \tag{4.134}$$

Hence the vectors $\boldsymbol{g} = \mathbf{1}$ and $\boldsymbol{w} = \mathbf{1}$ fulfill the side condition of Eq.4.125. Due to the Cauchy-Schwarz inequality r is less than, or equal to unity,

$$r = \frac{\boldsymbol{g}'\boldsymbol{M}_{xy}\boldsymbol{w}}{\sqrt{\boldsymbol{g}'\boldsymbol{M}_{xx}\boldsymbol{g}\;\boldsymbol{w}'\boldsymbol{M}_{yy}\boldsymbol{w}}} \leq 1, \tag{4.135}$$

for any real vectors $\boldsymbol{g}$ and $\boldsymbol{w}$. With $\boldsymbol{g} = \mathbf{1}$ and $\boldsymbol{w} = \mathbf{1}$ the maximum $r = 1$ is always obtained. (That this automatic maximum is not unique in real data is as likely as a unity correlation coefficient in real data.) Eq.4.129 proved.

$$Y = \sum_k w_k \ddot{y}_k = \sum_k 1\ddot{y}_k = \sum_k \ddot{y}_k = 1 \qquad \text{Eq.4.130 proved.} \tag{4.136}$$

$$p_k = \mathrm{E}(YY_k) = \mathrm{E}(1\ddot{y}_k) = \mathrm{E}(\ddot{y}_k) = m_k \qquad \text{Eq.4.131 proved.} \tag{4.137}$$

$$\begin{aligned}\mathrm{E}(\boldsymbol{e}\boldsymbol{d}') &= \mathrm{E}(\ddot{\boldsymbol{y}} - \boldsymbol{p}Y)(\ddot{\boldsymbol{x}} - \boldsymbol{t}X)' = \mathrm{E}(\ddot{\boldsymbol{y}} - m_yY)(\ddot{\boldsymbol{x}} - m_xX)'\\ &= \mathrm{E}(\ddot{\boldsymbol{y}} - m_y1)(\ddot{\boldsymbol{x}} - m_x1)' = \boldsymbol{M}_{yx} - m_xm_y' \qquad = \boldsymbol{S}_{yx},\end{aligned} \tag{4.138}$$

and similarly $\mathrm{E}(\boldsymbol{dd}') = \boldsymbol{S}_{xx}$ and $\mathrm{E}(\boldsymbol{ee}') = \boldsymbol{S}_{yy}$, thus completing the proof of the theorem.

Remark 4⋆1 The first canonical association coefficient $r_{1(M)}$ of $\boldsymbol{g}'\boldsymbol{M}_{xy}\boldsymbol{w}$ is not a correlation coefficient, whereas all following coefficients are, because, after the extraction of the first dimensions, the MVs have zero mean. Guttman (1941c) called the first solution "extraneous" because he defined a canonical correlation model, not a canonical association model on the contingency table. Others term the first solution "trivial".

Remark 4⋆2 The second canonical association coefficient of $\boldsymbol{M}_{yx}$, $r_{2(M)} = \boldsymbol{g}_2'\boldsymbol{M}_{xy}\boldsymbol{w}_2$, is identical to the first canonical correlation coefficient of $\boldsymbol{S}_{yx}$, $r_{1(S)} = \boldsymbol{g}'\boldsymbol{S}_{xy}\boldsymbol{w}$, because the residual moment of $\boldsymbol{M}_{yx}$, after extraction of the first dimension, is $\boldsymbol{S}_{yx}$ (see Eq.4.132), and in general

$$r_{a(S)} = r_{a+1(M)} \tag{4.139}$$

where $a = 1 \ldots A$ and A is the rank of $\boldsymbol{S}_{xy}$. The rank of the $(L \times K)$ matrix $\boldsymbol{M}_{yx}$ is $\min(L, K)$; the residual matrix $\boldsymbol{S}_{yx} = \boldsymbol{M}_{yx} - \boldsymbol{m}_y \boldsymbol{m}_x'$ has rank $A = \min(L, K) - 1$.

Remark 4⋆3 The sum of the squared nontrivial canonical correlation coefficients is

$$\varphi^2 = \sum_{a=1}^{A} r_{a(S)}^2 = \sum_{a=2}^{A+1} r_{a(M)}^2 \leq A = \min(L, K) - 1 \tag{4.140}$$

As proved by Lancaster (1969), the φ^2 in Eq.4.121 and in Eq.4.140 are identical. That the maximum of φ^2 is A has already been proved by Cramér (1946:282), which has lead to the construction of the coefficient Eq.4.122.

Remark 4⋆4 The canonical model is scale-free, i.e. if x_ℓ is multiplied by some constant $1/c$ and g_ℓ is multiplied by c, the linear form $X = \boldsymbol{g}'\boldsymbol{x}$ is not changed in any way. Hence, the CA coefficients computed from the product matrices $\boldsymbol{S}, \boldsymbol{R}, \boldsymbol{D}$ are identical, and those of $\boldsymbol{M}$ and $\boldsymbol{Q}$ are identical, and Eq.4.139 is enlarged:

$$r_{a(S)} = r_{a(D)} = r_{a(R)} = r_{a+1(M)} = r_{a+1(Q)}. \tag{4.141}$$

The relationships between the weights, and the loadings, of the differently scaled products are:

$$\begin{aligned} w_{k,a,(S)} &= w_{k,a+1,(M)} \\ &= w_{k,a+1,(Q)} / \sqrt{m_k} \\ &= w_{k,a,(D)} / \sqrt{m_k} \\ &= w_{k,a,(R)} / \sqrt{m_k - m_k^2} \end{aligned} \tag{4.142}$$

$$\begin{aligned} p_{k,a,(S)} &= p_{k,a+1,(M)} \\ &= p_{k,a+1,(Q)} \cdot \sqrt{m_k} \\ &= p_{k,a,(D)} \cdot \sqrt{m_k} \\ &= p_{k,a,(R)} \cdot \sqrt{m_k - m_k^2}. \end{aligned} \tag{4.143}$$

Remark 4⋆5 Another application of the Canonical Association model is found in the procedures for comparing two factor loading matrices, say $\boldsymbol{X}_{[K \times N]}$

Table 4.6: Canonical association and principal component analysis of products $\boldsymbol{R}$ (correlation) and $\boldsymbol{Q}$ (normed product), Tb.4.2

	Canonical correlation weights						Principal component loadings					
	$\boldsymbol{Q}$			$\boldsymbol{R}$			$\boldsymbol{Q}$			$\boldsymbol{R}$		
	LV_1	LV_2	LV_3	LV_1	LV_2	LV_3	LV_1	LV_2	LV_3	LV_1	LV_2	LV_3
$\ddot{x}_1$	50	–15	85	87	–97	**	50	–14	71	–10	81	57
$\ddot{x}_2$	71	–50	–50	121	21	**	71	–47	–41	–70	–60	–33
$\ddot{x}_3$	50	85	–15	*	*	**	50	80	–12	91	–12	–19
$\ddot{y}_4$	71	–50	50	121	–21	**	71	–47	41	–70	60	–33
$\ddot{y}_5$	50	–15	–85	87	87	**	50	–14	–71	–10	–81	56
$\ddot{y}_6$	50	85	15	*	*	**	50	80	12	91	12	–19

Canonical association coeff.			Eigenvalues		
ρ_1	1.00000	.76569	λ_1	2.00000	2.66667
ρ_2	.76569	.36569	λ_2	1.76569	2.05375
ρ_3	.36569	**	λ_3	1.36569	.93333
			λ_4	.63431	.34625
			λ_5	.23431	**
			λ_6	**	**

Two stars indicate automatic zeros.

One star indicates an arbitrarily fixed zero, both due to insufficient rank of R.

and $\boldsymbol{Y}_{[L\times N]}$, with $K \leq L$. Transformations $\boldsymbol{G}_{[K\times K]}$ and $\boldsymbol{W}_{[K\times L]}$ are sought which maximize the columnwise association of $\boldsymbol{X}^* = \boldsymbol{G}\boldsymbol{X}$ and $\boldsymbol{Y}^* = \boldsymbol{W}\boldsymbol{Y}$ and make the "congruence matrix" $\boldsymbol{R} = \boldsymbol{X}^*\boldsymbol{Y}^{*\prime}$ diagonal. For this application, no single point of the theorem Eqs.4.129,..., 4.132 holds true, because $\boldsymbol{X}\boldsymbol{X}'$ and $\boldsymbol{Y}\boldsymbol{Y}'$ are not diagonal.

Example. Table 4.6 shows the canonical association analysis of the product matrices $\boldsymbol{Q}$ and $\boldsymbol{R}$, taken from Tb.4.5. It demonstrates the correspondence between the coefficients $r_{(Q)}$ and $r_{(R)}$, as stated in Eq.4.141 and the differences in the weights, as is to be expected from Eq.4.142. The mean squared contingency is $\varphi^2 = 0.76569^2 + 0.36569^2 = 0.72$. Multiplying by 100 (cases) one arrives at the χ^2-value; Cramér's contingency coefficient is $C_{\text{Cramér}} = \sqrt{.72/2} = .60$, both identical to the values in Tb.4.3.

4.1.3 Principal Components of Contingency Tables

Let two categorical variables $\dot{x}$ and $\dot{y}$ be replaced by two sets of dichotomous variables $\ddot{\boldsymbol{x}}_{(Q)}$ and $\ddot{\boldsymbol{y}}_{(Q)}$ with scale unit such that $\text{E}(\ddot{\boldsymbol{x}}_{(Q)}\ddot{\boldsymbol{x}}'_{(Q)}) = \text{E}(\ddot{\boldsymbol{y}}_{(Q)}\ddot{\boldsymbol{y}}'_{(Q)}) = \mathbf{I}$ and $\text{E}(\ddot{\boldsymbol{x}}_{(Q)}\ddot{\boldsymbol{y}}'_{(Q)}) = \boldsymbol{Q}_{xy}$, as defined in Tb.4.4. Let the two sets be joined into one vector $\boldsymbol{u} = (\ldots, \ddot{\boldsymbol{x}}_{\ell,(Q)}, \ldots, \ddot{\boldsymbol{y}}_{k,(Q)}, \ldots)$ with product matrix

$$\mathrm{E}(\boldsymbol{u}\boldsymbol{u}') = \boldsymbol{Q} = \begin{bmatrix} \mathbf{I} & \boldsymbol{Q}_{xy} \\ \boldsymbol{Q}_{yx} & \mathbf{I} \end{bmatrix}. \tag{4.144}$$

The principal component model, then, is written as:

$$Z = \boldsymbol{w}'\boldsymbol{u}, \qquad \mathrm{E}(ZZ') = \boldsymbol{w}'\boldsymbol{Q}\boldsymbol{w} = 1 \tag{4.145}$$

$$\boldsymbol{u} = \boldsymbol{p}Z + \boldsymbol{e}, \qquad \mathrm{E}(\boldsymbol{u}|Z) = \boldsymbol{p}Z. \tag{4.146}$$

The component Z is split up into two pieces, X and Y, along the split of $\boldsymbol{u}$ into $\ddot{\boldsymbol{x}}$ and $\ddot{\boldsymbol{y}}$,

$$X = \sum_{\ell} w_\ell \ddot{x}_\ell = \boldsymbol{w}_x' \ddot{\boldsymbol{x}} \qquad Y = \sum_k w_k \ddot{y}_k = \boldsymbol{w}_y' \ddot{\boldsymbol{y}} \tag{4.147}$$

with variances and covariances

$$\mathrm{E}(XX') = \boldsymbol{w}_x' \mathbf{I} \boldsymbol{w}_x = g^2 \tag{4.148}$$

$$\mathrm{E}(YY') = \boldsymbol{w}_y' \mathbf{I} \boldsymbol{w}_y = h^2 \tag{4.149}$$

$$\mathrm{E}(XY') = \boldsymbol{w}_y' \boldsymbol{Q}_{yx} \boldsymbol{w}_x = g \cdot r \cdot h. \tag{4.150}$$

According to the principle of constant proportionality (Sc.3.5.4), the regression coefficient of X on the rest of Z (which is Y), i.e.:

$$b_{X|Y} = \frac{\mathrm{E}(XY')}{\mathrm{E}(YY')} = \frac{grh}{h^2} \tag{4.151}$$

and the regression coefficient of Y on the rest of Z (which is X)

$$b_{Y|X} = \frac{\mathrm{E}(XY')}{\mathrm{E}(XX')} = \frac{grh}{g^2} \tag{4.152}$$

are equal to the eigenvalue $\lambda - 1$, and thus numerically equal:

$$\frac{grh}{g^2} = \lambda - 1 = \frac{grh}{h^2}. \tag{4.153}$$

Hence the variances of X and Y are equal

$$g^2 = h^2 \tag{4.154}$$

and the correlation (more generally: the association) of X and Y is related to the eigenvalue of the PC analysis by:

$$r = \lambda - 1 \tag{4.155}$$

Because λ is maximized in the PC analysis, r is maximized at the same time; hence r is a canonical correlation (or association) coefficient, and X and Y are — appropriately rescaled to give $\mathrm{E}(XX') = \mathrm{E}(YY') = 1$ — the canonical variates, as described in Sc.4.1.2.

Summing up, a principal components analysis and a canonical association analysis of $\boldsymbol{Q}$ lead to essentially the same results. Some more remarks:

Remark 4⋆6 The relationship between PCA and CAA holds good for all dimensions. Because the first CAA coefficient $r_{1,(Q)} = 1$ is trivial, the first eigenvalue $\lambda_{1,(Q)} = r_{1,(Q)} + 1 = 2$ is trivial, too. Because $\sum_\ell \ddot{x}_\ell = \sum_k \ddot{y}_k = 1$, $\boldsymbol{Q}$ is singular and the last eigenvalue of $\boldsymbol{Q}$ is zero.

Remark 4⋆7 In contrast to CAA, the PCA is not scale-free. Only the PCA of the products $\boldsymbol{Q}$ and $\boldsymbol{D}$ can be translated into a CCA, but not the PCA of $\boldsymbol{M}, \boldsymbol{S}$ or $\boldsymbol{R}$. (In PCA, weights $\boldsymbol{w}$ and loadings $\boldsymbol{p}$ are proportional; in CAA, the weights $\boldsymbol{w}_y$ are proportional to $\mathrm{E}(\ddot{\boldsymbol{y}}\ddot{\boldsymbol{y}}')^{-1}\boldsymbol{p}$; only if $\mathrm{E}(\ddot{\boldsymbol{y}}\ddot{\boldsymbol{y}}') = \mathbf{I}$, this difference between PCA and CCA vanishes.)

Remark 4⋆8 For the sake of simplicity, the theorem Eqs.4.129,..., 4.132 was stated for the product matrices $\boldsymbol{M}$ and $\boldsymbol{S}$. With appropriate changes, the properties carry over to $\boldsymbol{Q}$ and $\boldsymbol{D}$:

$$w_{1,k,(Q)} = \sqrt{m_k} \qquad \text{the weights} \qquad (4.156)$$

$$p_{1,k,(Q)} = \sqrt{m_k} \qquad \text{the loadings} \qquad (4.157)$$

$$\boldsymbol{D}_{yx} = \mathrm{E}(\boldsymbol{e}\boldsymbol{d}'). \qquad \text{the residuals} \qquad (4.158)$$

Many authors (e.g. Benzécri 1976, Lebart, Morineau & Tabard 1978) recommend using $\boldsymbol{D}$ instead of $\boldsymbol{M}$, because then the trivial first solution of $\boldsymbol{M}$ is avoided.

Example. The right-hand part of Tb.4.6 shows the PC analysis of the product matrix $\boldsymbol{Q}$ of Tb.4.5. Observe the correspondence between the eigenvalues and the canonical correlations, as stated in Eq.4.154. The weights of CAA and PCA of $\boldsymbol{Q}$ are different, due to the different scalings of the LVs (in CCA they are scaled so that $\mathrm{var}(X) = \mathrm{var}(Y) = 1$, in PCA as $\mathrm{var}(X+Y) = 1$). However, the weights for the matrix $\boldsymbol{Q}$ are proportional to one other. The PCA of the correlation matrix $\boldsymbol{R}$ has no similarity, neither to the PCA of $\boldsymbol{Q}$, nor to the CAA of $\boldsymbol{R}$ or $\boldsymbol{Q}$. The example demonstrates that only the product $\boldsymbol{Q}$ (or $\boldsymbol{D}$) is appropriate for the PCA as an alternative method for the CAA.

4.1.4 Categorical Scaling

If $\dot{y}$ is a categorical variable, replaced by a set of Boolean variables $\ddot{\boldsymbol{y}} = [\ddot{y}_k, k = 1 \dots K]$, then the weights w_k, found by a canonical association analysis (Sc.4.1.2) or by a principal component analysis (Sc.4.1.3), provide a scaling of the categories of $\dot{y}$. Only one of the Boolean variables has the value 1 ($\ddot{y}_k = 1 \iff \dot{y} = k$) while all the others are zero. Hence the score on the LV is identical to the weight coefficient w_k corresponding to the category k, in symbols:

$$\begin{aligned} Y_n &= \sum_k w_k \ddot{y}_k \\ &= w_k \iff \dot{y} = k. \end{aligned} \qquad (4.159)$$

Thus, the estimated LV is a re-scaled, interval-scaled version of the categorical MV. If there are two categorical MVs and if the weights are found by CAA or PCA,

Table 4.7: Augsburg Longitudinal Study, school marks for German in grade 4

Timepoint	Scaling			Contingency table $M \times 10000$											
Scores	m_k	p_k	w_k												
mid-grade															
1 = sehr gut	3%	.31	1.79	339											
2 = gut	30%	.59	1.12	.	2983										
3 = befriedigend	32%	.10	.84	.	.	3169									
4 = ausreichend	24%	−.42	−.42	.	.	.	2419								
5 = mangelhaft	10%	−.56	−1.69	.	.	.	.	1000							
6 = ungenügend	1%	−.23	−2.30	.	.	.	.	.	89						
end of year															
1 = sehr gut	4%	.32	1.60	246	136	4	0	0	0	386					
2 = gut	29%	.58	1.12	93	2386	377	0	0	0	.	2856				
3 = befriedigend	33%	.11	.19	0	462	2415	373	4	0	.	.	3256			
4 = ausreichend	24%	−.41	−.84	0	0	369	1814	178	0	.	.	.	2360		
5 = mangelhaft	10%	−.55	−1.66	0	0	4	229	754	21	.	.	.	.	1008	
6 = ungenügend	1%	−.27	−1.91	0	0	0	4	64	68	.	.	.	.	.	136

Structural zeros are written as dots.
m_k = relative univariate frequencies.
p_k = loadings.
$w_k = p_k/m_k$ = scale value.
$\rho = 0.90$ = correlation(mid-grade,end).

then the re-scaled variables will have the maximum correlation. (Admittedly, the term latent variable is not totally appropriate for a rescaled MV; the term LV here is maintained in order to emphasize the similarity to the procedures with interval-scaled variables.)

The procedure for scaling categorical variables can be extended to more than two variables forming a super contingency table with 2^2 submatrices. If there are $J = 2, 3, 4 \ldots$ variables, the super contingency table will contain $J^2 = 2^2, 3^2, 4^2 \ldots$ sub-tables. The super contingency table $\boldsymbol{M}$ is transformed into $\boldsymbol{Q}$; the PC weights $\boldsymbol{w}_{(Q)}$ of $\boldsymbol{Q}$ are computed and back-transformed to apply to $\boldsymbol{M}$ by

$$w_{k,(M)} = w_{k,(Q)}/\sqrt{m_k}. \tag{4.160}$$

The first PC of $\boldsymbol{Q}$ is the trivial solution; therefor $\boldsymbol{D}$ may be used instead.

The Categorical Scaling procedure for two variables dates back to R. A. Fisher (1940), Hirschfeld (1935) and Horst (1935). For more than two variables the Categorical Scaling was initialized by Guttman (1941c, 1950a, 1950b), reinvented by Burt (1950) and developed by the French school of statistics (*"l'analyse des correspondances"*, Benzécri (1976)). McKeon (1966) and McDonald (1968) integrated the Categorical Scaling into the general framework of Multi-Set Factor analysis and optimal weighting.

New interest in Categorical Scaling arose in the seventies when the area of multidimensional scaling was treated by PLS algorithms, termed Alternating Least Squares (ALS) (cf. Young, de Leeuw & Takane 1976; van der Burg & de Leeuw 1980; van Rijckevorsel, Bettonvil & de Leeuw 1980; von Frenckell 1980; Young 1981). Several Applications can be found in Lohmöller & Oerter 1979:76, 83, 147, 153.

Application. Table 4.7 shows an application of Categorical Scaling to educational data. School marks (a six-point rating scale) often are suspected of not being interval-scaled, or even not constituting an ordinal scale, because the first rank in the one class might be the second in another. The Categorical Scaling procedure needs no assumptions on the ordering or spacing of the categories but discovers those scale values which optimally linearize the relationships between the variables.

The data used here are taken from the Augsburg Longitudinal Study (Hanke, Lohmöller & Mandl 1980). The variables are the marks for German in the 4th grade, given in the winter (mid-grade) and the summer (end of the school-year). The marks are given on a six-point scale, starting with *sehr gut* (very good). Tb.4.7 shows the lower part of the symmetric super contingency table $\boldsymbol{M}$, together with the means $\boldsymbol{m}$, the loadings $\boldsymbol{p}$ and the scale values $\boldsymbol{w}$. The correlation between the optimally scaled marks is $r = 0.90$. Using the unscaled marks (i.e. taking $1, 2, \ldots, 6$ as scales values) gives the correlation $r = 0.88$. (This was the biggest difference found in the 28 pairwise comparisons of the 8 ratings in the four years of primary school.) The difference is not at all important, so that the unscaled values may be used as well. The considerations regarding the scale quality of school marks have little empirical significance, as long as the correlation between the marks is taken as the criterion.

4.1.5 LV Path Analysis of Super Contingency Tables

In the Categorical Scaling procedure only the correlations between the LVs (the re-scaled MVs) are considered, and the procedure maximizes a weighted sum of all correlations, $f = \sum_i \sum_j v_i v_j r_{ji}$. The maximization criterion is expressed as a closed, differentiable function and an analytical solution — which is an LS solution — is possible (cf. Sc.3.4.2, Sc.4.1.4, cf. Horst 1965:chapter 24).

In LVP models the correlations between the LVs are explained and modeled out further by a system of linear relations, i.e. by the path model. The basic properties of PLS carry over to contingency tables, because a contingency table is a product matrix. The PLS approach to super-contingency tables has been outlined by Wold and Bertholet (1982a, 1982a, 1982c), Bertholet and Wold (1984, 1985) and Bertholet (1987). The extension of the PLS method presented in Sc.4.2 compromises the case of categorical variables as a special case, so no example will be given in the present section. Lohmöller (in press a) reports a LV path model with 10 categorical indicators, and he compares the results under four different scalings, namely a "raw scaling", an optimal scaling in a premodeling step, an optimal scaling as part of the PLS algorithm, and an analysis of polychoric correlations as produced by LISREL.

The only question we shall address in this section is the meaning of the PLS ModeA vs. ModeB in relation to the contingency table analysis. This is related to the question of which of the five product matrices $\boldsymbol{M}, \boldsymbol{Q}, \boldsymbol{S}, \boldsymbol{D}, \boldsymbol{R}$ introduced in Tb.4.4 is to be used for the analysis.

The distinction between ModeA and ModeB has its bearing inside the PLS estimation procedure when it comes to the estimation of the weights. The weights are estimated by single regression, which produces $\boldsymbol{w}_j \propto \text{cov}(\ddot{\boldsymbol{y}}_j; \tilde{Y})$ in a ModeA block, or by multiple regression, which produces $\boldsymbol{w}_j \propto \boldsymbol{S}_{jj}^{-1}\,\text{cov}(\ddot{\boldsymbol{y}}_j, \tilde{Y})$. However, $\boldsymbol{S}_{jj}, \boldsymbol{D}_{jj}$ and $\boldsymbol{R}_{jj}$ are singular, and the regular inverses do not exist. On the other hand, these three product matrices do not include the trivial solution (as do $\boldsymbol{M}_{jj}$ and $\boldsymbol{Q}_{jj}$), because the means are removed.

Blockwise orthonormalization. As shortly mentioned by Lohmöller (1979b) and in Eq.3.532, the distinction between ModeA and ModeB estimation can be placed outside the estimation procedure. All ModeB blocks are orthonormalized before and transformed back after the iterations phase (all ModeA blocks are transformed by an identity), so that within the iteration no distinction between ModeA and ModeB blocks needs to be made.

Let $\breve{\boldsymbol{y}}_j$ denote the original (correlated) variables in interpretation metric, $\boldsymbol{H}_j$ the transformation matrix, and $\boldsymbol{y}_j = \boldsymbol{H}_j\breve{\boldsymbol{y}}_j$ the orthonormalized variables in estimation metric, with $\text{E}(\boldsymbol{y}_j\boldsymbol{y}_j') = \mathbf{I}$. The estimated weights $\boldsymbol{w}_j$ apply to $\boldsymbol{y}_j, Y_j = \boldsymbol{w}_j'\boldsymbol{y}_j$. After the estimation phase, the weights are transformed back by $\breve{\boldsymbol{w}}_j = \boldsymbol{H}_j^{-1}\boldsymbol{w}_j$ to be applicable to the original variables, $Y_j = \breve{\boldsymbol{w}}_j'\breve{\boldsymbol{y}}_j$. A useful and interpretable transformation $\boldsymbol{H}_j$ is the symmetric transformation, the semi-inverse $\boldsymbol{H}_j = \breve{\boldsymbol{S}}_j^{-1/2}$:

$$\boldsymbol{H}_j = \breve{\boldsymbol{V}}_j\breve{\boldsymbol{\Lambda}}_j^{-1/2}\breve{\boldsymbol{V}}_j' \quad \text{with} \quad \breve{\boldsymbol{S}}_{jj} = \text{cov}(\breve{\boldsymbol{y}}_j) = \breve{\boldsymbol{V}}_j\breve{\boldsymbol{\Lambda}}_j\breve{\boldsymbol{V}}_j', \tag{4.161}$$

where $\breve{\boldsymbol{\Lambda}}_j$ is the diagonal matrix with the eigenvalues and $\breve{\boldsymbol{V}}_j$ the normed eigenvector matrix. Thus the transformed variables are orthogonal and the LS approximations to their originals (Johnson 1966, Kaiser 1967). When there are two sets of MVs $\breve{\boldsymbol{x}}$ and $\breve{\boldsymbol{y}}$, the correlations within the sets $\breve{\boldsymbol{S}}_{xx}$ and $\breve{\boldsymbol{S}}_{yy}$ "overshadow" the correlations $\breve{\boldsymbol{S}}_{yx}$ between the sets. The correlation $\boldsymbol{S}_{yx}$ between the orthogonalized, "purified" variables $\boldsymbol{x}$ and $\boldsymbol{y}$ give so to speak the net relationships (cf. Hanke, Lohmöller, Mandl 1980:166).

Analyse product matrix $\boldsymbol{D}$. Computation of the half inverses for super contingency tables is a very simple task. The product moment $\boldsymbol{M}_{yy}$ within a set is a diagonal matrix with elements m_k, the inverse $\boldsymbol{M}_{yy}^{-1}$ is a diagonal matrix with elements $1/m_k$ and the half inverse $\boldsymbol{M}_{yy}^{-1/2}$ is a diagonal matrix with elements $1/\sqrt{m_k}$. The blockwise orthonormalized super contingency table has the elements

$$q_{k\ell} = \frac{m_{kl}}{m_k m_\ell} \tag{4.162}$$

which is the definition for product matrix $\boldsymbol{Q}$, as given in Tb.4.4. The matrices $\boldsymbol{Q}_{jj}$ in the diagonal of $\boldsymbol{Q}$ are identities; the weights and loadings in an LVP model on $\boldsymbol{Q}$ are identical, just the same way as correlations and multiple regression coefficients are identical when the predictors are standardized and uncorrelated. The weights and loadings on $\boldsymbol{Q}$ are transformed back to weights and loadings on $\boldsymbol{M}$ by

$$w_{k,1,(M)} = w_{k,1,(Q)}/\sqrt{m_k} \tag{4.163}$$

$$p_{k,1,(M)} = p_{k,1,(Q)} \cdot \sqrt{m_k}. \tag{4.164}$$

The first dimension of LVs of $\boldsymbol{M}$ is trivial because the loadings are identical to the means, $p_{k,1,(M)} = m_k$, and the weights are unity, $w_{k,1,(M)} = 1$; weights and loadings of the first dimension of $\boldsymbol{Q}$ are $p_{k,1,(Q)} = w_{k,1,(Q)} = m_k$. The trivial dimension can be avoided by starting with the covariance matrix $\boldsymbol{S}$, where the means are removed and cannot blur the solution. However, the within-block covariances $\boldsymbol{S}_{jj}$ do not have full rank, and the inverses $\boldsymbol{S}_{jj}^{-1}$ do not exist. Some sort of generalized inverse could be applied, or the product $\boldsymbol{Q}$ with the trivial solution removed. Product $\boldsymbol{D}$ is both: it can be understood as $d_{k\ell} = q_{k\ell} - p_{k,1,(Q)}p_{\ell,1,(Q)}$ (see Eq.4.158), or as transformed by a generalized inverse (cf. McDonald, Torii & Nishisato 1979, Rao 1966). The product matrix $\boldsymbol{D}$ seems to be the right choice for the input to a LVP analysis of super contingency tables. The trivial solution is avoided and the loadings and weights are identical during the estimation phase. This is exactly the same matrix as used by Benzécri 1976 and his collaborators for the l'analyse des correspondances.

One could use the product $\boldsymbol{S}$ (i.e. the identical transformation) for some blocks and the product $\boldsymbol{D}$ (the orthonormal transformation) for others. This, then, would give the difference between ModeB and ModeA blocks. The variance of a variable in a ModeA block is part of the loss function to be minimized in the estimation. The variances of Boolean variables are directly related to the means, $s_{kk}^2 = m_k - m_k^2$. Thus, the greater the mean m_k (the bigger the group k), the higher the influence of category k on the loss function. This is not the case in ModeB blocks, where all $q_{kk} = 1$. In this way, something similar to the Principal Predictor model (Redundancy model) emerges.

As mentioned before, the Categorical Scaling of product matrix $\boldsymbol{M}$ leads to a trivial first dimension, where all weights and scores are unity and loadings identical to the means (i.e. univariate relative frequencies). Depending on the choice of the inner weighting scheme (see Sc.2.3.3), an LVP analysis with an unrestricted path model may come very close to, or exactly equal, the trivial solution of the Categorical Scaling method. If the first dimension of the LVPA of $\boldsymbol{M}$ misses the trivial solution, not one of the favorable properties stated in the theorem Eqs.4.129,..., 4.132 on contingency tables holds true. This argument confirms yet again that $\boldsymbol{D}$ or $\boldsymbol{S}$ are the right choices for the LVP analysis of super contingency tables. The separation between "location" information (means, first moments) and "association" (regression, correlation, contingency, second moments) is done automatically, when the conditions of theorem Eqs.4.129,..., 4.132 are met. If not, the researcher has to separate the two types of information "by hand" in a premodeling step.

4.2 Mixture of Categorical and Interval-Scaled Variables

Linear analysis of interval-scaled variables is a well-developed domain in statistics and includes methods like correlation, regression, factor analysis, principal component analysis, and path analysis. The association between a categorical and an interval-scaled variable is displayed by conditional averages: the categorical variable is considered as explanatory, the interval-scaled variable as target (dependent) variable, and the mean for each category of the explanatory variable is computed.

Analysis of variance (Anova) is used to test the relationships. The Anova was developed from the thirties to the sixties of this century and has brought to the fore a huge array of different models and computational schemes (cf. Winer 1971). In the last two decades regression theory has been laid down as a general and unifying theory that includes Anova as a special case (cf. Bock 1975). However, Anova testing procedures require the assumption of independent observations, which is not needed for the Stone-Geisser test. Standard textbooks on regression analysis (e.g. Wonnacott & Wonnacott 1981; Schönfeld 1969:I,85-91) now include chapters on Anova performed as regression analysis with dummy variables.

As a coefficient of association the proportion of explained variance R^2, or the adjusted coefficient of determination R^{*2} (Bock 1975:380), is used. The conditional means are used to provide a scaling of the categories. Prediction involving categorical variables is discussed, among others, by Guttman (1941b), Andrews and Messenger (1975), Hildebrand, Laing and Rosenthal (1977).

Tb.4.2 contains data for a variable z, which will be considered as interval-scaled. The total mean is $\bar{z} = 0$, the variance $s_z^2 = 1$. If the total sample of 20 cases is split up into three sub-samples, according to the categories of $\dot{x}$, then three conditional means $\bar{z}_1 = \text{mean}(y|\dot{x} = 1) = 6/5$, $\bar{z}_2 = 0$ and $\bar{z}_3 = -6/5$ are found. If the total sample is split up according to $\dot{y}$, the conditional means are $\bar{z}_4 = \text{mean}(z|\dot{y} = 4) = 7/10$, $\bar{z}_5 = -3/5$ and $\bar{z}_6 = -4/5$.

This chapter Sc.4.2 makes four preparatory steps before it comes to the treatment of multivariate models for mixed-scales variables in Sc.4.2.5. In Sc.4.2.1 the mixed product moment matrix is introduced which includes the products of interval-scaled and Boolean variables. Then three simple prediction models are treated, in Sc.4.2.2 with one categorical predictor, in Sc.4.2.3 with two categorical predictors and in Sc.4.2.4 with one categorical predictand.

4.2.1 The Mixed Product Moment Matrix

In order to attain a unified treatment for categorical and interval-scaled variables, the variable z is included in the moment matrices $\boldsymbol{N}$ and $\boldsymbol{M}$ defined in Sc.4.1.1. The raw product matrix of the vector variable

$$\boldsymbol{u} = (\ddot{\boldsymbol{x}}', \ddot{\boldsymbol{y}}', z)' = (\ddot{x}_1, \ddot{x}_2, \ddot{x}_3, \ddot{y}_4, \ddot{y}_5, \ddot{y}_6, z)' \tag{4.201}$$

Table 4.8: Raw product $\boldsymbol{N}$ and relative product $\boldsymbol{M}$ for data from Tb.4.2

	Raw product moment $\boldsymbol{N}$							Relative product moment $\boldsymbol{M}$						
	$\ddot{x}_1$	$\ddot{x}_2$	$\ddot{x}_3$	$\ddot{y}_4$	$\ddot{y}_5$	$\ddot{y}_6$	z	$\ddot{x}_1$	$\ddot{x}_2$	$\ddot{x}_3$	$\ddot{y}_4$	$\ddot{y}_5$	$\ddot{y}_6$	z
$\ddot{x}_1$	5						sym.	25						sym.
$\ddot{x}_2$	.	10						.	50					
$\ddot{x}_3$	.	.	5					.	.	25				
$\ddot{y}_4$	4	6	0	10				20	30	0	50			
$\ddot{y}_5$	0	4	1	.	5			0	20	5	.	25		
$\ddot{y}_6$	1	0	4	.	.	5		5	0	20	.	.	25	
z	6	0	−6	7	−3	−4	20	30	0	−30	35	−15	−20	100

Structural zeros are written as zeros.
Elements of $\boldsymbol{M}$ are multiplied by 100.

is

$$\boldsymbol{N} = \boldsymbol{u}\boldsymbol{u}' = \begin{bmatrix} \boldsymbol{N}_{xx} & & \text{sym.} \\ \boldsymbol{N}_{yx} & \boldsymbol{N}_{yy} & \\ \boldsymbol{N}_{zx} & \boldsymbol{N}_{zy} & \boldsymbol{N}_{zz} \end{bmatrix} = \begin{bmatrix} [n_{\ell\underline{\ell}}] & & \text{sym.} \\ [n_{z\ell}] & [n_{k\underline{k}}] & \\ [n_{z\ell}] & [n_{zk}] & [n_{zz}] \end{bmatrix} \tag{4.202}$$

and the relative product matrix is

$$\boldsymbol{M} = \frac{1}{N}\boldsymbol{u}\boldsymbol{u}' = \mathrm{E}(\boldsymbol{u}\boldsymbol{u}'). \tag{4.203}$$

Table 4.8 contains the products $\boldsymbol{N}$ and $\boldsymbol{M}$ for the artificial data in Table 4.2. The submatrices $\boldsymbol{N}_{xx}, \boldsymbol{N}_{yy}, \boldsymbol{N}_{yx}$ are absolute frequencies, submatrices $\boldsymbol{N}_{zx}, \boldsymbol{N}_{zy}$ contain the conditional raw sums for the sub-samples with $\ddot{x}_\ell = 1$ and $\ddot{y}_k = 1$. The conditional sums, divided by the sample size, give the conditional means:

$$\boldsymbol{B}_{z|x} = \boldsymbol{N}_{zx}\boldsymbol{N}_{xx}^{-1} = \boldsymbol{M}_{zx}\boldsymbol{M}_{xx}^{-1} \tag{4.204}$$

$$\boldsymbol{B}_{z|y} = \boldsymbol{N}_{zy}\boldsymbol{N}_{yy}^{-1} = \boldsymbol{M}_{zy}\boldsymbol{M}_{yy}^{-1}. \tag{4.205}$$

Thus, the product matrices $\boldsymbol{N}$ and $\boldsymbol{M}$ contain all the information that is required for an Anova or any other linear analysis of association.

The formulae in Tb.4.4, last column, can be used to compute the covariance matrix $\boldsymbol{S}$, the correlation matrix $\boldsymbol{R}$ and the scaled products $\boldsymbol{Q}$ and $\boldsymbol{D}$. That the first moment equals the second moment, $m_k = m_{kk}$, holds good only for Boolean variables. It is useful to give the interval-scaled variables zero mean and unit variance before the raw product is formed, as it is done in Tb.4.2. Only with this scaling do the relationships $m_{kz} = s_{kz}$ and $q_{kz} = d_{kz}$ (see Tb.4.5) hold true.

4.2.2 One Categorical Predictor

Analysis of variance, developed as a technique of its own, can be treated as a special case of regression analysis (see Bock 1975, Draper & Smith 1981:chapter 9). We will introduce LVs to show that Anova can be treated as a special case of PLS

Table 4.9: One categorical predictor

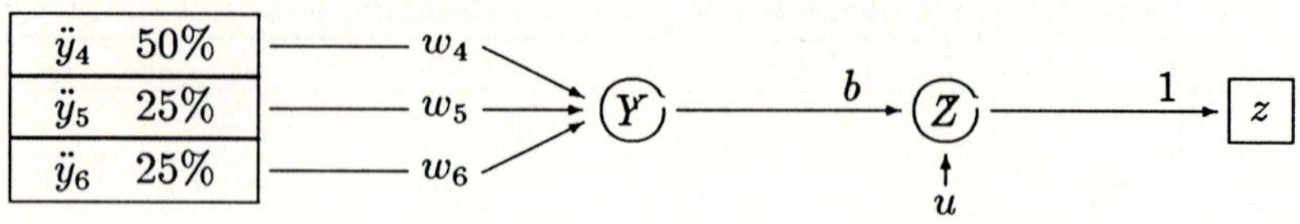

	m_{kk}	m_{zk}	$\bar{z}_k = \tilde{w}_k$	$w_{k,M}$	$p_{k,M}$	$w_{k,Q} = p_{k,Q}$
$\ddot{y}_4$	.50	.35	.70	.9949	.4975	.7035
$\ddot{y}_5$	.25	−.15	−.60	−.8528	−.2132	−.5683
$\ddot{y}_6$	.25	−.20	−.80	−1.1737	−.2843	−.5683

$s^2_{z|y} = 0.4950$
$Z = 0.7034Y + u$

path modeling. Thus, the assumptions leading to causal-predictive interpretation will be emphasized.

In the regressional analysis of an Anova model, the independent (categorical) variable $\dot{y}$ is replaced by Boolean variables $\ddot{\boldsymbol{y}}$ which are, then, transformed by a so-called linear contrast, $\boldsymbol{Y} = \boldsymbol{W}'\ddot{\boldsymbol{y}}$. The contrast matrix is orthonormal, $\boldsymbol{W}'\boldsymbol{W} = \mathbf{I}$, and because $\mathrm{E}(\ddot{\boldsymbol{y}}\ddot{\boldsymbol{y}}') = \boldsymbol{M}_{yy}$ is diagonal, the predictors $\boldsymbol{Y}$ are uncorrelated: $\mathrm{E}(\boldsymbol{Y}\boldsymbol{Y}') = \mathrm{E}(\boldsymbol{W}'\ddot{\boldsymbol{y}}\ddot{\boldsymbol{y}}'\boldsymbol{W}) = \boldsymbol{W}'\boldsymbol{M}_{yy}\boldsymbol{W} =$ diagonal. Note that the contrast matrix is known in advance and that the predictive power of $\boldsymbol{Y}$ is the surprise. It may turn out that some of the contrast vectors have no predictive value at all: $\mathrm{var}(\boldsymbol{w}'\ddot{\boldsymbol{y}}) = 0$.

In a descriptive approach one may wish to find the weight vector $\boldsymbol{w}$ (or $\tilde{\boldsymbol{w}} = b\boldsymbol{w}$) and the latent variable Y (or $\tilde{Y} = bY$):

$$\tilde{Y} = \tilde{\boldsymbol{w}}'\ddot{\boldsymbol{y}}, \qquad \mathrm{var}(\tilde{Y}) = \tilde{\boldsymbol{w}}'\boldsymbol{M}_{yy}\tilde{\boldsymbol{w}} = b^2 \tag{4.206}$$

$$Y = \boldsymbol{w}'\ddot{\boldsymbol{y}}, \qquad \mathrm{var}(Y) = \boldsymbol{w}'\boldsymbol{M}_{yy}\boldsymbol{w} = 1 \tag{4.207}$$

which explains as much as possible of the variance of the predictand z, leaving the variance of the residual u minimal:

$$z = \tilde{Y} + u = bY + u, \qquad \mathrm{E}(z\,|\,\ddot{\boldsymbol{y}}) \equiv \tilde{Y} = bY = b\boldsymbol{w}'\ddot{\boldsymbol{y}}. \tag{4.208}$$

Because $\dot{y}$ is polychotomous, any linear combination of its Boolean substitutes $\ddot{\boldsymbol{y}}$ and, consequently, $\tilde{Y} = bY = \mathrm{E}(z\,|\,\ddot{\boldsymbol{y}})$ have only a limited number of distinct values, denoted by $\ddot{y}_k, \tilde{Y}_k, \mathrm{E}(z\,|\,\ddot{y}_k)$. The conditional expectation of z for a given value $\ddot{y}_k$ is estimated by the subsample average:

$$\mathrm{E}(z\,|\,\ddot{y}_k) = \tilde{Y}_k = \frac{m_{zk}}{m_{kk}} \equiv \bar{z}_k. \tag{4.209}$$

As a consequence of to Eq.4.206 and the Boolean character of $\ddot{\boldsymbol{y}}$:

$$w_k = \bar{z}_k. \tag{4.210}$$

The variance of $\tilde{Y}$ is

$$s^2_{z|y} = \text{var}(\tilde{Y}) = \frac{\sum_k m_{zk}}{m_{kk}} \hat{=} \frac{1}{N}\sum_k n_k \bar{z}_k^2 \tag{4.211}$$

and consequently:

$$b = s_{z|y} = \text{cor}(z; Y). \tag{4.212}$$

The results for the data for $\ddot{\boldsymbol{y}}$ and z from Tb.4.2 are presented in Table 4.9. The correlation of the rescaled categorical variable Y and the interval-scaled variable z is $\rho = \sqrt{0.495} = 0.7034$ which often is called the nonlinear correlation coefficient η. The scaling of Y is chosen so as to maximize this correlation coefficient.

The covariance $p_k = \text{E}(\ddot{y}_k Y)$ can be understood as the LV loading and is easily computed by:

$$p_k = \text{cov}(\ddot{y}_k; Y) = w_k m_{kk}. \tag{4.213}$$

When the dichotomous variables $\ddot{\boldsymbol{y}}$ are scaled to have the product moment $\boldsymbol{Q}$, $\ddot{y}_{k,Q} = \ddot{y}_{k,M}/\sqrt{m_k}$, then the weights $w_{k,Q}$ and loadings $p_{k,Q}$ become identical:

$$p_{k,Q} = p_{k,M} \cdot \sqrt{m_k} = w_{k,Q} = w_{k,M}/\sqrt{m_k}. \tag{4.214}$$

Moreover, as long as the mean of z is zero,

$$p_{k,Q} = w_{k,Q} = w_{k,D} = p_{k,D}. \tag{4.215}$$

4.2.3 Two Categorical Predictors

Let us now suppose the problem is to predict the interval-scaled variable z by two trichotomous variables $\dot{x}$ and $\dot{y}$ (see Table 4.10, see also Bock 1975:381). This can be done by a usual two-way Anova. As a result for the data in Table 4.2, one gets a proportion of explained variance $\text{SS}_{\text{within}}/\text{SS}_{\text{total}} = R^2 = 15.929/20 = 0.7964$.

A multiple regression with six Boolean substitutes for $\ddot{\boldsymbol{x}}$ and $\ddot{\boldsymbol{y}}$ has to face the problem that the six dummies have only rank five, or, after standardization, rank four. So any one of the six dichotomous predictors may be deleted from the regression equation, the regression weights changing accordingly, and no interpretation of the weights is possible. One solution possible, where $w_2 = 0$ is fixed, is given in Tb.4.10 under the heading MR. The Anova gives essentially the same result.

The approach to be presented here avoids the ambiguity of the multiple regression approach by using latent variables. The model:

$$\boldsymbol{X} = \boldsymbol{V}'\ddot{\boldsymbol{x}}, \qquad \text{cov}(\boldsymbol{X}) = \boldsymbol{V}'\boldsymbol{M}_{xx}\boldsymbol{V} = \mathbf{I} \tag{4.216}$$

$$\boldsymbol{Y} = \boldsymbol{W}'\ddot{\boldsymbol{y}}, \qquad \text{cov}(\boldsymbol{Y}) = \boldsymbol{W}'\boldsymbol{M}_{yy}\boldsymbol{W} = \mathbf{I} \tag{4.217}$$

$$\begin{aligned} z &= \boldsymbol{g}'\boldsymbol{X} + \boldsymbol{b}'\boldsymbol{Y} + u, \\ \text{E}(z|\ddot{\boldsymbol{x}}, \ddot{\boldsymbol{y}}) &= \boldsymbol{g}'\boldsymbol{X} + \boldsymbol{b}'\boldsymbol{Y} = \boldsymbol{g}'\boldsymbol{V}'\ddot{\boldsymbol{x}} + \boldsymbol{b}'\boldsymbol{W}'\ddot{\boldsymbol{y}}. \end{aligned} \tag{4.218}$$

In a restricted version of the model Eqs.4.216,..., 4.218, one may use only the first dimension of the LVs X and Y; hence $\boldsymbol{V}$ and $\boldsymbol{W}$ are vectors, and $\boldsymbol{b}$ and $\boldsymbol{g}$ are scalars.

Table 4.10: Two categorical predictors

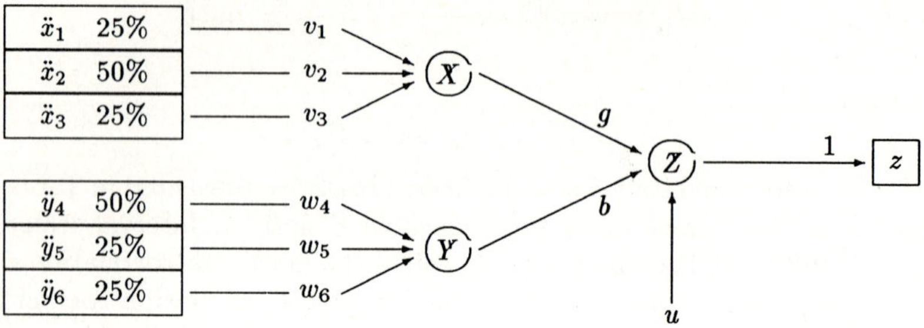

	MR	Split MR		PLS		PLS*	
	Z^*	X	Y	X	Y	X	Y
LV weights							
$\ddot{x}_1$	1.0357	1.5408	*	1.4142	*	.8593	*
$\ddot{x}_2$	*	*	*	.0000	*	-.5151	*
$\ddot{x}_3$	-.8571	-1.2751	*	-1.4142	*	-1.6525	*
$\ddot{y}_4$	.2857	*	.8516	*	.9949	*	1.4095
$\ddot{y}_5$	-.4286	*	-1.2774	*	-.8528	*	-.0845
$\ddot{y}_6$	-.3212	*	-.9580	*	-1.1371	*	.1396
Correlations and regressions							
X		1.0000		1.0000		1.0000	
Y		.5144	1.0000	.5829	1.0000	*	1.0000
Z		.8448	.6813	.8485	.7036	.7636	.4781
(g, b)	1.0000	.6722	.3355	.6640	.3165	.7536	.4764
R^2	.7964	.7964		.7861		.7964	

Stars denote fixed zeros.

One possible estimation procedure for Eq.4.218 makes use of the multiple regression solution (see Sc.3.5): the MR predictor Z^* is split up into the parts measured by the $\ddot{\boldsymbol{x}}$- and by the $\ddot{\boldsymbol{y}}$-variables. These two predictor LVs, then, are standardized, as shown in Tb.4.10 under the heading Split MR. The predictive power $R^2 = 0.7964$ of the two split MR predictors remains the same and the arbitrariness of fixing one coefficient to zero remains the same.

The other solution, presented in the right hand part of Tb.4.10, uses the PLS technique. As long as there is only one endogenous variable, the weights within each block are the same as for the model with one categorical predictor (compare the weights for $\ddot{\boldsymbol{y}}$ in Tb.4.9 and Tb.4.10). The squared multiple correlation is a little lower than in the MR approach: $R^2_{\text{PLS}} = 0.7861 < 0.7964 = R^2_{\text{MR}}$. This is to be expected from the arguments and experiments on Split MR and PLS in Sc.3.6. The PLS approach does not require fixing one weight coefficient to zero.

A second PLS solution, labeled PLS*, is given in Tb.4.10. Here the two LVs are constrained so as to be orthogonal and the Pattor rotation (see Sc.2.3.4) is applied. X and Y come out as totally uncorrelated, and the R^2 is the same as in the multiple regression estimation. All weights $\boldsymbol{v}$ and $\boldsymbol{w}$ have changed drastically as compared to the unrestricted PLS estimates. From the point of view of interpretation, the unrestricted solution offers a simple structure in the outer model, whereas the restricted solution is simpler with respect to the inner model.

The product matrices $\boldsymbol{M}$, $\boldsymbol{Q}$ and $\boldsymbol{D}$ may be used for the model in Tb.4.10 and give essentially the same results in the inner part of the model. In the outer part, however, the weights and loadings are different in the same way as shown in Tb.4.9. Matrices $\boldsymbol{M}$ and $\boldsymbol{Q}$ do not have the full rank, while $\boldsymbol{M}_{xx}$, $\boldsymbol{M}_{yx}$, $\boldsymbol{Q}_{xx}$ and $\boldsymbol{Q}_{yy}$ do. Therefore the PLS ModeB estimation can be chosen, which assures the scale-freeness in the inner model part. Matrices $\boldsymbol{D}_{xx}$ and $\boldsymbol{D}_{yy}$ are not invertible, and PLS ModeB estimation is the only possible.

The Anova procedure allows for interaction of the predictors. Writing the model with LVs, the interaction can be included in the inner part of the model:

$$Z = gX + bY + c(XY) + u, \tag{4.219}$$

or in the outer part of the model:

$$Z = g(\boldsymbol{v}'\ddot{\boldsymbol{x}}) + b(\boldsymbol{w}'\ddot{\boldsymbol{y}}) + c(\boldsymbol{g}'(\ddot{\boldsymbol{x}} \otimes \ddot{\boldsymbol{y}})) + u, \tag{4.220}$$

where $\ddot{\boldsymbol{x}} \otimes \ddot{\boldsymbol{y}}$ denotes the products of all pairs of elements of $\ddot{\boldsymbol{x}}$ and $\ddot{\boldsymbol{y}}$. (If $\ddot{\boldsymbol{x}}$ has L elements and $\ddot{\boldsymbol{y}}$ has K elements, $\ddot{\boldsymbol{x}} \otimes \ddot{\boldsymbol{y}}$ has $L \times K$ elements.)

The model Eq.4.220 is under-identified; the number of predictors is $L \times K$ and the number of weight coefficients is $L \times K + L + K$. (The three path coefficients g, b, c are made up for by the restriction for scale unambiguity (RSU) for the three LVs.) So $L + K - 1$ restrictions must be imposed on the weight vector $(\boldsymbol{v}'\boldsymbol{w}')'$. After omission of $L + K - 1$ of the product variables $\ddot{\boldsymbol{x}} \otimes \ddot{\boldsymbol{y}}$, the model Eq.4.220 can immediately be estimated by standard PLS programs.

4.2.4 One Categorical Predictand

Table 4.11 shows a prediction problem that is just the opposite of the problem in Tb.4.9. Here a polychotomous variable $\dot{y}$ is predicted by a continuous variable z. The classical method for solving this kind of problem is the discriminant analysis: A set of continuous variables $\boldsymbol{z}$ is used to construct discriminant variables $\boldsymbol{Z} = \boldsymbol{W}'\boldsymbol{z}$, or, to conform with the more general PLS model:

$$\boldsymbol{Y} = \boldsymbol{B}\boldsymbol{W}'\boldsymbol{z} + \boldsymbol{u} = \boldsymbol{B}\boldsymbol{Z} + \boldsymbol{u} = \tilde{\boldsymbol{Y}} + \boldsymbol{u}, \qquad \mathrm{E}(\boldsymbol{Y}|\boldsymbol{z}) = \boldsymbol{B}\boldsymbol{W}'\boldsymbol{z}. \tag{4.221}$$

The polychotomous variable is replaced by a set of Boolean variables $\ddot{\boldsymbol{y}}$ and related to the predictor $\boldsymbol{Y}$ by the model

$$\ddot{y}_k = p_{k0} + p_k Y + e_k, \qquad \mathrm{E}(\ddot{y}_k|Y) = p_{k0} + p_k Y, \tag{4.222}$$

where p_{k0} is the location parameter, p_k the loading coefficient and e the residual variable. In the example in Tb.4.2 and Tb.4.11 only one predictor z is used, hence

Table 4.11: One categorical predictand

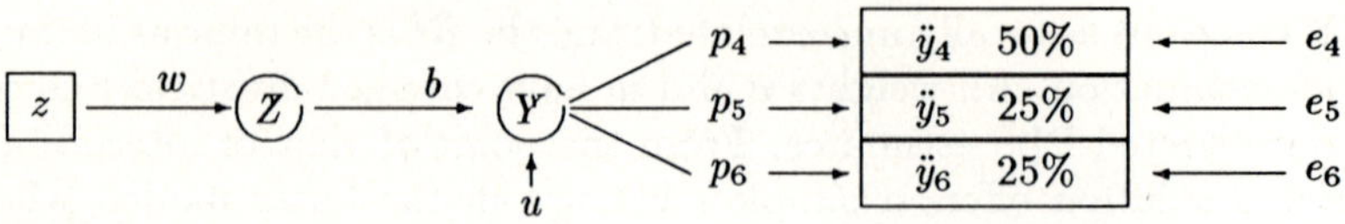

Predictor		True outcome				Estimated outcome			
$\tilde{Y}$	N_i	$\ddot{y}_4$	$\ddot{y}_5$	$\ddot{y}_6$	N_{ik}	$\tilde{y}_4$	$\tilde{y}_5$	$\tilde{y}_6$	$\sum_k$
						.50	.25	.25	$= p_{0k}$
						.35	−.15	−.20	$= p_k$
2	1	1	0	0	1	1.20*	−.05	−.15	1.00
1	6	1	0	0	5	.85*	.10	.05	1.00
		0	0	1	1				
0	6	1	0	0	4	.50*	.25	.25	1.00
		0	1	0	2				
−1	6	0	1	0	3	.15	.40	.45*	1.00
		0	0	1	3				
−2	1	0	0	1	1	−.20	.55	.65*	1.00
					average	.50	.25	.25	1.000
					variance	.1225	.0225	.0400	.1850

The star denotes the estimated category in case of a pick-the-highest-probability strategy.

$w = b = 1$, and $z = Z = \tilde{Y}$. z takes only five different values, denoted by $z_i = Z_i = \tilde{Y}_i$, with frequencies N_i.

Though $\ddot{y}_k$ is a zero-one-valued Boolean variable, the prediction $\tilde{y}_k = \ddot{y}_k - e_k$ is not Boolean and takes as many different values as $\tilde{Y}$ has. The values $\tilde{y}_k$ can be interpreted as the estimated probability of belonging to category k. However, the values of $\tilde{y}_k$ can lie outside the zero-one range and the values in Table 4.11 do not look like probabilities. The estimated scores sum up to unity for each case,

$$\sum_k \tilde{y}_{kn} = 1, \tag{4.223}$$

in just the same way as it holds true for a set of Boolean variables. However, the sum does not add up from one unity and several zeros, but from values like $1.20 - 0.15 - 0.05 = 1.00$. Three different approaches will be considered which assure that the estimated outcome $\tilde{y}_{kn}$ is inside the zero-one range or is even Boolean, the first two using nonlinear outer models, the third requiring distributional assumptions.

Logistic function. In the Rasch model and other similar logistic models the logistic function is applied instead of the linear model Eq.4.222:

$$\tilde{y}_{kn} = \Pr(y_{kn} = 1|Y_n) = \frac{\exp(p_k Y_n)}{1 + \exp(p_k Y_n)}. \tag{4.224}$$

The function on the right-hand side of Eq.4.224 assures that the expression $p_k Y_n$ is mapped into the zero-one range.

Threshold values. Threshold values τ_k are defined and by the decision rule:

$$\tilde{y}_{kn} = \begin{cases} 1 & \text{if } \tau_{k-1} \leq p_k Y_n < \tau_k \\ 0 & \text{otherwise} \end{cases}, \tag{4.225}$$

it is assured that the estimated outcome score is Boolean (see Muthén 1978, 1982, 1984, 1985). Both models Eq.4.224 and Eq.4.225 require difficult analytic treatment, including different treatment for categorical and continuous variables. Furthermore, they allow only for relatively small-sized models, whereas linear models can include more variables.

Classification. The problem of classification, usually treated as part of discriminant analysis, is as follows: Given a K-dimensional space spanned by $\tilde{y}_k$ and the centroids c_k of the classes $k = 1 \ldots K$ in this space, the problem is to allocate the points y_n to these centroids. This requires the knowledge of the distribution of the points around the centroids.

Let us suppose the point $\tilde{y}_n = (0, 0)$ is to be allocated to one of the centroids $c_1 = (1, 1)$ or $c_2 = (-2, -2)$ of a two-dimensional space. Without any knowledge about the distribution one would use the squared Euclidian distances $d^2(y_n, c_1) = (y_{1n} - c_{11})^2 + (y_{2n} - c_{12})^2 = (0 - 1)^2 + (0 - 1)^2 = 2$ and $d^2(y_n, c^2) = 4$ and take the case n as belonging to the class with centroid c_1. Knowing, however, that the variance around c_1 is $s_1^2 = 0.5$ and around c_2 is $s_2^2 = 8$, one finds that the distance of y_n from c_1 is twice the variance s_1^2, the distance from c_2 is half the variance s_2^2, and that one would better do to take y_n as belonging to c_2. As PLS needs no distributional assumptions for the estimation, it would be unfortunate to make them for a cosmetic fix-up of the results.

The prediction of polychotomies by linear models leaves many statisticians unconvinced and unsatisfied. Clearly, here is a choice between "puristic" abiding by theoretical specifications and constructions, and practical use of all empirical information available. Generalized linear models, for example, with features like Eq.4.224 and Eq.4.225 run out of computer time and space for even 10 or 20 variables, i.e. 50 categories. For a questionnaire survey, typically loaded with 100 and 200 questions, one would have to throw away 90% of the information available in a premodeling phase, in order to estimate a little, satisfying model. A linear model can handle a model that includes all variables.

4.2.5 More Variables

The formulae for the special cases treated in the preceding sections Sc.4.2.2 to Sc.4.2.4 are especially simple, because the interval-scaled variables are scaled to zero mean and unit variance and because no causal relationships between the LVs belonging to categorical variables are involved. The statistical properties of these special cases are well explored within the framework of Anova and discriminant analysis. They are least-squares methods and can be seen as special cases of the PLS method of Latent Variables Path (LVP) analysis. In fact, all examples are analysed by the PLS program (Lohmöller 1981a).

As the final example in Sc.4.3 will show, models with more than three LVs contain no difficulties and can be immediately analysed by standard PLS procedures. The only nonstandard step is the preparation of the input matrix. If the product matrix is to be used (program PLSC), then product matrix $\boldsymbol{D}$ or $\boldsymbol{S}$ must be used for the estimation, and the weights and loadings must be transformed back to the metric of $\boldsymbol{M}$ (Eqs.4.222, 4.143), i.e. to the metric of the Boolean variables, the metric of interpretation. (Program PLSCAL, part of PLS 1.8, performs the rescaling of the input product matrix.) If the data matrix is used as input (program PLSX), then the dummy variables must be scaled to the correct metric.

For a Maximum Likelihood approach to causal modelwith discrete variables one may refer to de Leeuw (1984). Israels (1984) develops a Redundancy model for qualitative variables, a ModeA approach in PLS terms.

Table 4.12: LV path diagram, including categorical and interval-scaled manifest variables

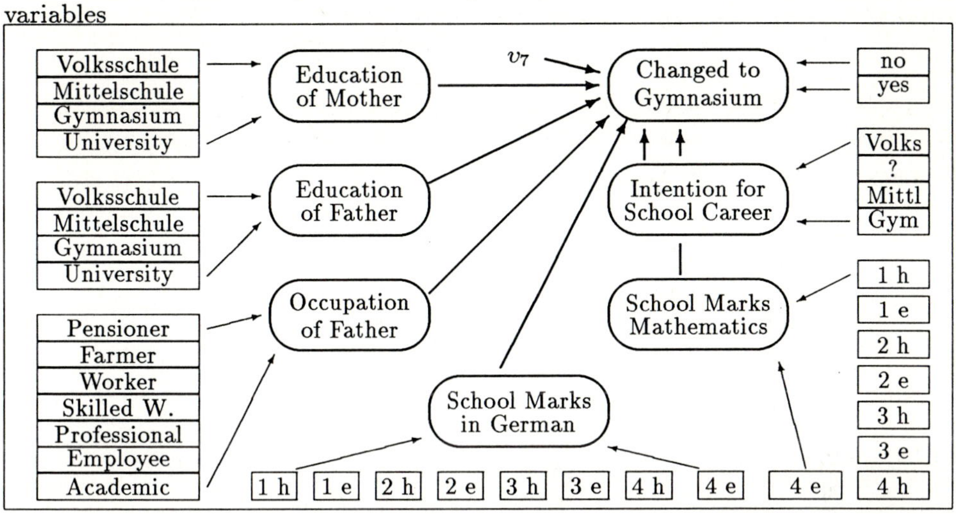

4.3 Application: SES and Educational Aspiration

Table 4.12 shows the diagram of a LV path model, where the manifest variables are partly considered as categorical, partly as interval-scaled. The model attempts to explain the school career decision that leads primary school children either to the Hauptschule or to the Gymnasium.

Data. The data for the model are taken from the Augsburg Longitudinal Study (Hanke, Lohmöller & Mandl 1980) on elementary school students. The sample contains all Augsburg children entering school in 1967. Due to home moving, 10% of the sample fluctuated in each year. From a total of 3500 children, only 1500 stayed the four years of primary school in the sample. This "longitudinal sample" is used here in order to avoid missing-data problems in parameter estimation.

Variables. The model starts with three categorical indicators for the social background of the child. Education of mother and education of father are tetrachotomous, and it would be no problem to assume even an ordinal measurement level. However, some problems are encountered with occupation of father. For example, farmers rank high if a socioeconomic index is to be constructed, and low, if the occupation variable is rescaled in order to constitute a socio-cultural scale.

As mediating variable the scholastic achievement of the student is introduced. Two LVs, achievement in German and in mathematics, are indicated by eight half-yearly marks in those subjects. As demonstrated in Tb.4.7, the marks may be assumed to be interval-scaled.

The variables to be predicted by the model (target variables) are the educa-

Table 4.13: School career decision model: Outer model parameters

m_k	$w_{k,D}$	$w_{k,(M)}$	$p_{k,(M)}$	Latent variable Manifest variable
				education of mother
76%	−.47	−.54	−.41	Volksschule (8 years)
20%	.67	1.49	.30	Mittelschule (10 Years)
2%	.37	2.43	.06	Gymnasium (13 years)
2%	.43	3.47	.05	university
				education of father
73%	−.48	−.56	−.41	Volksschule
15%	.40	1.04	.15	Mittelschule
4%	.24	1.26	.05	Gymnasium
8%	.74	2.62	.21	university
				occupation of father
.4%	−.15	−2.50	−.01	pensioner
.2%	−.08	−1.89	−.00	farmer
13%	−.42	−1.15	−.15	unskilled worker
37%	−.47	−.78	−.28	skilled worker, craftsman
6%	.11	.46	.03	professional
37%	.43	.71	.26	employee
7%	.62	2.32	.17	academic professions
				half-year school marks in German
		.13	.75	1st year, 1st half
		.13	.75	1st year, 2nd half
		.15	.85	2nd year, 1st half
		.14	.79	2nd year, 2nd half
		.16	.85	3rd year, 1st half
		.16	.81	3rd year, 2nd half
		.17	.86	4th year, 1st half
		.17	.81	4th year, 2nd half
				half-year school marks in mathematics
		.15	.87	1st year, 1st half
		.14	.82	1st year, 2nd half
		.15	.89	2nd year, 1st half
		.14	.84	2nd year, 2nd half
		.15	.89	3rd year, 1st half
		.14	.85	3rd year, 2nd half
		.15	.88	4th year, 1st half
		.15	.85	4th year, 2nd half
				educational aspiration of parents
38%	−.76	−1.23	−.47	no higher education, Volksschule
7%	−.04	−.15	−.01	(undecided)
14%	.26	.70	.10	Fach-, Real-, Mittelschule
40%	.60	2.32	.17	Gymnasium
				student's transition to Gymnasium
66%	−.58	−.71	−.47	no
34%	.81	1.40	.47	yes

Note: $w_{k,(M)} = w_{k,D}/\sqrt{m_k}$; $p_{k,(M)} = w_{k,D} \cdot \sqrt{m_k}$

Note: $\sum_k w_{k,(M)} \cdot m_k = 0$; $\sum_k w_{k,(M)} \cdot p_{k,(M)} = w^2_{k,(M)} \cdot m_k = 1$

tional aspiration of the parents, explored by a questionnaire, and whether the student entered the Gymnasium at the end of the fourth year or remained at the Volksschule (Hauptschule). These two variables, again, are categorical.

Outer model. Table 4.13 shows the coefficients for the outer part of the model. The first column (m_k) shows the means, i.e. the univariate frequencies of the categorical variables. The second column ($w_{k,D} = p_{k,D}$) gives the weights and the loadings in estimation metric, estimated from the distance product matrix $\boldsymbol{D}$ (cf. Sc.4.1.5). The third and fourth columns give the weights $w_{k,M}$ and the loadings $p_{k,M}$ relating the standardized LV to the zero-one scaled Boolean variables, i.e. in interpretation metric.

The scale values $w_{k,M}$ for all categories form a rank order with high values for good education, good occupation, high educational aspiration of the parents and for the transition to Gymnasium. Unexpectedly low ranks are found for pensioners ($N = 5$) and farmers ($N = 2$) and unexpectedly high ones for mothers with Gymnasium ($N = 34$) and university ($N = 21$) education. Due to the small frequencies of these categories, these values seem not to be reliable.

All loading coefficients of the school marks are very high and confirm the assumption of an underlying unidimensional latent variable. Three observations:

(i) The marks in German show increasing loadings over time, i.e. are indicative of increasing reliability of the marks.
(ii) The loadings of marks in mathematics are high from the beginning and higher throughout than in German.
(iii) The loadings of the marks at the end of the year are lower than in the middle of the year, possibly because they are influenced by other considerations than only the achievement of the student.

Notice that the LVs Achievement in German and Achievement in Mathematics are not determined as common factors of the marks, but to be good predictors for educational aspiration and transition to Gymnasium.

Inner model. Table 4.14 shows the relationships between the LVs. The path matrix, in general a full square matrix, is restricted by several technical and substantive assumptions:

(i) The diagonal is zero because there is no sense predicting a variable by itself.
(ii) The above-diagonal elements are zero, i.e. it is a recursive, loop-free model.
(iii) The coefficients between the first three LVs are fixed to zero, i.e. these variables are treated as exogenous and no attempt is made to explain the variation of these variables by the model.
(iv) Four more coefficients are fixed to zero because they turned out to be very close to zero.

When these restrictions have been imposed, no path coefficient is less than 0.05 and no residual covariance between endogenous variables is higher than 0.05.

Both LVs of scholastic achievement are related in the same way to the three social-background variables (correlation r ranges from 0.23 to 0.29) but the relationships to the two target variables are different. Achievement in mathematics

Table 4.14: School career decision model: Inner model parameters

LV-Correlation R

education of mother	100						
education of father	51	100					
occupation of father	40	61	100				
achievement, German	23	24	29	100			
achievement, mathematics	25	25	29	84	100		
educational aspiration	23	26	29	46	51	100	
student's transition to Gymnasium	30	32	33	58	65	52	100

Path coefficients B

	R^2							
education of mother	.	.	.	.	.	.	.	.
education of father	.	.	.	.	.	.	.	.
occupation of father	.	.	.	.	.	.	.	.
achievement, German	10	13	.	24	.	.	.	.
achievement, mathematics	71	6	.	.	83	.	.	.
educational aspiration	28	.	9	10	9	38	.	.
student's transition to Gymn.	50	8	7	5	5	45	23	.

Residual covariance Q

education of mother	100						
education of father	51	100					
occupation of father	40	61	100				
achievement, German	0	2	0	90			
achievement, mathematics	0	2	2	1	29		
educational aspiration	3	0	0	0	0	72	
student's transition to Gymnasium	0	0	0	0	0	0	50

All coefficients are multiplied by 100.

Note: $\boldsymbol{Q} = (\mathbf{I} - \boldsymbol{B})\boldsymbol{R}(\mathbf{I} - \boldsymbol{B})'$, $R^2 = \mathrm{Diag}(\mathbf{I} - \boldsymbol{Q})$ = squared multiple correlation

has higher influence on aspiration ($b = 0.38, r = 0.51$) and on the transition to Gymnasium ($b = 0.45, r = 0.65$) than achievement in German, which has only little additional influence on aspiration ($b = 0.09, r = 0.46$) and on the transition to Gymnasium ($b = 0.05, r = 0.58$). This again proves the higher reliability of the marks in mathematics.

The educational and occupational status of the parents influences not only the achievement of their child but has an additional, direct influence on the educational aspiration and the actual school career. If the school system (in 1971) really had provided equal opportunities to all gifted children, then no additional influence of parents on school career would have remained after the (reliably measured) scholastic achievement had been taken into account.

4.4 Different Slopes in Different Groups

When comparing different groups as described by a categorical variable $\dot{x}$, one may be interested not only in the location differences with regard to the predictand variables z, $\mathrm{E}(z|\dot{x})$, but also with regard to the specific influence of a moderator variable y on z, $\mathrm{E}(z|y,\dot{x})$ (cf. Jöreskog 1971, Lee & Tsui 1982). The obvious technique would be to do separate regression analyses for each group, followed by a comparison of the parameters. In LISREL it is possible to constrain some parameters to be equal across groups (cf. Sörbom 1974, 1978, 1982). This is also possible for PLS estimation but has not yet been implemented in a program.

Here, another approach is pursued that allows a generalization toward LV models. Products of interval- and categorical-scaled variables are introduced as predictands which permit, after some rearrangement, an interpretation as LV models. Sc.4.4.1 gives a preparatory introduction to MV models with product variables. Sc.4.4.2 then turns to LV models. These two sections are based on Lohmöller and Wold (1984).

4.4.1 MV Path Models with Product Variables

We consider the relationships between three variables $\dot{x}, y, z$ where $\dot{x}$ is dichotomous and replaced by two Boolean variables, $\ddot{x}_1$ and $\ddot{x}_2$, and where y and z are assumed to be interval-scaled. The regression model for the total sample is:

$$z = au + by + e, \qquad (4.401)$$

where u is a unit vector accounting for the location parameter as a "non-varying" variable. Notice that $u = \ddot{x}_1 + \ddot{x}_2 = 1$.

The regression model which considers the group differences can be written as a system of two equations, one for each group,

$$\begin{aligned} z &= a_1 u + b_1 y + e_1 && \text{for group } \dot{x} = 1 \\ z &= a_2 u + b_2 y + e_2 && \text{for group } \dot{x} = 2, \end{aligned} \qquad (4.402)$$

or as one equation,

$$z = (a_1 u + b_1 y)\ddot{x}_1 + (a_2 u + b_2 y)\ddot{x}_2 + e_3, \qquad (4.403)$$

where the first term of the sum is operative when $\ddot{x}_1 = 1$, i.e. in the first group, and where the second term is operative only in the second group. Eq.4.403 can be estimated by multiple regression after it is reformulated into

$$z = a_1\ddot{x}_1 + b_1(y\ddot{x}_1) + a_2\ddot{x}_2 + b_2(y\ddot{x}_2) + e_3, \qquad (4.404)$$

where products of a Boolean and an interval-scaled variable are used as predictors. Eq.4.404 may be reordered to attain:

$$z = (a_1\ddot{x}_1 + a_2\ddot{x}_2)u + (b_1\ddot{x}_1 + b_2\ddot{x}_2)y + e_3, \qquad (4.405)$$

Table 4.15: Data, product matrix, and scatter diagram for six cases

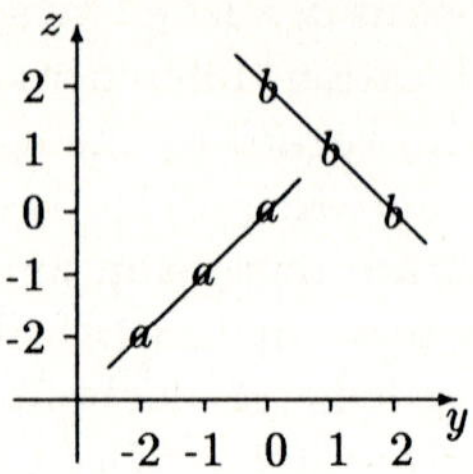

Group	Case	u	y	x_1	y_1	x_2	y_2	z
a	1	1	-2	1	-2	.	.	-2
a	2	1	-1	1	-1	.	.	-1
a	3	1	0	1	0	.	.	0
b	4	1	0	.	.	1	0	2
b	5	1	1	.	.	1	1	1
b	6	1	2	.	.	1	2	0
Sums:		6	0	3	-3	3	3	0
Product sums:								
u		6	†					
y		0	10					
x_1		3	-3	3	‡			
y_1		-3	5	-3	5			
x_2		3	3	0	0	3	*	
y_2		3	5	0	0	3	5	
z		0	6	3	-3	3	3	0

† Product of total sample. ‡ Products in first group. * Products in second group

where, in contrast to Eq.4.401, the constant parameters a and b of Eq.4.401 are replaced by the varying expressions in parentheses, which can be understood as regressions themselves, with $\ddot{x}_1$ and $\ddot{x}_2$ as predictors. The first term in the first parenthesis of Eq.4.405 tells us that the intercept a_1 is applied in the first group and a_2 in the second. Similarly, the slope b_1 is operative in the first group and the slope b_2 in the second. To make the interpretation of "regression coefficient regressed on grouping variables" more explicit, we rewrite Eq.4.405 as

$$z = (a + \acute{a}_1\ddot{x}_1 + \acute{a}_2\ddot{x}_2)u + (b + \acute{b}_1\ddot{x}_1 + \acute{b}_2\ddot{x}_2)y + e_3. \qquad (4.406)$$

This equation has 6 parameters; the estimable form of this equation, though, is Eq.4.404, which has only 4 predictor variables; thus Eq.4.406 is not estimable without further restrictions. The first type of restriction assumes that the parameters

a and b are those of Eq.4.401 and that the new parameters $\acute{a}_1, \acute{a}_2, \acute{b}_1, \acute{b}_2$ are the differences, e.g. $\acute{a}_1 = a_1 - a$. Thus, it follows that $\mathrm{E}(\acute{a}_1\ddot{x}_1 + \acute{a}_2\ddot{x}_2) = \mathrm{E}(\acute{b}_1\ddot{x}_1 + \acute{b}_2\ddot{x}_2) = 0$, that only four out of the six parameters of Eq.4.406 are independent, and that Eq.4.406 is identified.

The second type of restriction deletes redundant variables, and one may have the model

$$z = au + by + (\grave{a}_2 + \grave{b}_2 y)\ddot{x}_2 + e_3, \tag{4.407}$$

where the first group is used as a reference group and the deviation of the second group's parameters from the reference group's parameters are included as special parameters $\grave{a}$ and $\grave{b}$.

Table 4.15 contains a small data set with six observations on x, y, z and the unit vector u. The product variables $y_1 = y\ddot{x}_1$ and $y_2 = y\ddot{x}_2$ are included, as well as the product matrix $\boldsymbol{M}$ of all variables. Notice that the products of variables for different groups are zero, indicating that the groups are non-overlapping and independent. If we denote the moment matrix of predictors as $\boldsymbol{M}_{pp}$, of predictors with the criterion as $\boldsymbol{M}_{cp}$ and the vector of regression parameters as $\boldsymbol{B}_{c|p}$, the parameter for model Eq.4.401 are $a = 0$ and $b = 1$, giving a squared multiple correlation $R^2 = 0.60$.

$$\begin{aligned}
\boldsymbol{M}_{pp} &= \begin{bmatrix} 6 & 0 \\ 0 & 10 \end{bmatrix} & \boldsymbol{M}_{pp}^{-1} &= \tfrac{1}{60}\begin{bmatrix} 10 & 0 \\ 0 & 6 \end{bmatrix} \\
\boldsymbol{M}_{cp} &= \begin{bmatrix} 0 & 6 \end{bmatrix} \\
\boldsymbol{B}_{c|p} &= \begin{bmatrix} a & b \end{bmatrix} \\
&= \begin{bmatrix} 0 & 1 \end{bmatrix} & R^2 &= 0.60
\end{aligned} \tag{4.408}$$

The parameters for the first group in Eq.4.402 are

$$\begin{aligned}
\boldsymbol{M}_{pp} &= \begin{bmatrix} 3 & -3 \\ -3 & 5 \end{bmatrix} & \boldsymbol{M}_{pp}^{-1} &= \tfrac{1}{6}\begin{bmatrix} 5 & 3 \\ 3 & 3 \end{bmatrix} \\
\boldsymbol{M}_{cp} &= \begin{bmatrix} -3 & 5 \end{bmatrix} \\
\boldsymbol{B}_{c|p} &= \begin{bmatrix} a_1 & b_1 \end{bmatrix} \\
&= \begin{bmatrix} 0 & 1 \end{bmatrix} & R^2 &= 1.0
\end{aligned} \tag{4.409}$$

and for the second group

$$\begin{aligned}
\boldsymbol{M}_{pp} &= \begin{bmatrix} 3 & 3 \\ 3 & 5 \end{bmatrix} & \boldsymbol{M}_{pp}^{-1} &= \tfrac{1}{6}\begin{bmatrix} 5 & -3 \\ -3 & 3 \end{bmatrix} \\
\boldsymbol{M}_{cp} &= \begin{bmatrix} 3 & 1 \end{bmatrix} \\
\boldsymbol{B}_{c|p} &= \begin{bmatrix} a_2 & b_2 \end{bmatrix} \\
&= \begin{bmatrix} 2 & -1 \end{bmatrix} & R^2 &= 1.0.
\end{aligned} \tag{4.410}$$

The slope in the first group is positive, $b_1 = 1$, in the second group negative, $b_1 = -1$. This effect is covert in model Eq.4.401, with $b = 1$.

The regression for model Eq.4.404 gives

$$\begin{aligned}
\boldsymbol{M}_{pp} &= \begin{bmatrix} 3 & -3 & 0 & 0 \\ -3 & 5 & 0 & 0 \\ 0 & 0 & 3 & 3 \\ 2 & 0 & 3 & 5 \end{bmatrix} \qquad \boldsymbol{M}_{pp}^{-1} = \tfrac{1}{36}\begin{bmatrix} 5 & 3 & 0 & 0 \\ 3 & 3 & 0 & 0 \\ 0 & 0 & 5 & -3 \\ 0 & 0 & -3 & 3 \end{bmatrix} \\
\boldsymbol{M}_{cp} &= \begin{bmatrix} -3 & 5 & 3 & 1 \end{bmatrix} \\
\boldsymbol{B}_{c|p} &= \begin{bmatrix} a_1 & a_2 & b_1 & b_2 \end{bmatrix} \\
&= \begin{bmatrix} 0 & 1 & 2 & -1 \end{bmatrix} \qquad R^2 = 1.0.
\end{aligned} \tag{4.411}$$

The results of the simultaneous estimation in Eq.4.411 are identical to the separate estimation in Eqs.4.409, 4.410; the squared multiple regression is $R^2 = 1$. Notice that $\boldsymbol{M}_{pp}$ (Eq.4.411) would be singular if the means had been subtracted, i.e. if the covariances had been analysed.

The regression for model Eq.4.407 gives the same results for the total sample, $a = 0$ and $b = 1$, which at the same time are the parameters for the first group:

$$\begin{aligned}
\boldsymbol{M}_{pp} &= \begin{bmatrix} 6 & 0 & 3 & 3 \\ 0 & 10 & 3 & 5 \\ 3 & 3 & 3 & 3 \\ 3 & 5 & 3 & 5 \end{bmatrix} \qquad \boldsymbol{M}_{pp}^{-1} = \begin{bmatrix} 5 & 3 & -5 & -3 \\ 3 & 3 & -3 & -3 \\ -5 & -3 & 10 & 0 \\ -3 & -3 & 0 & 6 \end{bmatrix} \\
\boldsymbol{M}_{cp} &= \begin{bmatrix} 0 & 6 & 3 & 1 \end{bmatrix} \\
\boldsymbol{B}_{c|p} &= \begin{bmatrix} a & b & \grave{a}_2 & \grave{b}_2 \end{bmatrix} \\
&= \begin{bmatrix} 0 & 1 & 2 & -2 \end{bmatrix} \qquad R^2 = 1.0.
\end{aligned} \tag{4.412}$$

The parameters for the second group are $a_2 = a + \grave{a}_2 = 0 + 2 = 2$ and $b_2 = b + \grave{b}_2 = 1 - 2 = -1$; they are identical to the results in Eq.4.411, even if they do not appear so at first glance. They can easily be transformed into each other.

Two truly restricted models may now be considered. Model Eq.4.413 assumes equal intercepts for both groups but with different slopes

$$\begin{aligned}
z &= au + b_1\ddot{x}_1 y + b_2\ddot{x}_2 y + e_4 \\
&= 1u + 1.6\ddot{x}_1 y - 1.6\ddot{x}_2 y + e_4, \qquad R^2 = 0.64
\end{aligned} \tag{4.413}$$

and model Eq.4.414 assumes equal slopes but allows for different intercepts:

$$\begin{aligned}
z &= a_1\ddot{x}_1 + a_2\ddot{x}_2 + by + e_5 \\
&= -1\ddot{x}_1 + 1\ddot{x}_2 + 0y + e_5, \qquad R^2 = 0.80.
\end{aligned} \tag{4.414}$$

The explanatory power, measured by $r^{2\prime} = 1 - \mathrm{SSQ}(e_5)/\,\mathrm{SSQ}(z)$ where $\mathrm{SSQ}(\cdot)$ is the sum of squares, is $r^{2\prime} = 0.64$ for Eq.4.413 and $r^{2\prime} = 0.80$ for Eq.4.414 — clearly less than the $r^{2\prime} = 1.0$ for the unrestricted models Eqs.4.402,..., 4.407 but more than the more restricted model Eq.4.401 with $r^{2\prime} = 0.60$.

4.4.2 LV Path Models with Product Variables

The step from Eq.4.404 to Eq.4.405 was made in order to facilitate the interpretation of group-specific regression coefficients. In this section, a step in another direction will be taken, Eq.4.404 being developed in order to allow for an interpretation in terms of LVs. The model is:

$$\begin{aligned} z &= c(w_1\ddot{x}_1 + w_2\ddot{x}_2) + d(w_3\ddot{x}_1 y + w_4\ddot{x}_2 y) + e_3 \\ &= cY_1 + dY_2 + e_3. \end{aligned} \tag{4.415}$$

The terms in parentheses in Eq.4.415 are weighted aggregates of the MVs and are written as LVs in Eq.4.415. The first LV, Y_1, accounts for the group-specific location differences in z and the second LV, Y_1, accounts for the group-specific influences of y on z.

Eq.4.415 has 6 parameters but there are only 4 predictor variables. Thus, the model is underidentified. The usual restriction var$(Y_1) = 1$ and var$(Y_2) = 1$ may not be justified because $\ddot{x}_1$ and $\ddot{x}_2$ are Boolean with means different from zero. However, the restrictions $\mathrm{E}(Y_1^2) = 1$ and $\mathrm{E}(Y_2^2) = 1$ lead to identifiability as well. A shorthand parameter estimation for Eq.4.415 replaces the restriction for scale unambiguity (RSU) by the assumption $\breve{c} = \breve{d} = 1$ and uses the estimates for Eq.4.404,

$$z = 1(0\ddot{x}_1 + 2\ddot{x}_2) + 1(1\ddot{x}_1 y - 1\ddot{x}_2 y) + e_3. \tag{4.416}$$

The matrix of sums of squares and cross product is then:

$$\mathrm{SSP}\begin{bmatrix} \breve{Y}_1 \\ \breve{Y}_2 \\ z \end{bmatrix} = \begin{bmatrix} 12 & & \\ -6 & 10 & \\ 6 & 4 & 10 \end{bmatrix}, \tag{4.417}$$

which is used to standardize the solution for Eq.4.416 by $c = \sqrt{12}$, $w_2 = 2/\sqrt{12}$, $d = \sqrt{10}$, $w_3 = 1/\sqrt{10}$, $w_4 = -1/\sqrt{10}$.

Notice that in Eq.4.417 the two predictor LVs Y_1 and Y_2 are highly correlated. The same holds true for the MV predictors,

$$\mathrm{SSP}\begin{bmatrix} \ddot{x}_1 \\ \ddot{x}_2 \\ \ddot{x}_1 y \\ \ddot{x}_2 y \\ z \end{bmatrix} = \begin{bmatrix} 3 & & & & \\ 0 & 3 & & & \\ -3 & 0 & 5 & & \\ 0 & 3 & 0 & 5 & \\ -3 & 3 & 5 & 1 & 10 \end{bmatrix}, \tag{4.418}$$

where the association between the two blocks is stronger than within blocks.

Eq.4.403 is the basis for another LV path model:

$$\begin{aligned} z &= f(w_5\ddot{x}_1 + w_6\ddot{x}_1 y) + g(w_7\ddot{x}_2 + w_8\ddot{x}_2 y) + e_4 \\ &= fY_3 + gY_4 + e_4. \end{aligned} \tag{4.419}$$

The substantive interpretation of this model is not so elegant and straightforward: Y_3 is measured and operative only in the first group, and f indicates the influence in this group; Y_4 has no meaning in the first group. The predictor LVs establish no joint framework for the prediction of the outcome variables z. The reordering of the SSP matrix Eq.4.419 according to the structure of this model produces:

$$\mathrm{SSP}\begin{bmatrix}\ddot{x}_1\\ \ddot{x}_1 y\\ \ddot{x}_2\\ \ddot{x}_2 y\\ z\end{bmatrix}=\begin{bmatrix}3 & & & & \\ -3 & 5 & & & \\ 0 & 0 & 3 & & \\ 0 & 0 & 3 & 5 & \\ -3 & 5 & 3 & 1 & 10\end{bmatrix}, \tag{4.420}$$

where the LVs Y_3 and Y_4 are computed using $f = g = 1$ and the coefficients in Eq.4.411 as weights. The association between the blocks of MV predictors is zero, and consequently between the LV predictors also.

Drawing from Sc.3.4 it is to be expected that a PLS parameter estimation for a model like Eq.4.420 with two predictor blocks gives the same R^2 as a multiple regression with one predictor block, because it is here that this very special case is encountered that the association between blocks is zero (cf. Sc.2.4).

For model Eq.4.415 may be expected that the results are suboptimal compared with Eq.4.420, because the association between the blocks are far stronger than within the blocks. This is confirmed by an PLS analysis which gives the estimates:

$$\begin{aligned} z = &.529987(-.408248\ddot{x}_1 + .408248\ddot{x}_2) + \\ &+.379532(.438529\ddot{x}_1 y + .087706\ddot{x}_2 y) + e_4, \end{aligned} \tag{4.421}$$

with $r^{2\prime} = 1 - \mathrm{SSQ}(e_4)/\,\mathrm{SSQ}(z) = 0.684211$. If the Pattor rotation of the extended method (Sc.2.3.4) is used and the two LVs are forced to be uncorrelated, the estimates for solution #1 (cf. Sc.2.3.6) are:

$$\begin{aligned} z = &.834353(-.149718\ddot{x}_1 + .557600\ddot{x}_2) + \\ &+.713805(.378381\ddot{x}_1 y - .238385\ddot{x}_2 y) + e_4, \end{aligned} \tag{4.422}$$

with $r^{2\prime} = .933110$, which is clearly a higher explanatory power than in Eq.4.421 but, nevertheless, lower than in Eq.4.404 or Eq.4.419. If we take solution #3, the explained variance $r^{2\prime} = 0.995789$ approaches unity but the weights no longer form a block-diagonal matrix.

The difference $R^2_{\mathrm{Eq.4.419}} - R^2_{\mathrm{Eq.4.421}} = 1.0 - 0.684211 = 0.315789$ or $R^2_{\mathrm{Eq.4.419}} - R^2_{\mathrm{Eq.4.422}} = 0.066890$ seems to be the price that has to be paid for the interpretability of the LVs. If one wants to avoid this loss of predictive power, one could use a model like Eq.4.419 for the estimation and then reorganize and rescale the model to the form of Eq.4.415.

4.4.3 Metric of Product Variables

In least squares regression analysis first order information (means, location parameters) and second order information (variances, correlation, covariances, slopes) are usually separated in advance, the analysis being done separately. In some models, however, the estimation method can separate the first order from the second order information — for example in multiple regression analysis, when a constant "variable" is included in the set of predictors, and in categorical scaling (cf. Sc.4.1) where the first LV turns out to be a constant "variable", and the loadings of the first LV are identical to (or a function of) the means.

The problem of the metric — more specifically the problem of the scale origin — in second order models is whether the means have to be removed from the second order moment matrix, or whether the estimation method can separate the first and second order information. Computer programs for covariance and correlation computation usually first compute the mean vector $\boldsymbol{m} = \frac{1}{N}\boldsymbol{x}\mathbf{1}$ and the relative raw moment matrix $\boldsymbol{M} = \frac{1}{N}\boldsymbol{x}\boldsymbol{x}'$ and then correct for the means to get the covariance $\boldsymbol{S} = \boldsymbol{M} - \boldsymbol{m}\boldsymbol{m}'$.

The correction made afterwards for the means is impossible when product variables are included. The second order moment, then, includes product sums of three variables, i.e. third order moments. Take a model like:

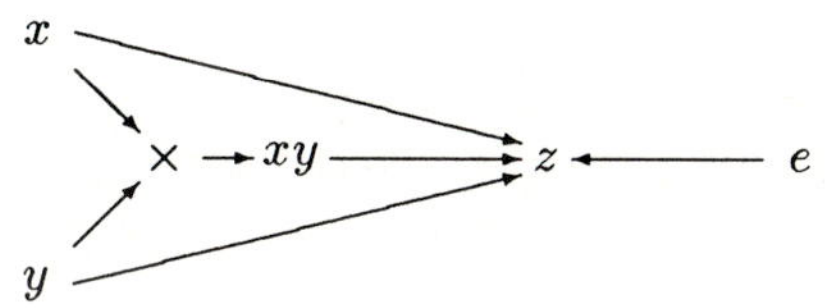

(4.423)

where x is a variable of any type, y a interval-scaled variable, and xy the product variable. The product matrix has the elements:

$$\boldsymbol{M} = \mathrm{E}\begin{bmatrix} xx & & & \\ xyx & xyxy & & \\ yx & yxy & yy & \\ zx & zxy & zy & zz \end{bmatrix} \tag{4.424}$$

Let us take as an example the element $\mathrm{E}(xyz)$. The means of the variables being denoted by $\bar{x}, \bar{y}, \bar{z}$, the expectation of the triple product of the centered variables is:

$$\begin{aligned} s_{xyz} &= \mathrm{E}((x-\bar{x})(y-\bar{y})(z-\bar{z})) \\ &= \mathrm{E}(xyz - \bar{x}yz - x\bar{y}z - xy\bar{z} + \bar{x}\bar{y}z + \bar{x}y\bar{z} + x\bar{y}\bar{z} - \bar{x}\bar{y}\bar{z}) \\ &= \mathrm{E}(xyz) + \bar{x}\bar{y}\bar{z} - \bar{z}(\mathrm{E}(xy) - \bar{x}\bar{y}) - \bar{y}(\mathrm{E}(xz) - \bar{x}\bar{z}) - \bar{x}(\mathrm{E}(yz) - \bar{y}\bar{z}) \\ &= m_{xyz} + \bar{x}\bar{y}\bar{z} - \bar{z}s_{xy} - \bar{y}s_{xz} - \bar{x}s_{yz} \end{aligned} \tag{4.425}$$

being generally not identical to the covariance of the product variable xy with z,

$$s_{(xy)z} = \mathrm{E}(xyz) - \mathrm{E}(xy)\,\mathrm{E}(z), \tag{4.426}$$

or something like:

$$\breve{s} = \mathrm{E}(xyz) - \bar{x}\bar{y}\bar{z}. \tag{4.427}$$

Eq.4.426 and Eq.4.427 are identical when $\mathrm{E}(z) = 0$. Eqs.4.425,..., 4.427 are identical when all means are zero, or when all pairwise covariances are zero, or when an appropriate three-element sample of means and covariances cancel out each other (see last row of Eq.4.425).

Table 4.16: Squared multiple correlations for model Eq.4.4xx and data Tb.4.16, with different scale means and estimation methods.

(a) Product matrix and PLS Mode	(b) DESIGN r_{12}	(c) Raw data means of y and z		
		$y = 0$, $z = 0$	$y = 1$, $z = 0$	$= 1$, $= 1$
$\boldsymbol{D}$, ModeA	orthogonal	1.0000	1.0000	1.0000
	correlated	.7231	.7231	.7231
$\boldsymbol{M}$, ModeB	orthogonal	1.0000	.9331	.8736
	correlated	.7231	.6842	.7516
$\boldsymbol{M}$,ModeA	orthogonal	1.0000	.9331	.9721
	correlated	.8941	.6842	.8683
$\boldsymbol{M}$, multiple regression		1.0000	1.0000	1.0000

In product matrix $\boldsymbol{D}$, the means of $\boldsymbol{x}$, xy and z are removed.
In product matrix $\boldsymbol{M}$, means are as given by column heads.

Thus, there is a difference as to whether the variables are centered to zero mean before the computations of the expectations (or product sums), or corrected for the means afterwards. As may be seen from Eq.4.427, all pairwise relations are removed from the triple product when the means are removed. This guarantees that the simple effect and the interaction effect in model Eq.4.423 are well separated.

Let us suppose we introduce LVs for x and xy in model Eq.4.423. Using split multiple regression estimation, the effect of x on z would be correctly assigned to the LV of the first LV; the LV of xy would account for the different slopes. Using PLS estimation, this would not be guaranteed. It could happen that the iteration oscillates between assigning the effect of x on z either to the first or to the second LV (because the indicators of both LVs contain the related effect), or that the iteration stops at a suboptimal point. The orthogonalization procedure (Sc.2.3.4) is sometimes but not always a remedy.

This is demonstrated in Table 4.16, using the data on six cases from Tb.4.15. If we introduce two LVs, the model is Eq.4.415. We already know that the true squared multiple correlation is $R^2 = 1$. Notice that y and z have zero mean, so two of the conditions for Eq.4.425 and Eq.4.427 to be identical are met. The scaling experiment with the six-case data varies three conditions:

(a) x is centered or not; PLS estimation with ModeA or ModeB predictor blocks
- **(a1)** x is centered: matrix $\boldsymbol{D}$; only ModeA possible
- **(a2)** x is not centered: matrix $\boldsymbol{M}$; ModeB
- **(a3)** x is not centered: matrix $\boldsymbol{M}$; ModeA

(b) the two predictor LVs are
- **(b1)** forced to orthogonality (Pattor rotation) or
- **(b2)** left correlated.

(c) the product variable xy is
- **(c1)** centered to zero mean,

(c2) original mean,

(c3) all variables are de-centered, $y := y + 1, z := z + 1$.

The entries in Tb.4.16 are the squared multiple correlations R^2 and show marked differences. First of all, the orthogonalization seems to work in most cases where the correlated LVs fail to reach $R^2 = 1$. However, this is a consequence of the very peculiar data set, the like not being observed in other data sets, the 20-case data set from Tb.4.2, for instance.

Condition (a1) together with the orthogonalization (b1) always reaches the $R^2 = 1$. Here the means $\bar{y}$ and $\bar{z}$ are zero from the beginning, the means $\bar{x}$ and $\bar{xy}$ being removed afterwards.

Condition (a2) together with (c1) gives the same result as (a1). Results would look worse had $\bar{y}$ not been equal to zero. In this case, the estimation runs into oscillations.

Condition (c3) with mean of xy different from zero gives the worst results, as is to be expected from the discussion of Eq.4.427.

Conclusion. It is recommended that centred data be used, i.e. matrix $\boldsymbol{D}$ or $\boldsymbol{S}$ from Tb.4.4. The super-contingency table $\boldsymbol{M}$ may be helpful when inspecting the data but it is not the right table to use for the PLS analysis. If groups have different sizes, the choice of the scale unit (the standard deviation), and thus that of $\boldsymbol{D}$ and $\boldsymbol{S}$, becomes crucial.

4.5 Application: TV Consumption and Fear of Crime

The methods developed in the previous sections will be subsequently applied to a substantive area where the basic hypotheses are not bivariate but trivariate, i.e. include an interaction term. Gerbner's (1984) basic hypothesis is *not* that the more ($y =$) people watch TV, the more ($z =$) they fear becoming victims of crime, $z = f(y)$. His hypothesis is rather that the effect of TV watching is different for ($x =$) men and women, for upper and lower classes, for young people and older: $z = f(x, y, xy)$. Gross (1984) has elaborated on this and distinguishes two types of interaction which he calls the correspondence between TV message and TV reaction: if TV shows lower class people as victims of crime, lower class people will show more fear of crime the more they watch TV.

The data to be used for the example in this and the following chapter are taken from Violence Profile No.11 (Gerbner, Gross, Morgan & Signorielli 1980, 1982; Morgan 1984).[2] The data were collected in March 1979 through personal interviews; the sampling design was (U.S.) national, full probability, stratified by geographic area and clustered at the neighborhood level.

Table 4.17 shows the LV path model on TV consumption and Fear of Crime (taken from Lohmöller and Wold 1984). The target variable of the model is

[2] I am indebted to Michael Morgan for providing the data in Tb.4.18 and for the permission to use it.

Table 4.17: TV Consumption and Fear of Crime: path diagram

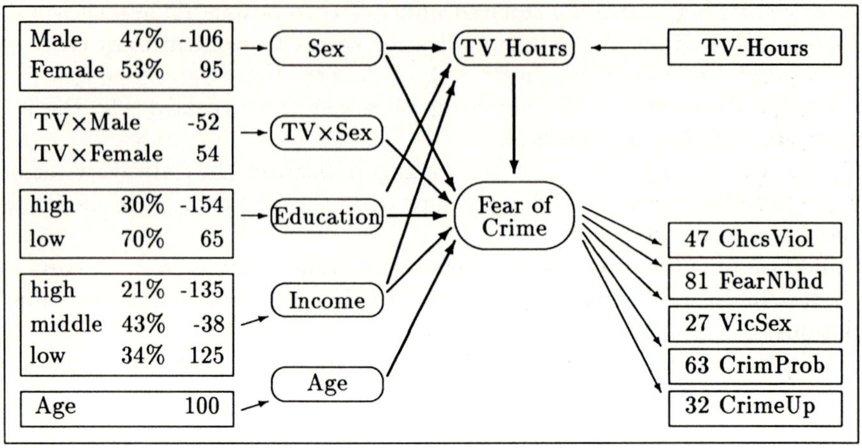

the Fear of Crime, modeled as a latent variable and measured by 5 indicators. Four background variables are introduced: the dichotomous variables Sex and Education, the trichotomous variable Income, and the interval-scaled Age in years ($m = 43.13, s = 18.03$). The moderator variable is TV watching, (hours spent daily, scaled to zero mean and unit variance). As a second moderator variable, the product of the Boolean variable Sex and of TV is introduced. This variable is designed to account for different influences of TV Watching on Fear, i.e. different for males and females.

The product data, presented in Table 4.18, are defined to be sum of products, $s_{xy} = \bar{\sum}_n x_n y_n$. Depending on the two variables x and y involved, the symbol s_{xy} denotes a bivariate frequency, a correlation coefficient, or a conditional mean.

(i) If both x and y are Boolean variables, then s_{xy} denotes a bivariate relative frequency and is an entry from a contingency table. Thus, the product data for the three categorical variables consists of three contingency tables (Sex×Education, Sex×Income, Education×Income) and three diagonal tables with univariate relative frequencies).

(ii) If both x and y are interval-scaled variables with zero mean and unit variance, then s_{xy} is a correlation coefficient. Thus, the product data for the last seven MVs in Tb.4.18 are correlations.

(iii) If x_ℓ is a Boolean variable and y is interval-scaled, then $N s_{xy}$ is the sum of y within the group of cases defined by $x_\ell = 1$, and s_{xy}/s_{xx} is the average of y in the group $x_\ell = 1$. Thus, the average of TV watching (measured on a standardized scale) for high-education people is $-0.07/0.30 = -0.23$, for low-income people $0.07/0.70 = 0.10$.

Four different analyses were tried out and compared. Firstly, the categorical variables were scaled so as to give either the product matrix $\boldsymbol{M}$ or $\boldsymbol{D}$. For $\boldsymbol{D}$ the ModeA estimation has to be chosen for all categorical variables, whereas for $\boldsymbol{M}$ the ModeB estimation was chosen. Secondly, both the $\boldsymbol{D}$-ModeB and the $\boldsymbol{M}$-

Table 4.18: TV Consumption and Fear of Crime: product data

Sex,Male	47															
Sex,Female	.	53														
TV, Male	44	-44	100													
TV, Female	-44	44	-78	100												
Education high	16	14	1	-5	30											
Education low	32	39	-1	5	.	70										
Income high	10	11	-2	-2	9	12	21									
Income middle	20	23	0	0	13	30	.	43								
Income low	17	19	2	2	8	28	.	.	36							
Age	-1	1	-2	4	-9	9	-1	-12	13	100						
TV hours	-3	3	25	41	-7	7	-6	0	6	4	100					
ChcsViol	-7	7	-10	15	-5	5	-5	-2	6	-9	9	100				
FearNbhd	-13	13	-22	22	-6	6	-3	-6	8	25	1	13	100			
VicSex	-3	3	-3	8	-1	1	-1	0	2	6	8	1	9	100		
CrimeProb	-3	3	-5	10	-7	7	-5	-3	8	10	8	13	30	8	100	
Crimeup	-3	3	-4	9	-4	4	-2	-1	3	4	7	13	5	7	7	100

ModeA models were estimated (a) with Sex and Sex×TV as separated blocks and (b) with Sex and Sex×TV as one joint block. The four estimates show differences only in the third decimal, concerning the squared multiple correlation of the target variable Fear of Crime. The estimates are:

- Model = $\boldsymbol{D}$-ModeA, split: $R^2 = 0.147744$
- Model = $\boldsymbol{M}$-ModeB, split: $R^2 = 0.147746$
- Model = $\boldsymbol{D}$-ModeA, joint: $R^2 = 0.146325$
- Model = $\boldsymbol{M}$-ModeB, joint: $R^2 = 0.147869$.

Thus the differences are markedly smaller than those between the results for the artificial data reported in Tb.4.16.

The weights, part of the outer model parameters shown in Table 4.19, provide a rescaling of the categorical variables. The Sex variable is scored with -1.06 for males and 0.95 for females. Low income and low education are scored as positive. With this scaling, Sex, Education, Income and Age are positively correlated to the amount of TV watching and Fear (Tb.4.20). The loadings of the five indicators on the LV Fear of Crime are moderate.

The path coefficients, part of the inner model parameters shown in Table 4.20, show a weak connection between TV-watching and the background variables Sex, Education, and Income, the multiple squared correlation R^2 being 4%.

$$\text{TV} = .05\text{Sex} + .13\text{Education} + .13\text{Income} + U. \tag{4.501}$$

Now let all variables of the model be used to predict for Fear of Crime:

Table 4.19: TV Consumption and Fear of Crime: Outer model parameters

Mean	Loading	Weight	LV	MV
47%	-50	-106	SEX	male [1 = male, 0 = female]
53%	50	95		female [1 = female, 0 = male]
7.18	-94	-52	SEX×TV	TV watching, men only
7.53	94	54		TV watching, women only
30%	-46	-154	EDUCATION	some college
70%	46	65		no college
21%	-28	-135	INCOME	over \$25'000
43%	-16	-38		\$10'000 – \$25'000
34%	45	125		under \$10'000
43.13	100	100	AGE	Age [in years]
7.36	100	100	TV	TV watching [hours per day]
73	47	31	FEAR	Chance of involvement in violence
20	81	63		Fear, neighborhood not save
78	27	16		Women more likely to be victim
24	63	38		Crime is a very serious problem
96	32	21		Crime is rising

All numbers without decimal point or without percentage sign are multiplied by 100.
For categorical variables, the means are written as percentages.

Table 4.20: TV Consumption and Fear of Crime: Inner model parameters

LV	LV-Correlation							Path coefficients							R^2
Sex	100														
Sex×TV	94	100													
Low education	6	7	100												
Low income	-1	0	17	100											
Age	2	3	21	22	100										
TV-Hours	6	9	15	15	4	100		5	.	13	13	.	.	.	4%
Fear	24	24	20	23	18	10	100	16	8	12	18	11	3	.	15%
	S	×	E	I	A	TV	F	S	×	E	I	A	TV	F	

All coefficients multiplied by 100. Dots indicate fixed zeros.

$$\begin{aligned}
\text{Fear} &= .16 \cdot \text{Sex} + .08 \cdot (\text{TV} \times \text{Sex}) + .03 \cdot \text{TV} \\
&\quad + .12 \cdot \text{Education} + .18 \cdot \text{Income} + .11 \cdot \text{Age} + U &(4.502)\\
&= .16 \cdot \left\{ \begin{array}{rcl} -1.06 & \cdot & \text{Male} \\ .95 & \cdot & \text{Female} \end{array} \right\} \\
&\quad + .08 \cdot \left\{ \begin{array}{rcl} -.52 & \cdot & \text{Male} \times \text{TV} \\ .54 & \cdot & \text{Female} \times \text{TV} \end{array} \right\} + .03 \cdot \text{TV} + \ldots + U &(4.503)\\
&= \left\{ \begin{array}{rl} -.17 & \text{if male} \\ .15 & \text{if female} \end{array} \right\} + \left\{ \begin{array}{rl} -.01 & \text{if male} \\ +.07 & \text{if female} \end{array} \right\} \cdot \text{TV} + \ldots + U &(4.504)
\end{aligned}$$

The Fear of Crime higher is for women ($b = 0.16$), the elderly ($b = 0.11$), low income (b $= 0.18$) and low education people ($b = 0.12$), and is fairly independent of TV-watching ($b = 0.03$). But if the different reactions of men and women to TV are taken into account ($b = 0.08$), then the connection becomes practically zero for men ($b = 0.03 + (-0.52 \cdot 0.08) = -0.01$) and higher for women ($b = 0.03 + (0.54 \cdot 0.08) = 0.07$). All the associations in this model are not very strong, but with $N = 2060$ cases one can assume that they are different from zero. The predictive power of TV consumption is reduced when Income, Education and Age are introduced as additional predictors.

This example shows that a mixture of categorical- and interval-scaled variables may be treated in one LV-path model. The categorical variables are rescaled (scored) so that they provide an optimal fit to the model. In our example no unexpected scale value (cf. weights in Tb.4.2) was found. Moreover, the example shows how differences between groups can be included in a path model. Differences in the means are included by forming optimal contrasts between the groups and the contrast values are provided by the scaling procedure. Differences in the associations (different correlations, different path coefficients) are covered by forming a product variable from group-describing Boolean variables (e.g. Sex) and interval-scaled variables (e.g. TV-watching).

4.6 Conclusion

The methods presented in Sc.4 enlarge the standard PLS modeling technique and make it applicable to categorical variables and products of categorical and interval-scaled variables. As a routine procedure, categorical variables are replaced by as many Boolean variables as the variable has categories. This procedure introduces a singularity into the analysis, because the Booleans always sum up to unity. This singularity propagates through the analysis and shows up at several points, thereby causing the singularity of certain product matrices.

The introduction of product variables adds to the problems when the original variables x and y, as well as their product xy, are included in the model. Traditionally Anova and Ancova methods take care of the singularities by reparametrization of the model. Three techniques for reaching correct and satisfying solutions have been tried in this chapter: namely, the split MR (join blocks for estimation, split them for interpretation), the reordering into blocks so that estimation is made easier, and the restriction by Pattor rotation. Results appeared to differ and were not always encouraging. It requires a thorough understanding of the singularities included in the model to find out what remedial device could possibly help.

The choice of the product matrices $\boldsymbol{R}, \boldsymbol{S}, \boldsymbol{M}, \boldsymbol{D}, \boldsymbol{Q}$ turns out to be crucial for the choice of the ModeA/ModeB of model estimation, because $\boldsymbol{R}, \boldsymbol{S}, \boldsymbol{D}$ are singular and allow only for the ModeA estimation.

Chapter 5

Predictive vs. Structural Modeling: PLS vs. ML

The Latent Variables Path (LVP) model and also the first viable estimation technique were developed by Karl Jöreskog (from 1973 onwards). Right from the beginning his LISREL method was of the utmost elegance and perfection, as regards the deductive properties of the model. The LISREL program has since been gradually improved (LISREL III: Jöreskog & van Thillo 1973; LISREL VII: Jöreskog & Sörbom 1987), as regards convenience for the user, additional estimation criteria and model diagnostics. However, one inconvenient side of LISREL, well known to all users, is incurable: LISREL tends to produce improper solutions characterized by negative variance estimates and LV correlations greater than unity (van Driel 1978, Jöreskog 1981, Fornell & Bookstein 1982, Rindskopf 1983, 1984).

Why, then, does LISREL give improper solutions and why are PLS estimates biased? When does LISREL give proper solutions and when are PLS estimates consistent? Also, why do LISREL and PLS come to numerically different estimates, when both start with the same data matrix and same path diagram?

The answer to these questions is not only a matter of least squares vs. maximum likelihood, a matter of different statistical estimation principles. Despite the similarity of the path diagram, PLS and LISREL modeling are basically different as regards the question of what the relevant information for the model is:

- LISREL is a covariance structure model, PLS is a data structure model.
- LISREL is a common factor approach to LVs, PLS a component approach.
- LISREL models the relationships between the variables (structural modeling), PLS models additionally the relationships between the cases, thereby reconstructing / predicting the data matrix (predictive modeling).

What is said here about the LISREL model, i.e. the LVP model with covariance reconstructing estimation methods, holds for other modeling approaches as well. Covariance structure modeling represents at the present time the mainstream in psychometrics, as is indicated in articles by Bentler (1976, 1980, 1982, 1983, 1986), Bentler & Weeks (1979, 1980, 1982), Bentler & Lee (1983), McArdle (1979, 1980, 1981, 1982), McArdle & McDonald (1984), as well as all those of Jöreskog. The programs which implement these techniques are COSAN (Fraser 1980), RAM (McArdle & McDonald 1984) and EQS (Bentler 1985).

5.1 Covariance vs. Data Structure Models

In this section, the arguments of Sc.1.3 will be expanded and several distinctions introduced, which will help to understand better the difference between the ML methods and the PLS methods of latent variables path analysis. The distinctions concern the basic entities of the model (data points vs. random variables), the type of the model (data structure vs. covariance structure), the type of parameters (structural vs. incidental), and the type of application (general vs. applied psychology, e.g.). The type of the model determines the type of the estimation procedure and also the differences in the results.

Known vs. anonymous arrays. The principal component analysis of a real matrix $\boldsymbol{y}_{[K \times N]}$ is symmetric (cf. Lohmöller & Wold 1980). The eigenvalues of $\boldsymbol{y}$ and of $\boldsymbol{y}'$ are identical. It is simply a matter of scaling, as to whether the left or right hand eigenvector matrix of $\boldsymbol{y}$ becomes the loading matrix $\boldsymbol{\Pi}$, or the score matrix $\boldsymbol{\eta}$. This symmetry does not hold for the common factor model, which implies a genuine difference between variables and observational units.

Let us consider the data matrix $\boldsymbol{y}$ as a real function of the Cartesian product of two sets of entities $\mathcal{K}$ and $\mathcal{N}$,

$$\boldsymbol{y} = y(\mathcal{K} \times \mathcal{N}) \tag{5.1}$$

where $\mathcal{N} = \{n, \underline{n}, \underline{\underline{n}}, \ldots\}$ is the array (set, mode, coordinate) of cases and $\mathcal{K} = \{k, \underline{k}, \underline{\underline{k}}, \ldots\}$ is the array of attributes of these cases. The arrays are called "modes" by Tucker (1966), "coordinates" by Cattell (1966:78) and "facets" by Guttman. In the terminology of Anova a difference is made between fixed and random arrays. An array ("factor") is called *random*, if the elements at hand are a sample which is meant as representative for a population of elements; the single elements of this array remain exchangeable and anonymous. An array is called *fixed*, if it includes all the elements of substantive interest; every single element of this array is "known" to, and specified by, the researcher. To put it another way: an array is *anonymous* (or random or "sampled"), if the substantive interest goes beyond the elements at hand. The distinction "known vs. anonymous" expresses in subject-matter related terms what the distinction "fixed vs. random" expresses in statistical terms.

Abstract vs. applied science. In a general-psychology study the attribute array is fixed and the case array is sampled. The substantive interest is in the structure of intelligence, for example, or in the general relationship between intelligence and achievement. The structural statements on intelligence are assumed to hold not only for the sample of cases from which they were gained, but also for a population greater than this sample. Contrary to this situation, in a sociometric test the case array (the member of the group under study) is specified, whereas the attribute array (the sociometric criteria) must be chosen so as to be representative for the relationships between the group members and so as to be generalizable beyond the used criteria.

Applied psychology (psychometry, individual or differential or diagnostic psychology) is a blend of both types. The main interest lies in the individuals and the focus is on the differences and similarities between the individuals concerned. The attributes are designed to constitute a fairly general and abstract framework for the comparison of the individuals. Hence, in both arrays, the cases and the attributes are considered as non-sampled. However, the difference between general and applied is not an "either-or", but a "more-or-less" distinction, because every applied scientific research (in psychology, education, marketing research, or whatever) attempts to construct insights which are valid and useful beyond the data at hand.

The statistical variables. Pursuing the idealized difference between abstract (general) and applied studies further, the basic entities of the models have either one or two indices:

$$\boldsymbol{y} = [y_k], \qquad k = 1 \dots K \qquad \text{for abstract models} \tag{5.2}$$
$$\boldsymbol{y} = [y_{kn}], \qquad k = 1 \dots K, n = 1 \dots N \qquad \text{for applied models.} \tag{5.3}$$

In the statistical formulation of the model the basic entities are considered as random variables. The model parameters, together with the model relations, generate the joint distribution of the variables $\boldsymbol{y}$. Notice that in the general model there are K statistical variables, i.e. as many as attributes, and that in the applied model there are $K \times N$ statistical variables.

The abstract LV model. The linear LV model for the variables defined in Eq.5.2 (the general LV model) is written as:

$$y_k = \pi_{k0} + \sum_j \pi_{kj}\eta_j + \epsilon_k, \qquad \forall k. \tag{5.4}$$

All quantities on the right-hand side of Eq.5.4 are unknowns; η and ϵ denote random variables, and π_{k0} and π_{kj} parameters, i.e. algebraic variables whose fixed but unknown values have to be estimated. To make the model estimable, the random variables have to be replaced by parameters. In a moment structure model it is assumed that only the first and second moments of $\boldsymbol{y}$ are of interest, all higher moments being zero. The parameters of the first moment $\mathrm{E}(\boldsymbol{y})$ of Eq.5.4 are the first-moment parameters,

$$\pi_{k0}, \quad \rho_{j0} = \mathrm{E}(\eta_j), \quad \theta_{k0} = \mathrm{E}(\epsilon_k), \tag{5.5}$$

which generate the first-moment model (i.e. on the averages)

$$\mathrm{E}(y_k) = \pi_{k0} + \sum_j \pi_{j0}\rho_{j0} + \theta_{k0}, \qquad \forall k. \tag{5.6}$$

The parameters for the second (central) moment are the second-moment parameters

$$\pi_{kj}, \quad \rho_{ji} = \mathrm{E}(\eta_j\eta_i'), \quad \theta_{k\ell} = \mathrm{E}(\epsilon_k\epsilon_\ell'), \quad \gamma_{kj} = \mathrm{E}(\eta_k\epsilon_j'), \tag{5.7}$$

producing the model (i.e. on the covariances)

$$\boldsymbol{\Sigma} = \mathrm{E}[(\boldsymbol{y} - \mathrm{E}(\boldsymbol{y}))(\boldsymbol{y} - \mathrm{E}(\boldsymbol{y}))'] = \boldsymbol{\Pi}\mathbf{P}\boldsymbol{\Pi}' + \boldsymbol{\Theta} + \boldsymbol{\Gamma} + \boldsymbol{\Gamma}'. \tag{5.8}$$

Once the parameters Eqs.5.5, 5.7 are known or estimated, the first and the second moments on the left-hand side of Eqs.5.6, 5.8 can be deduced (cf. Sc.2.2.2). If the first-order parameters (Eq.5.5) are restricted to zero, the model is called a covariance structure model, if not, a moment structure model. With the additional restrictions

$$\gamma_{kj} = 0, \qquad \theta_{k\ell} = 0 \quad \text{if} \quad k \neq \ell, \tag{5.9}$$

we have the common factor model.

The applied LV model. The linear model for the variables defined in Eq.5.3 (the applied LV model) is written as:

$$y_{kn} = \pi_{k0} + \sum_j \pi_{kj}\eta_{jn} + \epsilon_{kn}, \qquad \forall k, \forall n. \tag{5.10}$$

Next, a quite different technique of parameterizing the random variables will be employed. A second relation is defined, the weight relation,

$$\eta_{jn} = \omega_{j0} + \sum_k \omega_{kj} y_{kn} + \delta_{jn}, \qquad \forall k, \forall n. \tag{5.11}$$

with predictor specification

$$\mathrm{E}(\eta_{jn}|y_{kn}) \equiv \tilde{\eta}_{jn} = \omega_{j0} + \sum_k \omega_{kj} y_{kn}. \tag{5.12}$$

The "truly" latent variables η_{jn} in Eq.5.10 are replaced by their conditional expectation $\tilde{\eta}_{jn}$ (defined in Eq.5.12), and the relation Eq.5.10 is now written:

$$y_{kn} = \tilde{\pi}_{k0} + \sum_j \tilde{\pi}_{kj}\tilde{\eta}_{jn} + \tilde{\epsilon}_{kn}, \qquad \forall k, \forall n. \tag{5.13}$$

The model Eqs.5.12, 5.13 has two types of parameters:

(i) The structural parameters $\tilde{\pi}_{k0}, \tilde{\pi}_{kj}, \omega_{k0}, \omega_{kj}$: These are the same for all cases and related only to the attributes in the model.

(ii) The incidental parameters[1] $\tilde{\eta}_{jn}$ and $\tilde{\epsilon}_{kn}$: The index n reveals that these are case values of the variables.[2]

None of the structural parameters carries the index n for the cases; the case array may be completely neglected while estimating the weights and the loadings. The means $m_k = \text{mean}(y_{kn})$ and the covariances $s_{k\ell} = \text{cov}(y_k; y_\ell)$ are sufficient for the estimation of the structural parameters, which may require an iterative algorithm. Once the structural parameters are known, the incidental parameters are immediately computed, the LV scores $\tilde{\eta}_{jn}$ by Eq.5.12, and the residual scores by $\tilde{\epsilon}_{kn} = y_{kn} - \tilde{\pi}_{k0} - \sum_j \tilde{\pi}_{kj}\tilde{\eta}_{jn}$.

[1]The distinction between structural and incidental parameters is due to Zellner.

[2]No matter whether the $\tilde{\eta}_{jn}$ and $\tilde{\epsilon}_{kn}$ are interpreted as parameters (i.e. fixed but unknown constants) or as random variables, both interpretations are consistent with the model Eqs.5.12, 5.13. If $\tilde{\eta}$ and $\tilde{\epsilon}$ are random variables, the observations $y = f(\tilde{\eta}, \tilde{\epsilon})$ are random and, consequently, $\tilde{\eta} = f(y)$ and $\tilde{\epsilon} = f(y, \tilde{\eta})$ are random. If $\tilde{\eta}$ and $\tilde{\epsilon}$ are interpreted as fixed values, the observations come out as fixed values by Eq.5.13 and, consequently, $\tilde{\eta}$ and $\tilde{\epsilon}$ are fixed values.

First recapitulation. The operative use of a model may be oriented either more towards an abstract (theoretical) or an applied goal. The more abstract and general model is focussed only on the structure of the attributes (a fixed array of entities) and establishes a framework that can be applied to a population greater than the observations at hand (a random array of entities), which are nothing else than a sample and may be replaced by any other representative sample. The applied model is focussed on the cases at hand (a fixed array) and provides incidental parameters for the description of the cases, for comparisons, differences and similarities. The observed means and covariances form a sufficient data basis for the estimation of the structural parameters.

When referring to their relevant basis of information, the two types of models will be called *covariance structure models* (or moment structure, or even correlation structure), or *data structure models.* For the covariance structure model the covariance matrix $\boldsymbol{S} = \mathrm{E}(\boldsymbol{y}\boldsymbol{y}')$ contains all the information which is of interest, whereas in the data structure model it is only a vehicle for simplifying the numerical workload of estimation. The differences between covariance and data structure models influence the choice of estimation criteria for the models.

Data structure estimation. For the loading relation Eq.5.13 predictor specification is assumed:

$$\mathrm{E}(y_{kn}|\tilde{\eta}_{jn}) = \tilde{y}_{kn} = \sum_j \tilde{\pi}_{kj}\tilde{\eta}_{jn}, \quad \forall k, \forall n. \tag{5.14}$$

The conditional expectations Eq.5.14 imply that the discrepancies between the observed values y_{kn} and the systematic part $\tilde{y}_{kn}$ are minimized:

$$\sum_k \overline{\sum_n} \mathrm{E}(y_{kn} - \tilde{y}_{kn})^2 = \min. \tag{5.15}$$

Defining $\tilde{\boldsymbol{\Sigma}}_{[K\times K]}$ as the covariance of $\tilde{y}_{[K\times N]}$,

$$\tilde{\sigma}_{k\ell} = \overline{\sum_n} \mathrm{E}[(\tilde{y}_{kn} - \mathrm{E}(\tilde{y}_k))(\tilde{y}_{\ell n} - \mathrm{E}(\tilde{y}_\ell))], \tag{5.16}$$

the property Eq.5.15 can be rewritten as

$$\sum_k (s_{kk}^2 - \tilde{\sigma}_{kk}^2) = \mathrm{trace}(\boldsymbol{S} - \tilde{\boldsymbol{\Sigma}}) = \min. \tag{5.17}$$

We replace the theoretical items $\tilde{\eta}$ and $\tilde{\pi}$ by their estimated counterparts $\hat{\eta}$ and p and rewrite Eqs.5.14,..., 5.17:

$$y_{kn} = \sum_j p_{kj}\hat{\eta}_{jn} + e_{kn} = \hat{y}_{kn} + e_{kn} \tag{5.18}$$

$$\hat{\sigma}_{kk} = \overline{\sum_n}(\hat{y}_{kn} - \overline{\sum_n}\hat{y}_{kn})^2 \equiv \hat{\sigma}_k^2 \tag{5.19}$$

$$\begin{aligned} \mathrm{OLS} &= \sum_k \overline{\sum_n}(y_{kn} - \hat{y}_{kn})^2 \stackrel{!}{=} \min \\ &= \sum_k (s_{kk} - \hat{\sigma}_{kk}) = \mathrm{trace}(\boldsymbol{S} - \hat{\boldsymbol{\Sigma}}) \stackrel{!}{=} \min. \end{aligned} \tag{5.20}$$

The Ordinary Least Squares (OLS) estimation criterion is formulated in terms of the data points y_{kn} and minimizes the discrepancies between observed and estimated values. The OLS criterion is applied only to the diagonal elements of the covariance matrix: "Maximize the explained variance diag($\hat{\mathbf{\Sigma}}$)."

Covariance structure estimation. The covariance structure model in estimated form is written:

$$\hat{\mathbf{\Sigma}} = \mathbf{\Sigma}(\hat{\mathbf{\Pi}}, \hat{\mathbf{P}}, \hat{\mathbf{\Theta}}, \hat{\mathbf{\Gamma}}) = \hat{\mathbf{\Pi}}\hat{\mathbf{P}}\hat{\mathbf{\Pi}}' + \hat{\mathbf{\Theta}} + \hat{\mathbf{\Gamma}} + \hat{\mathbf{\Gamma}}'. \tag{5.21}$$

The Unweighted Least Squares (ULS) criterion, the Generalized Least Squares (GLS), and the Maximum Likelihood (ML) criterion are covariance oriented and minimize the discrepancies between the empirical covariance matrix $\boldsymbol{S}$ and the estimated theoretical covariance $\hat{\mathbf{\Sigma}}$ (cf. Jöreskog 1978):

$$\text{ULS} = \text{trace}(\boldsymbol{S} - \hat{\mathbf{\Sigma}})^2 = \sum_k \sum_\ell (s_{k\ell} - \hat{\sigma}_{k\ell})^2 \stackrel{!}{=} \min \tag{5.22}$$

$$\text{GLS} = \text{trace}(\mathbf{I} - \boldsymbol{S}^{-1}\hat{\mathbf{\Sigma}}) \stackrel{!}{=} \min \tag{5.23}$$

$$\text{ML} = \text{trace}(\hat{\mathbf{\Sigma}}^{-1}\boldsymbol{S}) - \log\left|\hat{\mathbf{\Sigma}}^{-1}\right|\boldsymbol{S} \stackrel{!}{=} \min. \tag{5.24}$$

In most applications of the covariance structure model Eq.5.21 it is assumed that $\mathbf{\Gamma} = \mathbf{0}$ and $\mathbf{\Theta}$ = diagonal, whence it follows that a perfect fit to the diagonal of $\boldsymbol{S}$ is always possible and that the remaining parameters have to be estimated so that the off-diagonal elements of $\boldsymbol{S}$ are optimally reproduced: "Maximize the explained covariance".

Second recapitulation. When applied to the covariance matrix, the estimation criteria of data structure models minimize the residual variances, whereas the criteria of the covariance structure models minimize the residual covariances. Hence it is to be expected that a component model like PLS will look better than a factor model, like LISREL, when only the fit to the diagonal of $\boldsymbol{S}$ is taken into account, worse if off-diagonal fit measures are applied. How big the differences are will be demonstrated in Sc.5.4, following some systematic results on one-block (Sc.5.2) and two-block (Sc.5.3) models with artificial data.

5.2 Scored and Unscored LVs

The factor model. Factor analysis, as presented in textbooks, proceeds in two steps. First, the "factors" are estimated, and this means estimating the loading matrix $\mathbf{\Pi}$, often simply called the factor matrix. For the first step, an estimation criterion like ML is adopted, which serves for estimation of the parameters (loadings, LV covariance, residual covariance), their standard deviations as well as for the testing of the model against the data.

Secondly, the "factors" are estimated, this, though, now means the estimation of the factor scores η_{jn}. If in the first step ML was used to estimate the unscored factors, a new criterion must now be established, say the LS criterion as discussed in Sc.3.3.4, to solve for the scores.

As a third step, one can use the scored factors to recalculate the factor covariance, loading pattern and loading structure coefficients. It becomes apparent that there may be differences between the unscored (step one) and scored (step two) factors regarding the MV-LV relationships. This will be shown in more detail.

For the demonstration of the problems of LV scores, the most simple model, the unrestricted one-factor model will be used. The model and the related covariance matrices are defined by

$$\boldsymbol{y} = \boldsymbol{\pi}\eta + \boldsymbol{\epsilon} \tag{5.25}$$

$$\boldsymbol{\tau} = \mathrm{E}(\boldsymbol{y}\eta) \qquad = \text{LV structure} \tag{5.26}$$

$$\rho^2 = \mathrm{E}(\eta\eta) = 1 \qquad = \text{LV variance} \tag{5.27}$$

$$\boldsymbol{\Theta} = \mathrm{E}(\boldsymbol{\epsilon}\boldsymbol{\epsilon}') \qquad = \text{residual covariance} \tag{5.28}$$

$$\boldsymbol{\Sigma} = \boldsymbol{\pi}\rho^2\boldsymbol{\pi}' + \boldsymbol{\Theta} \qquad = \text{MV covariance,} \tag{5.29}$$

and the estimates are denoted by $\boldsymbol{p} \mathrel{\hat{=}} \boldsymbol{\pi}$, $\boldsymbol{t} \mathrel{\hat{=}} \boldsymbol{\tau}$, $r^2 \mathrel{\hat{=}} \rho^2$, and $\boldsymbol{C} \mathrel{\hat{=}} \boldsymbol{\Theta}$.

First step. As an example, the MV covariance

$$\boldsymbol{S} = \begin{bmatrix} 1.0 & .64 & .64 \\ .64 & 1.0 & .64 \\ .64 & .64 & 1.0 \end{bmatrix} \tag{5.30}$$

is used, with the obvious solution

$$\boldsymbol{p} = \boldsymbol{\pi} = \boldsymbol{t} = \boldsymbol{\tau} = (.80, .80, .80)', \tag{5.31}$$

which fulfills any covariance-structural criterion Eqs.5.22,..., 5.24 perfectly.

Second step. If the LV scores are to be estimated, the weight equation

$$\tilde{\eta} = \boldsymbol{w}'\boldsymbol{y} \tag{5.32}$$

is introduced with the solution $\boldsymbol{w} = \boldsymbol{S}^{-1}\boldsymbol{t}$ (cf. Sc.3.3.2, ModeB weights). The weights for the example Eq.5.30 are

$$\boldsymbol{w} = \begin{bmatrix} .35 \\ .35 \\ .35 \end{bmatrix}, \qquad \boldsymbol{S}^{-1} = \begin{bmatrix} 2.00 & -.78 & -.78 \\ -.78 & 2.00 & -.78 \\ -.78 & -.78 & 2.00 \end{bmatrix}. \tag{5.33}$$

Third step, solution #1 Using the LV scores $\tilde{\eta}$ as defined in Eq.5.32, the covariances defined in Eqs.5.26,..., 5.28 are recalculated:

$$\tilde{r}^2 = \mathrm{var}(\tilde{\eta}) = \boldsymbol{w}'\boldsymbol{S}\boldsymbol{w} = \quad 0.84 \quad \neq \rho^2 \tag{5.34}$$

$$\tilde{\boldsymbol{t}} = \mathrm{cov}(y; \tilde{\eta}) = (.80, .80, .80)' = \boldsymbol{\tau} \tag{5.35}$$

$$\tilde{\boldsymbol{p}} = (.87, .87, .87)' = \boldsymbol{\pi}/\tilde{r}. \tag{5.36}$$

The variance of $\tilde{\eta}$ is less than the variance of η; and the LV pattern coefficients $\tilde{\boldsymbol{p}}$ must be adjusted to this variance.

Third step, solution #2 The weights are rescaled to obtain a LV $\acute{\eta}$ with unit variance:

$$\acute{\boldsymbol{w}} = f\boldsymbol{w} = \frac{\boldsymbol{w}}{\sqrt{\boldsymbol{w}'\boldsymbol{S}\boldsymbol{w}}} = (.38, .38, .38)' \tag{5.37}$$

$$\acute{r}^2 = \text{var}(\acute{\eta}) = \acute{\boldsymbol{w}}'\boldsymbol{S}\acute{\boldsymbol{w}} = \quad 1 \quad = \rho^2 \tag{5.38}$$

$$\acute{\boldsymbol{t}} = \text{cov}(y, \acute{\eta}) = (.87, .87, .87)' = \boldsymbol{\tau}/\tilde{r} \tag{5.39}$$

$$\acute{\boldsymbol{p}} = (.80, .80, .80)' = \boldsymbol{\pi}. \tag{5.40}$$

Here the pattern coefficients $\boldsymbol{p}$ are correct but the structures $\acute{\boldsymbol{t}}$ are different from the covariance-structural solution Eq.5.31.

Third step, solution #3 Schönemann (1981) enlarges Eq.5.32 by an arbitrary variable δ, which is uncorrelated with $\boldsymbol{y}$ and $\tilde{\eta}$ and has variance $1 - \tilde{r}^2$:

$$\eta_\delta = \boldsymbol{w}'\boldsymbol{y} + \delta = \tilde{\eta} + \delta. \tag{5.41}$$

Hence it follows that:

$$r_\delta^2 = \rho^2, \quad \boldsymbol{t}_\delta = \boldsymbol{\tau}, \quad \boldsymbol{p}_\delta = \boldsymbol{\pi}. \tag{5.42}$$

To compute the intelligence factor score for a sample of 40 cases, Schönemann uses as variable δ the date of Easter Sunday (March 20 is coded as 3.20) in the last 40 years, appropriately transforms δ to $\text{cov}(\delta; \tilde{\eta}) = 0$ and $\text{var}(\delta) = 1 - \tilde{r}^2$, then adding it to $\tilde{\eta}$; he can then predict perfectly the date of Easter Sunday by the intelligence scores. The scores $\tilde{\eta}_\delta$ of Eq.5.41 are indeterminate, due to the indeterminateness and arbitrariness of δ, and can be used for any joke and any nonsense.

Comparison. Comparing the three solutions for factor scores, it is noted that

- $\tilde{\eta}$ and $\acute{\eta}$ in solution #1 and #2 have other relations (pattern, structure coefficients) with the MVs than the covariance-structural solution;
- η_δ in solution #3 includes the determination of an indeterminate and arbitrary variable δ, so that the scores on $\tilde{\eta}_\delta$ are also indeterminate;
- No determinate factor scores exist which have the same relations as the covariance-structural solution.

As long as the estimation of loadings and scores is done successively, no coherent solution exists. The inconsistency may be moved to the scores (factor indeterminacy), the LV correlation, LV pattern, LV structure, or even the residual covariance (see Bentler 1982:109). The only way to reach a coherent set of estimates for all structural and incidental parameters is to include the scores or the weight matrix in the basic set of parameters, which are estimated simultaneously. The indeterminacy of factor scores has led to some discussion regarding the factor-analytic tradition in psychometrics (cf. Sc.3.3.5;cf. Schönemann 1981, Steiger 1979a, 1979b, Rozeboom 1982, Greene 1978, Mulaik & McDonald 1978, Williams 1978, Mulaik 1981).

The component model. The Principal Component (PC) model is the data-structural counterpart to the factor model Eq.5.25. The PC parameter estimates for the covariance matrix Eq.5.30 are found easily, because all off-diagonal elements $s_{k\ell}$ of $\boldsymbol{S}$ are equal. Weight coefficients must be equal, and one may use $\breve{\boldsymbol{w}} = (1,1,1)'$ as a first approximation. The variance of the equally weighted sum of the MVs is:

$$\begin{aligned}\breve{r}^2 &= \breve{\boldsymbol{w}}'\boldsymbol{S}\breve{\boldsymbol{w}} = \sum_k\sum_\ell w_k s_{k\ell} w_\ell = \sum_k\sum_\ell 1 s_{k\ell} 1 \\ &= (K^2-K)s + K = K^2 s + K(1-s) \\ &= 9\times 0.64 + 3(1-0.64) = 6.84,\end{aligned} \tag{5.43}$$

where K is the number of MVs. The properly scaled weights which produce a standardized LV are found by:

$$w_k = \frac{\breve{w}_k}{\breve{r}} = \frac{1}{\breve{r}} = \frac{1}{\sqrt{K^2 s + K(1-s)}} = 0.38. \tag{5.44}$$

The covariance of MV and LV is

$$\begin{aligned}t_k &= \operatorname{cov}(y_k;\tilde{\eta}) = \operatorname{cov}(y_k, \boldsymbol{w}'\boldsymbol{y}) = \boldsymbol{S}\boldsymbol{w} = \sum_\ell w_k s_{k\ell} \\ &= w_k(K^2 s + K(1-s)) \\ &= \frac{K^2 s + K(1-s)}{\sqrt{K^2 s + K(1-s)}} = \sqrt{s + (1-s)/K} \\ &= \sqrt{0.64 + (1-0.64)/3} = 0.87.\end{aligned} \tag{5.45}$$

Because both MVs and LV are standardized, the structure coefficients $\boldsymbol{t}$ are at the same time the pattern coefficients $\boldsymbol{p}$.

$$\boldsymbol{p} = \boldsymbol{t} = (.87, .87, .87)'. \tag{5.46}$$

Bias and consistency. The estimates Eq.5.46 for the component model are clearly different from the estimates Eq.5.30 for the factor model. It being stipulated that the factor model is the true model, the PC estimates are distorted by the bias factor:

$$\text{bias(p)} = \frac{p_{\text{PC}}}{p_{\text{FA}}} = \sqrt{\frac{s + (1-s)/K}{s}}. \tag{5.47}$$

The bias factor depends primarily on the value of s (the higher the true loadings, the smaller the bias can be) and on K, i.e. the number of MVs. The higher the number of indicators used, the closer the PC estimate comes to the FA estimate.

This, then, gives the theorem of *consistency at large* (Lyttkens 1973, Wold 1982d:25s): When the number of MVs is increased to infinity, the bias factor tends to unit:

$$\lim_{K\to\infty} \text{bias} = 1. \tag{5.48}$$

For the most simple case, when $\boldsymbol{S}$ is a correlation with all off-diagonal elements being equally $s_{k\ell} = s = 0.64$, Eq.5.45 is used to demonstrate that:

$$\begin{aligned} K &= 3, \quad 6, \quad 12, \quad 24, \quad 48, \quad 96, \quad \ldots, \quad \infty \\ p &= .87, \quad .84, \quad .82, \quad .8093, \quad .8047, \quad .8023, \quad \ldots, \quad .8000. \end{aligned} \tag{5.49}$$

The more general form of the theorem of consistency at large argues that the estimate $\tilde{\eta}$ of a LV η,

$$\tilde{\eta} = \mathrm{est}(\eta) = \boldsymbol{\omega}'y = \boldsymbol{\omega}'(\boldsymbol{\pi}\eta + \boldsymbol{\epsilon}) = \boldsymbol{\omega}'\boldsymbol{\pi}\eta + \boldsymbol{\omega}'\boldsymbol{\epsilon}, \tag{5.50}$$

consists of two parts, namely the reliable part $\boldsymbol{\omega}'\boldsymbol{\pi}\eta$ and the impurities $\boldsymbol{\omega}'\boldsymbol{\epsilon}$, which are inevitably included in the weighted aggregate $\tilde{\eta}$. Assuming $\mathrm{E}(\eta\boldsymbol{\epsilon}') = 0$ and $\mathrm{E}(\tilde{\eta}) = 0$, the variance of $\tilde{\eta}$ is decomposed into the reliable and nonreliable parts

$$r = \mathrm{E}(\tilde{\eta}\tilde{\eta}') = \boldsymbol{\omega}'\boldsymbol{\pi}1\boldsymbol{\pi}'\omega + \boldsymbol{\omega}'\boldsymbol{\Theta}\boldsymbol{\omega} \tag{5.51}$$

with $\mathrm{E}(\eta\eta') = 1$ and $\boldsymbol{\Theta} = \mathrm{E}(\boldsymbol{\epsilon}\boldsymbol{\epsilon}')$. If the number of indicators is increased and the loadings of these additional indicators do not tend to zero, then the first summand in Eq.5.51 increases in quadratic fashion; if at the same time the residual covariance for these additional indicators tends to zero, the second summand in Eq.5.51 increases in linear progression. Hence the proportion of reliable variance in Eq.5.51 will increase, and the LV estimate $\tilde{\eta}$ will show higher correlation to the "true" LV η. (For a more formal proof see Wold 1982c:26.)

Residual covariances. Consistency at large means that the LV estimate $\tilde{\eta}$ approaches the true LV η as the number of indicators increases. As a consequence of determining a better estimate of the LV, the scores, loadings and patterns are improved and the residual variables are estimated more correctly also. In particular, if the residuals are assumed to be uncorrelated, $\boldsymbol{\Theta} = \mathrm{E}(\boldsymbol{\epsilon}\boldsymbol{\epsilon}') =$ diagonal, the empirical residual covariance will come closer to diagonality. With fewer MVs, the residuals will be biased and "overextraction" will be generally observed.

Assuming again the most simple example with all off-diagonal elements of $\boldsymbol{S}$ being identically $s_{k\ell} = s = 0.64$, the residual covariances are computed by $\boldsymbol{\Theta} \hat{=} \boldsymbol{C} = \boldsymbol{S} - \boldsymbol{p}\boldsymbol{p}'$, which for the PC estimates of the loadings gives:

$$\begin{aligned} c_{kk} &= s_{kk} - p^2 = 1 - (s + (1-s)/K) \\ &= 1 - (.64 + (1 - .64)/3) = 0.24 \\ c_{k\ell} &= s_{k\ell} - p^2 = s - s + (1-s)/K) = \frac{1-s}{K} \\ &= -.12. \end{aligned} \tag{5.52}$$

The comparison of the PC and the FA estimate of the residual covariance:

$$\boldsymbol{C}_{\mathrm{PC}} = \begin{bmatrix} .24 & -.12 & -.12 \\ -.12 & .24 & -.12 \\ -.12 & -.12 & .24 \end{bmatrix} \qquad \boldsymbol{C}_{\mathrm{FA}} = \begin{bmatrix} .36 & 0 & 0 \\ 0 & .36 & 0 \\ 0 & 0 & .36 \end{bmatrix} \tag{5.53}$$

shows that the FA estimate is correctly diagonal, whereas the PC estimates of the residual variables are negatively correlated. (When, after the extraction of the LV, positive MV covariances turn into negative residual covariances, it is termed *overextraction*). The bias of the residual covariances (here defined as a difference) is the square of the bias of the loadings (at least in this simplified model)

$$\text{bias}(c_{k\ell}) = \frac{1-s}{K}. \tag{5.54}$$

Clearly the bias of the residual covariance (i.e. the overextraction) decreases as the number K of MVs increases. With reference to Sc.3.2.1, the rank of the residual matrix $\boldsymbol{\Theta}$ is rank$(\boldsymbol{S}) - 1$, i.e. equal to or less than $K - 1$. With only one indicator, the LV estimate is equal to the MV and there is no residual, $\boldsymbol{\Theta} = \mathbf{0}$. With two indicators, the rank of $\boldsymbol{\Theta}$ is one, and the two residual variables are perfectly (negatively or positively) correlated. Notice that in Eq.5.53 all row sums and all column sums of $\boldsymbol{\Theta}_{\text{PC}}$ are zero: $\mathbf{1}'\boldsymbol{\Theta}\mathbf{1} = 0$; the general expression $\boldsymbol{\omega}'\boldsymbol{\Theta}\boldsymbol{\omega} = 0$ was already presented in Eq.3.229, and from Eq.3.229 it also follows that:

$$\boldsymbol{\Theta}\boldsymbol{\omega} = \text{E}(\boldsymbol{\epsilon}\boldsymbol{\epsilon}'\boldsymbol{\omega}) = 0 = \omega_k\theta_{kk} + \sum_{\ell \neq k} \omega_\ell\theta_{k\ell}. \tag{5.55}$$

Without loss of generality, one may assume only positive weights ω_k. Then the first summand in Eq.5.55 is necessarily positive and the second summand has the same absolute value, with negative sign. Hence the off-diagonal elements of $\boldsymbol{\Theta}$ must be dominantly negative, in order to balance the diagonal element, while the more off-diagonal elements there are in one row (i.e. the larger $K - 1$), the closer to zero the single elements can be, and the less the bias and overextraction are.

5.3 Consistency and Bias in a Two-Block Model

The preceding Sc.5.2 showed that PLS estimates of loadings are biased, the degree of the bias being greater the smaller the set of indicators is. In this section we now turn our attention to two-block models and we will see that PLS is not always biased. The example in this section is taken from Lohmöller (1982a), the theorem from Areskoug (1982).

One essential difference between the Maximum Likelihood (ML) method and the Partial Least Squares (PLS) method of parameter estimation for LVP models can be demonstrated by a very small model which contains only two latent variables, each indicated by a set of three manifest variables. PLS versions of this model are the well-known Canonical Correlation model (CC) and Tucker's (1958) Interbattery Factor model. A small artificial correlation matrix will suffice to demonstrate the main aspects of the difference. A small real world data example will be given later.

Let us suppose a sample correlation matrix $\boldsymbol{S} = \text{cor}(\boldsymbol{x}, \boldsymbol{y})$ is given for two blocks of standardized variables $x_\ell, \ell = 1 \ldots L$ and $y_k, k = 1 \ldots K$ with $L = K = 3$,

$$S = \frac{1}{100} \begin{bmatrix} 100 & & & & & \\ 10 & 100 & & & & \\ 10 & 10 & 100 & & & \\ 9 & 9 & 9 & 100 & & \\ 9 & 9 & 9 & 10 & 100 & \\ 9 & 9 & 9 & 10 & 10 & 100 \end{bmatrix}. \tag{5.56}$$

All correlation coefficients are very small, the correlations between blocks ($s_{k\ell} = 0.09$) being somewhat smaller than the correlations within blocks ($s_{k\underline{k}} = s_{\ell\underline{\ell}} = 0.10$). Depending on sample size, it would in most cases make no sense at all to analyse the association between the two blocks, because the correlation cannot be securely distinguished from a zero correlation.

However, let us assume a model with a latent variable ξ underlying the manifest variables $\boldsymbol{x}$ and a latent variable η for the MVs $\boldsymbol{y}$. The parameter of interest is the correlation coefficient $\rho = \text{cor}(\xi; \eta)$. A canonical correlation analysis gives the estimate:

$$r_{\text{CC}} = 0.225, \tag{5.57}$$

while the ML method of LISREL gives:

$$r_{\text{ML}} = 0.900. \tag{5.58}$$

The LISREL result appears unbelievable and extremely exaggerated, because the estimated association $r_{\text{ML}} = 0.900$ is ten times as high as the correlation $s_{k\ell} = 0.090$ between the observed variables. The bias of the estimated correlation is

$$\text{bias}(r) = \frac{r_{\text{CC}}}{r_{\text{ML}}} = \frac{0.225}{0.900} = 0.25. \tag{5.59}$$

The factor model. If we assume a two-block factor model with LVs ξ and η, where all variables are standardized, the loading coefficients (pattern coefficients) being denoted by π, the path coefficient between ξ and η by ρ and the variance of the outer residual variables by θ^2, then the covariance matrix of the manifest variables has the theoretical structure:

$$\boldsymbol{\Sigma} = \begin{bmatrix} \pi\pi + \theta^2 & & & & & \\ \pi\pi & \pi\pi + \theta^2 & & & & \\ \pi\pi & \pi\pi & \pi\pi + \theta^2 & & & \\ \pi\rho\pi & \pi\rho\pi & \pi\rho\pi & \pi\pi + \theta^2 & & \\ \pi\rho\pi & \pi\rho\pi & \pi\rho\pi & \pi\pi & \pi\pi + \theta^2 & \\ \pi\rho\pi & \pi\rho\pi & \pi\rho\pi & \pi\pi & \pi\pi & \pi\pi + \theta^2 \end{bmatrix}. \tag{5.60}$$

Given the sample correlation matrix $\boldsymbol{S}$ in Eq.5.56 and $\boldsymbol{\Sigma}$, as defined in Eq.5.60, the following set of estimates gives a perfect fit:

$$\pi \mathrel{\hat{=}} p = \sqrt{0.10}, \qquad \rho \mathrel{\hat{=}} r = 0.90, \qquad \theta^2 \mathrel{\hat{=}} c^2 = 0.90. \tag{5.61}$$

The path coefficient r is be found in this simple case by the ratio of between-block correlation to within-block correlation:

$$\rho_{\mathrm{FA}} = \frac{\pi\rho\pi}{\pi\pi} = \frac{\sigma_{kk}}{\sigma_{k\ell}} \hat{=} \frac{0.09}{0.10} = 0.90. \tag{5.62}$$

The solution Eqs.5.61, 5.62 fulfills at the same time the Generalized Least Squares (GLS), the Maximum Likelihood (ML), and the Unweighted Least Squares (ULS) criterion perfectly.

The component model. The two-block component model prescribes to estimate the LVs by standardized weighted aggregates of the manifest variables, $X = \boldsymbol{g}'\boldsymbol{x}$ and $Y = \boldsymbol{w}'\boldsymbol{y}$. In case of the Canonical Correlation model (CC) the correlation between these two aggregates is maximized:

$$r_{\mathrm{CC}} = \mathrm{cor}(X;Y) = \mathrm{cor}(\boldsymbol{g}'\boldsymbol{x};\boldsymbol{w}'\boldsymbol{y}) \stackrel{!}{=} \max. \tag{5.63}$$

Ignoring the conventional standardization $\mathrm{var}(X) = \mathrm{var}(Y) = 1$ for a moment, we chose the weights for the sample correlation matrix $\boldsymbol{S}$ in Eq.5.56 to be all equal, $\breve{g}_\ell = \breve{w}_k = 1$. Then:

$$\mathrm{var}(\breve{X}) = \breve{\boldsymbol{g}}'\boldsymbol{S}_{xx}\breve{\boldsymbol{g}} = \sum_{\ell\underline{\ell}} s_{\ell\underline{\ell}} = L^2 s_{\ell\underline{\ell}} + J(1 - s_{\ell\underline{\ell}}) = 3.60 \tag{5.64}$$

$$\mathrm{var}(\breve{Y}) = \breve{\boldsymbol{w}}'\boldsymbol{\Sigma}_{yy}\breve{\boldsymbol{w}} = \sum_{k\underline{k}} s_{k\underline{k}} = K^2 s_{k\underline{k}} + K(1 - s_{k\underline{k}}) = 3.60 \tag{5.65}$$

$$\mathrm{cov}(\breve{X};\breve{Y}) = \breve{\boldsymbol{g}}'\boldsymbol{S}_{yx}\breve{\boldsymbol{w}} = \sum_k\sum_\ell s_{k\ell} = KL s_{k\ell} = 0.81 \tag{5.66}$$

$$\mathrm{cov}(\breve{X},\breve{Y}) = \begin{bmatrix} 3.60 & 0.81 \\ 0.81 & 3.60 \end{bmatrix}, \tag{5.67}$$

and the correlation coefficient is

$$r_{\mathrm{CC}} = \frac{\mathrm{cov}(\breve{X};\breve{Y})}{\sqrt{\mathrm{var}(\breve{X})\,\mathrm{var}(\breve{Y})}} = \frac{0.81}{3.60} = 0.225. \tag{5.68}$$

This solution holds not only for the CC model, but for all PLS methods discussed in Sc.3.4. For $K = L$, all within-block correlations $s_{k\underline{k}} = s_{\ell\underline{\ell}}$ identical and all between-block correlations $s_{k\ell}$ identical, Eq.5.68 can be written:

$$r_{\mathrm{PLS}} = \frac{K^2 s_{k\ell}}{K^2 s_{k\underline{k}} + K(1 - s_{k\underline{k}})} = \frac{s_{k\ell}}{s_{k\underline{k}} + (1 - s_{k\underline{k}})/K}. \tag{5.69}$$

The results from Eq.5.62 and Eq.5.69 are combined to define the bias factor Eq.5.59 in terms of the MV correlations:

$$\mathrm{bias}(r) = \frac{r_{\mathrm{PLS}}}{r_{\mathrm{FA}}} = \frac{\frac{s_{k\ell}}{s_{k\underline{k}} + (1 - s_{k\underline{k}})/K}}{s_{k\ell}/s_{k\underline{k}}} = \frac{s_{k\underline{k}}}{s_{k\underline{k}} + (1 - s_{k\underline{k}})/K}. \tag{5.70}$$

Several comments are in order here:

Remark 5⋆1 The canonical correlation coefficient is consistent at large, with respect to a covariance-analytical model: Inasmuch as the number K of indicators for each block increases, the bias factor approaches unity:

$$\lim_{K\to\infty} \text{bias}(r) = 1. \tag{5.71}$$

This is demonstrated by increasing the block size K:

$$\begin{array}{rcl} K &=& 3,\quad 6,\quad 12,\quad 24,\quad 48,\quad 96,\quad \ldots,\quad \infty \\ r_{\text{PLS}} &=& .225,\ .360,\ .514,\ .655,\ .758,\ .823,\ \ldots,\ .900 \end{array} \tag{5.72}$$

Remark 5⋆2 The bias factor of r is the inverse square root of the loadings, Eq.5.47, $\text{bias}(r) = 1/\sqrt{\text{bias}(p)}$. Hence, from $p_{\text{PLS}} = p_{\text{FA}}\text{bias}(p)$ and $b_{\text{PLS}} = b_{\text{FA}}\text{bias}(b)$ it follows that the product of loadings and LV correlation is unbiased:

$$\begin{aligned} s_{k\ell} &= p_{\text{PLS}}\, r_{\text{PLS}}\, p_{\text{PLS}} = p_{\text{FA}}\text{bias}(p)\; r_{\text{FA}}\text{bias}(r)\; p_{\text{FA}}\text{bias}(p) \\ &= p_{\text{FA}}\, r_{\text{FA}}\, p_{\text{FA}}. \end{aligned} \tag{5.73}$$

This relations holds exactly only in the unrealistic case of all correlations being equal. For the more general case we may replace $r_{\text{Eq.5.68}}$ by the average of the inner correlations (r in Sc.2.4.3), and the squared loadings by the communality H^2 (Eq.2.109), and then compare

$$r_{\text{PLS}} H^2_{\text{PLS}} \stackrel{?}{=} r_{\text{FA}} H^2_{\text{FA}}. \tag{5.74}$$

Remark 5⋆3 Areskoug (1982) has proved the general theorem that the bias of the loadings and the bias of the canonical correlation coefficient cancel out each other, even for unequal block-sizes and unequal weight and loading coefficients. Also the prediction of the $\boldsymbol{y}$-variables by the $\boldsymbol{x}$-variables (after elimination of the LVs) is consistent at large. The inconsistency of the within-block parameters, the loadings and residual covariances, is inevitable. The bias factor is a function of the true loadings and "cannot be eliminated without additional information. This is the price to be paid for the unidentifiability inherent in the unconstrained residual structure" i.e. within the blocks (Areskoug 1982:105).

Remark 5⋆4 The first essential difference between the ML and LS method lies in the prescription, that in the LS method the LV must estimated as a linear combination of its indicators. Through this requirement, the LV is closely related to the observed and measured data. The multiple correlation of the LV on its MVs is perfect in the LS method but not in the ML method. We may rewrite the correlation matrix for the $\boldsymbol{x}$-variables, together with their correlations with the estimated X (the LV-structure coefficients), thus:

$$\text{cov}(\boldsymbol{x}, X)_{\text{PLS}} = \frac{1}{100}\begin{bmatrix} 100 & & & \\ 10 & 100 & & \\ 10 & 10 & 100 & \\ 63 & 63 & 63 & 100 \end{bmatrix}, \tag{5.75}$$

$$\text{cov}(\boldsymbol{x}, \xi)_{\text{ML}} = \frac{1}{100}\begin{bmatrix} 100 & & & \\ 10 & 100 & & \\ 10 & 10 & 100 & \\ 32 & 32 & 32 & 100 \end{bmatrix}. \tag{5.76}$$

The LV-structure coefficient for the ML estimate is $\sqrt{0.10} = 0.32$, for the LS estimate 0.63 (see Eq.5.45). The squared multiple correlation is for the LS case $R^2_{X|\boldsymbol{x}} = 1$ by prescription, and for the ML case $R^2_{\xi|\boldsymbol{x}} = 0.25$. Thus the overlap between the data and the theory, the MVs and the LVs, is only 25%. The LV in the ML solution is very far from being observable and the correlation $r = 0.90$ between these untouchable variables is even more distant from observation. The LS estimates of the LVs are bound completely to the observations and the estimate of the correlation between the LVs is conservative, on the secure side.

Conclusions. The mathematical considerations of the model show very clearly that the ML estimate $r_{\text{FA}} = 0.900$ is correct, i.e. consistent with the implied model. Thus it is up to the researcher whether he "believes" more in his data or in his model. If he is sure that the model is correct, that his MVs $\boldsymbol{x}$ and $\boldsymbol{y}$ are valid indicators of the LVs ξ and η and that it is only for practical, not theoretical reasons that no more MVs are available, then he may accept the ML estimate as the figure he is interested in. He should, however, be aware of the fact that his measure of the association between the two unobserved constructs ξ and η are very far from being observable, and that only 25% are validated, and 75% are postulated. Also, if the sample correlation matrix in Eq.5.56 is not gathered with sufficient precision and if his result is, for example, $s_{k\ell} = 0.11$ instead of $s_{k\ell} = 0.09$, then the correct LISREL estimate will turn out to be $r = 1.10$.

The LS estimate is on the conservative side, because it overestimates the loadings, thus linking the LVs more closely to the observations and underestimating the relationships between the LVs. It can never happen that the correlation r_{PLS} exceeds unity, or that the residual variance is negative (or other Heywood cases), because the LV correlation is the correlation between linear combinations.

Table 5.1 shows the real world data example that led to the construction of the more extreme artificial data in Eq.5.56. This example is taken from Weber (1982), who did not so much stress the very high correlation coefficient he had previously found by LISREL. Comparison with the CC and IBF estimates indicates that the estimate can depend largely on the statistical assumption only introduced to facilitate the estimation.

LS methods are closer to the data, more explorative and more data analytic. ML methods are closer to the model, more confirmative and more model analytic.

5.4 The Interpretation of Consistency at Large

In the previous section, reference was made to the limited number of manifest variables available as indicators for a construct variable. Is it true that the number is limited only for practical reasons? This is questionable, especially in the human sciences. For example, is it possible to find an arbitrarily large number of indicators for the socioeconomic status of the family of a student, after education and occupation of the parents have been taken as indicators? If we include the income of the father in the set of indicators, the meaning of the LV is likely to veer more towards stressing the socio-*economic* status of the father. If we include

Table 5.1: Association between German economic development and concern of wealth in the throne speeches of the Kaiser: LVP analysis with LISREL and PLS

Variable	PLS ModeA IBF model		PLS ModeB CC model		LISREL	
Loadings:						
Rye price (2nd diff.)	23.92	0	24.88	0	21.90	0
Wheat price (2nd diff.)	20.65	0	18.88	0	16.20	0
Wealth - others	0	1.89	0	1.23	0	1.00
Wealth - transactions	0	.53	0	.84	0	.64
Wealth - participants	0	.31	0	.37	0	.30
LV correlations						
Economic performance	1.00		1.00		1.00	
Wealth concerns	-.47	1.00	-.59	1.00	-.85	1.00

Economic performance: Indicators are the prices at the Berlin wholesale grain market from 1871 to 1912, more precisely the second differences, i.e. the speed of price changes. Scale unit: Mark.

Wealth concerns: Content analytic indicators, extracted from the speeches of the German Kaiser, 1871 - 1912. Scale unit: Number of words per 1000.

Negative correlation: "As the economy improves there is less concern with wealth, while a declining economy is associated with increasing wealth concerns."

Data from Weber (1982)

the number of books in the household as indicator, the LV will tend to become a socio-*cultural* indicator. It may well be for theoretical reasons that the number of indicators is limited.

Experience tells us that no empirical correlation matrix is like Eq.5.56. Each indicator behaves differently: a first variables is correlated with a second rather more than it is with a third, which goes to show that the first and second indicators are influenced by a doublet factor. And if the first three indicators are related to each other rather more than they are to a fourth indicator, these three can be understood as being influenced by a triplet factor. With each new indicator introduced in the set of indicators, it is possible that a whole bunch of new factors is introduced. A careful selection of indicators may help to reduce the influence of factors other than the unique factors and the common factor, it cannot eliminate them totally, though.

To put it another way: if only very few indicators are available, it is virtually impossible to prove the assumption that the indicators measure nothing other than the unidimensional LV and that no doublet, triplet, etc. factors are included. If, on the other hand, there are many indicators, the unidimensionality can be *confirmed* by the data. The common factor approach (LISREL etc.) and the component approach (PLS) draw different conclusions from this fact.

PLS. If there is not more than one indicator, the PLS estimate for the LV is identical with the MV, which, then, is a "non-estimate". If there are three indicators (cf. example in Sc.5.3), the estimate of the LV is too close to the data and the relations between the LVs are underestimated. The more indicators used, the more the estimates of the inner relationships approach the "true" parameters.

With an increasing number of indicators, the estimate of the LV correlation comes closer to the true value. This is what is stated in the theorem on consistency at large (Wold 1982d): The PLS estimates approach the true values proportionally, as the number of cases in the sample increases (i.e. the sample correlation matrix becomes a better estimate of the population correlation matrix) and as the number of indicators increases. The substantive interpretation and justification of this theorem is obvious: The more the "existence" of a LV is confirmed by the data, i.e. the more the model assumptions are justified by the data, the more does is make sense at all to estimate the parameters of this model, and the more consistent the parameter estimates become. Conversely: if the model assumptions are not supported by the data, it is better to stick closely to the data than to try to estimate the parameters of a postulated model. However, cf. Dijkstra (1981, 1983).

ML. ML estimates are always consistent with respect to the model postulated, no matter how poor the indicators and the data are. However, cf. Boomsma (1982), and Giegler (1983).

Target of consistency. Consistency of an estimator means that the estimate tends to the true value — as defined in the model — inasmuch as the number of cases increases indefinitely. Consistency at large of an LV means that the estimate of the LV tends to the true value inasmuch as both the number of cases and the number of indicators tend to infinity. Consistency at large is based on a double aggregation — over cases and over MVs; hence it implies a double invocation of the law of large numbers.

In Sc.5.1 the difference between "data structure" and "covariance structure" was elaborated. If the research problem is a data-structural one, all the considerations regarding consistency in Sc.5.2 and Sc.5.3 are of secondary relevance, the LS criterion assuring that the parameters reconstruct and optimally predict the data points.

A consistency argument is pointless when the target of consistency is not desirable. In Sc.5.2 and Sc.5.3 the convergence of PLS estimators on ML parameters was proved and implicitly it was more or less supposed that the PLS estimation is only a surrogate procedure for the ML estimation. The theorem of consistency at large applies to a situation when the PLS estimation is a surrogate procedure applied to a covariance-structural model.

Table 5.2: ML and PLS estimates for Lord's vocabulary data

	PLS model		LISREL	
	η_1	η_2	η_1	η_2
Loadings				
y_1	8.47	.	7.50	.
y_2	8.51	.	7.70	.
y_3	.	9.23	.	8.51
y_4	.	9.28	.	8.67
LV correlation				
η_1	1.00		1.00	
η_2	.75	1.00	.90	1.00
Diagonal fit indices				
H^2, communality	.86		.71	
$1 - c^2/s^2$	.86		.72	
Off-diagonal fit indices				
$1 - c/s$	.8		1.00	
$1 - \text{rms}(\boldsymbol{S} - \hat{\boldsymbol{\Sigma}})/s$	1.00		.99	
r_{BB}	.71		.95	
$\chi^2(df = 6, N = 101)$	65.		.1	

$s^2 = 91.94$ = average of MV variances

$s = 61.33$ = root mean square of MV covariances (diagonal excluded)

$\chi^2 = 2.26$ = under the hypotheses $\boldsymbol{\Sigma}$=diagonal

Explanation of fit indices see Sc.2.4

5.5 Some PLS - LISREL Comparisons

In Sc.5.5 several examples will be given showing the differences and the similarities of PLS and LISREL estimates for some selected examples. (The models are well-known LISREL examples, here cited from Jöreskog 1978; the PLS counterparts were first presented by Lohmöller and Wold 1981.)

Fit indices. Two diagonal-fit indices and four off-diagonal-fit indices (already defined in Sc.2.4.3) will be used to summarize the results of PLS and LISREL estimation. The diagonal-fit indices are the averaged communality H^2 (Eq.2.97, Eq.2.109) and the averaged proportion of explained variance "$1-c^2/s^2$" (Eq.2.132); both indices are unity in the case of perfect fit and zero where the model explains nothing of the variance. For standardized MVs both indices give identical values.

The first off-diagonal-fit index is the ratio of the root mean square (rms) of the residual covariances over the rms of the observed covariances, "$1-c/s$" (Eq.2.133). The other off-diagonal indices measure the discrepancy between the observed $\boldsymbol{S}$ and the reconstructed $\hat{\boldsymbol{\Sigma}}_1$ (Eq.2.128) which is considered explicitly in LISREL but not in PLS. For the sake of comparison, it is constructed in PLS as $\hat{\Sigma}_1 = \boldsymbol{PRP}' + \mathrm{diag}(\hat{\Theta})$. The other fit indices are the χ^2_{ML} (Eq.2.125), and the r_{BB}-index based on this χ^2_{ML} (Eq.2.131) .

Lord's vocabulary data. Table 5.2 shows the results of PLS and LISREL analysis of a simple data set used by Jöreskog (1978:452). There are 2 LVs, each indicated by 2 MVs. The first two MVs are vocabulary subtests, administered under liberal time limits; the other two MVs are high speed vocabulary tests. The MVs are unstandardized and the different variances of the MVs have a certain influence on the PLS estimation. The communality is $H^2 = 86\%$ in PLS and $H^2 = 71\%$ in LISREL. The correlation between the two LVs is $r = 0.75$ in PLS and $r = 0.90$ in LISREL. Notice that $(rH^2)_{\mathrm{PLS}} = 0.75 \times 0.86 = 0.65$ and $(rH^2)_{\mathrm{LISREL}} = 0.90 \times 0.71 = 0.64$.

The question is whether the two test batteries are congeneric ($\rho = 1$) or only parallel ($\rho < 1$). This research question implies clearly a covariance-structural model and, for an answer, the parameter ρ should be estimated as precisely as possible. Hence we conclude that $r = 0.90$ is the more precise answer to the hypothesis $\rho = 1$. But if we tried to predict the scores y_3 and y_4 on the speed tests from the scores on the slow speed tests y_1 and y_2, then a data-structural approach like the Principal Predictor model (Sc.3.4.1) with PLS estimation would be more appropriate.

Stability of alienation. Table 5.3 and Table 5.4 contain the results of 10 analyses in all. Tb.5.3 starts with three PLS models with different weight estimation modes, termed `D111`, `D100`, `D000`, and two LISREL counterparts, termed `NO-X` and `FI-X`. In Tb.5.3 standardized MVs are used, while in Tb.5.4 the analyses are repeated for unstandardized variables. For the inner part of the model the results for the standardized and unstandardized LISREL models are identical, confirming that the LISREL model is scale-free. Also the PLS-`D111` model (all blocks in ModeB) is scale-free, as to be expected from Dijkstra's (1981:54) remarks. When comparing the three standardized PLS models, one finds no big differences, at least in the inner model part.

In the unstandardized data set, though, the loss function is drastically influenced by the variances of the MVs, especially by s_2^2. The variance of y_2=SEI is much larger than the other variances, without, however, any interpretable relationship with the variances of the other variables. If the variances are taken into account by the loss function, the loading of SEI increases markedly. The explained variance $1 - c^2/s^2$ increases from 81% (standardized, Tb.5.2) to 97% (unstandardized, Tb.5.3), but the communality drops from 81% to 77%. Also the inner model parameters change and become in part more similar to the LISREL estimates. Also the off-diagonal fit measures improve.

The LISREL model corresponding to the PLS `D100` model is a fixed-X model. This means that the exogenous variables are taken as error-free and that the Γ_{LISREL} coefficients are in lieu of the PLS weights. A model like this would be used for the prediction of the endogenous variables, the changed estimates

Table 5.3: ML and PLS estimates for Stability of Alienation model, standardized manifest variables

	PLS model			LISREL	
	D111	D100	D000	NO-X	FI-X
Loadings					
η_1 = Background					
y_1 = Education	.96	.96	.91	.83	.79
y_2 = SEI	.75	.75	.84	.65	.32
η_2 = Alienation '67					
y_3 = '67 Anomia	.91	.91	.91	.81	.81
y_4 = '67 Powerlessness	.92	.91	.91	.81	.81
η_3 = Alienation '71					
y_5 = '71 Anomia	.93	.92	.92	.84	.84
y_6 = '71 Powerlessness	.89	.91	.91	.80	.80
Path coefficients and R^2					
$b_{2,1}$	-.44	-.44	-.43	-.57	-.49
$b_{3,1}$	-.18	-.18	-.18	-.15	-.12
$b_{3,2}$	.52	.52	.52	.66	.69
R_2^2, Alienation '67	.19	.19	.19	.32	.24
R_3^2, Alienation '71	.39	.39	.38	.58	.57
Diagonal fit indices					
H^2, communality	.80	.80	.81	.63	.63
$1 - c^2/s^2$	.80	.80	.81	.63	.63
Off-diagonal fit indices					
$1 - c/s$	.81	.82	.82	1.00	1.00
r_{BB}	.62	.62	.62	.97	.97
$\chi^2(df = 6, N = 101)$	87.	88.	102.	8.	8.

$s^2 = 1.00$ = average of MV variances

$s = 0.45$ = root mean square of MV covariances (diagonal excluded)

$\chi^2 = 229.0$ = under the hypotheses Σ =diagonal

PLS-D111 is a ModeB model, WeightMode=(ModeB,ModeB,ModeB)=(1,1,1)

PLS-D100 is a ModeC model, WeightMode=(ModeB,ModeA,ModeA)=(1,0,0)

PLS-D000 is a ModeA model, WeightMode=(ModeA,ModeA,ModeA)=(0,0,0)

LISREL NO-X is a model without x-variables, Θ=diagonal

LISREL FI-X is a fixed-X model, $\Theta_\epsilon = S_{xx}$, unrestricted

Explanation of fit indices see Sc.2.4

Wheaton, Muthén, Alwin and Summers (1977), Jöreskog (1978:452)

Table 5.4: ML and PLS estimates for Stability of Alienation model, unstandardized manifest variables

	PLS model			LISREL	
	D111	D100	D 000	NO-X	FI-X
Loadings					
η_1 = Background					
y_1 = Education	2.98	2.98	1.73	2.58	2.45
y_2 = SEI	15.90	15.98	21.21	13.76	6.79
η_2 = Alienation '67					
y_3 = '67 Anomia	3.11	3.20	3.21	2.80	2.80
y_4 = '67 Powerlessness	2.80	2.72	2.71	2.48	2.48
η_3 = Alienation '71					
y_5 = '71 Anomia	3.30	3.32	3.33	2.97	2.97
y_6 = '71 Powerlessness	2.82	2.80	2.79	2.52	2.59
Path coefficients and R^2					
$b_{2,1}$	-.44	-.43	-.33	-.57	-.49
$b_{3,1}$	-.18	-.18	-.13	-.15	-.12
$b_{3,2}$	.52	.52	.56	.66	.69
R_2^2, Alienation '67	.19	.19	.11	.32	.24
R_3^2, Alienation '71	.39	.39	.38	.58	.57
Diagonal fit indices					
H^2, communality	.80	.80	.77	.37	.44
$1 - c^2/s^2$	.41	.60	.97	.45	.16
Off-diagonal fit indices					
$1 - c/s$	.78	.78	.95	1.00	1.00
r_{BB}	.62	.63	.66	.97	.97
$\chi^2(df = 6, N = 101)$	87.	88.	102.	8.	8.

$s^2 = 83.94$ = average of MV variances

$s = 14.62$ = root mean square of MV covariances (diagonal excluded)

$\chi^2 = 229.0$ = under the hypotheses $\boldsymbol{\Sigma}$ =diagonal

PLS-D111 is a ModeB model, WeightMode=(ModeB,ModeB,ModeB)=(1,1,1)

PLS-D100 is a ModeC model, WeightMode=(ModeB,ModeA,ModeA)=(1,0,0)

PLS-D000 is a ModeA model, WeightMode=(ModeA,ModeA,ModeA)=(0,0,0)

LISREL NO-X is a model without x-variables, Θ=diagonal

LISREL FI-X is a fixed-X model, $\Theta_\epsilon = \boldsymbol{S}_{xx}$, unrestricted

Explanation of fit indices see Sc.2.4

Wheaton, Muthén, Alwin and Summers (1977), Jöreskog (1978:452)

reflecting the changed purpose of the model. The changes are more pronounced for the unstandardized MVs than they are for the standardized, as they are also for the PLS models.

Peer influence on aspiration. Duncan, Haller and Portes (1968) presented one of the first path models which included latent variables (not only unobserved variables, as in Blalock 1963). They treated their model on Peer Influence on Aspiration (model IV, p.234) with classical path-analytical tools by an ad-hoc Unweighted Least Squares (ULS) estimation method. Jöreskog and Sörbom (1978:31-39) applied LISREL estimation to the (nearly) unchanged model. The original Duncan-Haller-Portes model has two endogenous LVs (respondent's ambition and best friend's ambition) in a reciprocal relationship; the LVs are directly regressed on the exogenous MVs. Here in Table 5.5 the model is re-specified and two exogenous LVs are introduced: respondent's background and best friend's background. The indicators for the two background LVs overlap. The model, as presented in Tb.5.5, is adjusted to the capabilities of the program PLSC: The nonrecursive path model is replaced by the reduced form, so that it can be consistently estimated by OLS regression.

The LISREL estimates in Tb.5.5 are identical to the Jöreskog and Sörbom results, when the two exogenous LVs have been eliminated. The PLS estimates are quite different, the loadings being higher and the path coefficients smaller. The communality of the LISREL model is $H^2 = 63\%$, of the PLS model 82%. The off-diagonal fit indices are: $\chi^2_{\text{LISREL}} = 12(\text{df} = 16, N = 101)$, the $\chi^2_{\text{PLS}} = 129(\text{df} = 27, N = 101)$, which, when transformed into the reliability index, gives an $r_{\text{BB}} = 95\%$ fit for the LISREL estimates and an $r_{\text{BB}} = 51\%$ fit for PLS.

Summary. When measured by summary statistics related to the diagonal elements or to the off-diagonal elements of the MV covariance matrix, PLS and LISREL results show striking differences. For the small models presented in this section, the communality for the PLS models is, typically, 20 percentage points higher than for LISREL models. This figure shows to what extent PLS is better when the aim of the analysis is to reconstruct (predict) the data point from the latent variables and the other model parameters.

The off-diagonal fit, as measured by the reliability index, is for PLS models 20 percentage points lower than for LISREL models, or even 44 points as in Tb.5.5. This figure shows how much better suited LISREL is for a covariance structural approach.

The examples show the numerical implications of the decision regarding the research problem and the type of model. The dichotomies outlined in Sc.5.1 — general vs. applied, explanation vs. prediction, covariance structure vs. data structure — are not merely a theoretical "Glasperlenspiel" but influence the results considerably. The differences of the results may be even stronger than the decisiveness of the researcher regarding the type of his model.

The differences in the single parameters do not cancel out in a summary statistic like H^2 or r_{BB}, because the differences all tend systematically in the same

Table 5.5: ML and PLS estimates for Peer Influence model with overlapping exogenous indicators

	PLS model				LISREL			
	η_1	η_2	η_3	η_4	η_1	η_2	η_3	η_4
LV weights for exogenous variables								
y_1 Resp. parental aspiration	.24	.	.	.	.36	.	.	.
y_2 Respondent's IQ	.48	.	.	.	.56	.	.	.
y_3 Respondent's SES	.43	.37	.	.	.49	.11	.	.
y_4 Friend's SES	.44	.39	.	.	.16	.42	.	.
y_5 Friend's IQ	.	.57	.	.	.	.66	.	.
y_6 Friend's parental aspiration	.	.19	.	.	.	.28	.	.
LV loadings								
y_1 Resp. parental aspiration	.36	.	.	.	.36	.	.	.
y_2 Respondent's IQ	.70	.	.	.	.56	.	.	.
y_3 Respondent's SES	.47	.27	.	.	.49	.11	.	.
y_4 Friend's SES	.38	.36	.	.	.16	.42	.	.
y_5 Friend's IQ	.	.81	.	.	.	.66	.	.
y_6 Friend's parental aspiration	.	.33	.	.	.	.28	.	.
y_7 Resp. occupational aspiration	.	.	.90	.	.	.	.76	.
y_8 Resp. educational aspiration	.	.	.91	.	.	.	.82	.
y_9 Friend's educational aspirat.	.	.	.	.91	.	.	.	.84
y_{10} Friend's occupational aspirat.	.	.	.	.90	.	.	.	.76
LV correlation								
η_1 Respondent's background	1.00				1.00			
η_2 Friend's background	.71	1.00			.53	1.00		
η_3 Respondent's ambition	.59	.46	1.00		.68	.46	1.00	
η_4 Friend's ambition	.50	.66	.44	1.00	.49	.75	.40	1.00
Path coefficients								
η_3 Respondent's ambition	.55	.05	.	.	.61	.14	.	.
η_4 Friend's ambition	.02	.64	.	.	.13	.68	.	.
R^2	.	.	.35	.43	.	.	.48	.57
Diagonal fit indices								
H^2, communality (only endog.)	.82				.63			
$1 - c^2/s^2$	.82				.63			
Off-diagonal fit indices								
$1 - c/s$	.77				1.00			
$1 - \text{rms}(S - \hat{\Sigma})/s$	1.00				.88			
r_{BB}	.51				.95			
$\chi^2 : df, (N = 101)$	129:27				12:16			

$s^2 = 1.00$ = average of MV variances

$s = 0.30$ = root mean square of MV covariances (diagonal excluded)

$\chi^2 = 226.0$ = under the hypotheses Σ =diagonal ($N = 101, df = 45$)

Explanation of fit indices see Sc.2.4

Duncan, Haller, Portes (1968) Jöreskog (1978:432)

direction. In this respect the summary statistics function like a magnifying glass. However, they are not model parameters and not part of the model. Even if the PLS and ML estimates of a single parameter are systematically different, the researcher may have difficulties in telling which one is more in accordance with his theory and expectation, because both lie within a theoretically acceptable range.

If PLS is used for practical reasons as a surrogate estimation procedure for a covariance structural model, the high χ^2-statistic and the low reliability index make one aware of the bias of the single parameters. These summary statistics should not be used for a decision about the fit of the model, because these indices are not optimized in the estimation procedure. (More references for PLS-LISREL comparisons: Lohmöller & Wold 1981, Jöreskog & Wold 1982b, Fornell & Bookstein 1982, Lohmöller 1988a).

5.6 The PLS Solution of the Identification Problem

A question often difficult to discuss in connection with structural equation analysis is the identification status of the model, or, more exactly, the identifiability of the parameters of the model (Dupačová & Wold 1975, 1982, Wegge 1981). A model is identified, if only one set of parameters exists that generates the observations, given the hypothesized structure of the model. For a covariance structure model, this means that no other set of parameter values leads to the observed covariance matrix but the obtained set of parameters. Conversely, if an estimation criterion has been established, the parameter estimates are unique only if the model is identified. If a model is not identified, the model must be restricted in order to attain the status of identification. Three kinds of restrictions may be distinguished:

(i) restriction for scale unambiguity (RSU),

(ii) restrictions for identifiability (in a narrower sense) (cf. Seidel 1987, Stelzl 1986), and

(iii) additional restrictions leading to testability (cf. de Pijper & Saris 1982).

The first two kinds of restrictions will be discussed in some detail in this section, with emphasis on the differences between the PLS and the ML species of LV path models.

5.6.1 Restriction for Scale Unambiguity (RSU)

If a model includes latent variables, then the scale of the LV must be determined. Otherwise, the LV can be multiplied by some constant c, and if all regression coefficients, for which this LV is the regressor, are divided by c, and all regression coefficients, for which this LV is the regressand, are multiplied by c, then this new set of parameters will produce the same covariance matrix. Three ways to determine the scale of an LV will be considered.

RSU1, standardized LVs. The first, and in the psychometric tradition widely used, way is to standardize the LV to unit variance. In LISREL, however, this kind of scale-fixing can be applied only to the exogenous LVs, because their variances (the diagonal of $\mathbf{\Phi}$) are model parameters. The variances of the endogenous LVs are not model parameters but result as functions from other model parameters.[3] In PLS basic method, the LVs are always scaled to unit variance, at least during the estimation phase.

RSU2, fixed loadings. The second type of scale-fixing applies to the loadings. For each LV one loading coefficient is fixed, the scale of the LV thus being tied to the scale of an MV, after which a score on the LV can be interpreted "in the metric" of that MV. It is advisable to fix the (prospected) highest loading of each LV.

RSU3, standardized residuals. A third type of scale-fixing was tried out as a compromise between the first two. The scale of the endogenous LVs was fixed by fixing the scale of the inner residual variables (the path residuals), i.e. by fixing the diagonal of $\mathbf{\Psi}$.

Comparison. The different sorts of restriction for scale unambiguity should lead to essentially the same results. Essentially means identical, up to a rescaling of the LVs. This assumption was tested with LISREL IV, by comparing RSU2 and RSU3 on different models. Equivalent models, differing only in the type of scalefixing (RSU2 or RSU3) are expected to give the same χ^2 fit statistic and the same standardized solution. When PLS estimates are taken as starting (and final) values for the diagonal of $\mathbf{\Psi}$, the variance of the endogenous LVs are expected to be close to unity and differ from unity only in as much as PLS estimates differ from ML estimates in general.

However, the results are different from those expected. Table 5.6 shows a very small example.[4] The difference in the standardized solutions is on the second decimal place, and the fit statistics are $\chi^2 = 45.4996$ for the RSU2 estimates and $\chi^2 = 45.3502$ for the RSU3 estimates. Considering that both models are identified, and this means mathematically equivalent and unique, this difference is too great. Notice that scale-fixing by the "unconventional" restriction RSU3 (on the residual variances) indicates a slightly better fit.

In this example the difference between RSU2 and RSU3 is not substantial. This is due to the fact that this model has a small diameter (where the diameter is the number of arrows in the longest pathway across the path diagram). In models with big diameters the variances of the exogenous and the first endogenous LVs show reasonable values, whereas the variances of the last LVs in a causal chain become greater and greater with each iteration step and, finally, may reach absurd values; the LISREL estimates of the path coefficients diverge out to bizarre values and, sometimes, the state of empirical non-identifiability is reached.

This leads to the conclusion that the theoretically equivalent models, differing only in the "restrictions for scale unambiguity", may prove to be numerically non-

[3]In LISREL V it is possible to fix the scales of the LVs, Jöreskog and Sörbom 1981.

[4]Correlations are taken from the example in Sc.2.5

Table 5.6: Fixed loadings or fixed residual variances?

Parameter	PLS model	LISREL, fixed loadings			LISREL, fixed ψ_j^2		
		unstandardized		standardized	unstandardized		standardized
Loadings			θ^2			θ^2	
Educ mother	91 . .	91 . .	21	91 . .	89 . .	21	89 . .
Educ father	84 . .	73 . .	49	73 . .	71 . .	49	71 . .
Occu father	90 . .	*91 . .	22	91 . .	88 . .	23	88 . .
Press achvm	. 92 .	. *98 .	20	. 90 .	. 85 .	20	. 90 .
Press activ	. 90 .	. 91 .	31	. 84 .	. 79 .	31	. 83 .
Press intel	. 94 .	. 100 .	17	. 92 .	. 87 .	17	. 91 .
PMA verbal	. . 90	. . *92	28	. . 85	. . 69	28	. . 85
PMA number	. . 89	. . 85	38	. . 79	. . 64	38	. . 79
PMA space	. . 57	. . 45	83	. . 42	. . 34	83	. . 42
Path coefficients							
η_1	63 . .	62 . .		68 . .	70 . .		67 . .
η_2	2 66 .	-8 86 .		-9 86 .	-11 101 .		-9 86 .
Inner residual variance							
ψ_k^2	100 61 55	100 46 31		100 54 36	*100 *61 *55		100 55 36
LV covariance							
η_1	103 . .	100 . .		108 . .	100 . .		100 . .
η_2	63 100 .	62 85 .		68 100 .	70 110 .		67 100 .
η_3	43 57 100	45 68 86		49 80 100	60 104 153		48 80 100
χ^2/df		45.4996 / 25			45.3502 / 25		

equivalent. The estimates seem to be more reliable if the scales of the LVs are tied as closely as possible to the observed variables. Our experiment with RSU3 scale-fixing justifies the recommendation of Jöreskog and Sörbom (1978) to use type RSU2 scale-fixing. The problem is not the mathematical-statistical frame of LISREL but the numerical properties of the iterative algorithm of the LISREL program (Fletcher & Powell 1963, Fletcher & Reeves 1964).

The result with LISREL is that the RSU is non-trivial. The estimation procedure is influenced by which parameters are fixed, or not. If a parameter is free, the derivatives of $\boldsymbol{S}$ with respect to this parameter have to be minimized, and the estimate has a certain standard error; if it is fixed, for any reason whatever, the algorithm leaves the derivatives as they are, and the parameter has no standard error.

RSU1 is applied always applied in PLS. Any other scaling may be used. Here the RSU influences only the scales, not the total composition of the results. The theoretically meaningful distinction between restriction for scale unambiguity and restrictions for identifiability (in a narrower sense) and testability is strictly maintained in the PLS technique, whereas in the LISREL algorithm this distinction is blurred.

(Experience with RSU2 and RSU3 was gathered during the construction of program PLSC. PLSC includes a feature that creates a complete input deck for LISREL, in which the PLS estimates are used as starting values; the restriction

Table 5.7: Five equivalent loading matrices for a two-factor model

Solution:	#1	#2	#3	#4	#5
y_1	80 *0	70 -40	57 -57	40 -70	0 -80
y_2	70 40	80 *0	77 -20	70 -40	40 -70
y_3	57 57	77 20	80 *0	77 -20	57 -57
y_4	40 70	70 40	77 20	80 *0	70 -40
y_5	0 80	40 70	57 -57	70 40	80 *0

* = fixed zero

on the parameters are made in such a way as to maintain the intentions expressed in the PLS model and to yield an identified LISREL model. In PLSC version 1.4, technique RSU3 was used, which resulted in sometimes horrible LISREL results. Thus the translation rule had to be changed completely, so that PLSC 1.6 and newer versions use RSU2 technique.)

5.6.2 Restrictions for Identifiability

Identifiability of a model is, typically, achieved by restrictions on single parameters. For factor models this can be achieved by fixing one or several loading coefficients to zero. In this section it will be demonstrated that in some cases a restriction on a single parameter is equivalent to a rotation criterion, an orthonormality condition, or a residual analysis.

Table 5.7 shows a small factor model, which assumes that the five MVs can be explained by two standardized uncorrelated LVs. Five loading matrices are given, each one with a fixed zero in a different position. The parameters for each of the five models are thus identifiable. The five loading matrices are equivalent up to an orthogonal rotation, the first and the last one being Varimax rotation solutions, the middle one being a general-and-contrast-factor solution. This example shows that a rotational criterion has the same effect on the identification status as the restriction of a single parameter.

Let us consider solution #3 in Tb.5.7 in more detail. The second LV is a general factor and, had we used least squares, would be the first principal component (PC). The second PC can be determined by the deflation technique, meaning that we subtract the part of the MVs correlated with the first PC from the MVs and perform the PC analysis the residuals. However, the deflation technique is equivalent to an appropriate rotation of two or more LVs which are determined simultaneously (cf. Horst 1965:chapter 8). Thus, the deflation can also be considered as a kind of restriction serving to achieve the status of identifiability.

5.6.3 Identifiability in PLS Model

Herman Wold claims that in PLS no identifiability problems arise. To be more precise, the identifiability problems are solved in a very general and straightforward manner.

First of all, the restrictions for scale unambiguity are always realized by method

RSU1: i.e. the LVs are scaled to unit variance. With RSU2 one must be sure that the one fixed loading will come out different from zero; had one fixed a loading coefficient to unity, for example, where the true coefficient is zero, this would have no restrictive effect at all.

The second prescription warranting identifiability restricts the class of admitted models. It is prescribed that the total set of MVs is partitioned into exhaustive and non-overlapping subsets (or blocks) of MVs, each block being attached to only one LV. Furthermore, it is prescribed that the LVs form a recursive (loop-free) path model, a causal chain in a broader sense.[5] With these two prescription, every model — PLS as well as LISREL models — is identified.

Both prescriptions, as fas as non-overlapping subsets and loop-free path models are concerned, allow for some exceptions.

The variables of two different subsets need not be totally different. The same variable may be repeated in another block. This is usually so in hierarchical models (cf. Noonan & Wold 1983). In such a case the LV weights make sense but the loadings and the outer residuals have to be interpreted with care. The two loadings relating the same MV to two correlated LVs are LV structure coefficients, i.e. covariances, and not LV pattern coefficients, i.e. regressions of MVs on LVs. The LV patterns must be computed by different formulae (cf.Tb.2.4), and the residuals by using these LV pattern coefficients.

If the inner part of the model is not a loop-free path model, the outer part of the model can be estimated, i.e. the weights and the correlations of the LVs can be determined by Wold's basic method. But to estimate the path model, other estimation methods must be applied, for example, the Fix-Point method (Bergström 1981, 1986, Bergström & Wold 1980, 1982, Bodin 1970, 1974, Hui 1978, Hui & Wold 1982, Lyttkens 1973, Mosbaek & Wold 1970, Strotz & Wold 1960, Wold 1965, 1981c).

[5]In a narrower sense, the member of a chain are connected only to their direct neighbors.

Chapter 6

Latent Variables Three-Mode Path (LVP3) Analysis

Three-mode data y_{ktn} are indexed by elements of three different sets of indices. Thus, y_{ktn} may denote the score of case $n \in \mathcal{N}$ at time $t \in \mathcal{T}$ with respect to attribute $k \in \mathcal{K}$. The term "three-mode" as well as the first three-mode structural model, the three-mode factor model (FA3 model), was introduced by Tucker (1963, 1964, 1965, 1966, 1967, 1972). The three-mode path model with latent variables (LVP3 model), introduced by Lohmöller and Wold (1980), uses a three-mode factor model as outer and a three-mode path model as inner model. An application of LVP3 analysis was presented by Lohmöller and Wold (1982). The program for LVP3 analysis is called PLS3 and is written by Lohmöller (1981a:chapter 6).

Unlike the rectangular parameter matrices of the "two-mode" LV path model, loading and path matrices of the LVP3 model are specified as "Kronecker structured matrices", $\boldsymbol{P} = \boldsymbol{P}_1 \otimes \boldsymbol{P}_2$ and $\boldsymbol{B} = \boldsymbol{B}_1 \otimes \boldsymbol{B}_1$. The estimation relies on Lohmöller's theorem on Kronecker principal components, first presented by Lohmöller and Wold (1980) and repeated here in Sc.6.2, after some reflections on three-mode data in Sc.6.1.

6.1 Three-Way Data Models

Three-way data are ordered by three indices; for example, y_{ktr} may denote the election outcome of a party k at election time point t in a region r, and $x_{\ell tr}$ the number of employed, and unemployed, workers at time t in region r. If theory is concerned only with the influence of unemployment categories (x_ℓ) on voting outcome (y_k), time points or regions or both may be used as observational units, and we have a two-way ordering of observations. In this case, the indices k and ℓ refer to "specified" (individualized, known, see Sc.5.1) coordinates of the data array, and t and r refer to "unspecified" (anonymous) coordinates of replications.

It may, however, turn out that the time dimension taps a causal influence of its own, i.e. a certain segments of the electorate changes its voting behaviour with time in a systematic and interpretable way. Then the time must be specified theoretically, the model considers the variables $x_{\ell t}$ and y_{kt}, with regions as the only observational units, and we have a two-way ordering of variables. As a general method for modeling variables ordered in two ways and observed in one way (three-way or three-mode data), the three-mode path analysis with latent variables can be used.

The basic PLS design is concerned with two-ways data: One array of variables is recorded over one array of observational units. The array of variables may consists of attributes, or one attribute a different time points, or one attribute measured by different methods, etc. The variables are considered as "specified" and "known" (see Sc.5.1), and structural parameters account for the relationships between the variables. The array of observational units may consist of individuals, groups of individuals, time points, regions, countries, etc. The units are treated as "unspecified" and "random", and incidental parameters account for the relationships between the units.

Many sciences use cross-classified (two-way-ordered) variables. In psychological studies often an array of traits is considered, each on being measured by the same array of methods (multitrait-multimethod (MTMM) model), and this crossclassification of traits and methods is observed over an array of individuals. Three-mode path analysis puts the interaction of traits and methods into the structural parameters, and the incidental parameters (factor scores) locate the individuals on latent variables which are not pure trait variables or pure method variables but trait-by-method variables like the MVs.

If economic indicators, measured at a small number of significant time points, are observed over a large array of countries (or regions, or branches, or firms), then the cross-classification of indicators and time points are treated as structural, and the countries as incidental.

6.2 The Kronecker Principal Component (KPC) Model

The most conspicuous item of Tucker's three-mode factor model is the Kronecker product operator "$\otimes$". The Kronecker product[1] (or direct product) $\boldsymbol{A}$ of two matrices $\boldsymbol{B}$ and $\boldsymbol{C}$ is written in matrix notation as:

$$\boldsymbol{A} = \boldsymbol{B} \otimes \boldsymbol{C} \tag{6.1}$$

and in elementary notation as:

$$a_{(ji)(k\ell)} = b_{jk}c_{i\ell}. \tag{6.2}$$

The rows and the columns of $\boldsymbol{A}$ are indexed by combined indices (ji) and $(k\ell)$. A combined index consists of two indices ordered in a way that the first index, e.g. j, is kept constant temporarily, while the second index, e.g. i, is running.

[1]Concerning the term Kronecker product, Searle (1966:215) writes: "Although MacDuffee (1933) cites many references from the late nineteenth and early twentieth centuries in an extensive discussion of direct products, nowhere does he mention Kronecker in this regard. One may speculate, therefore, as to how his name has become associated with direct products."

The KPC model. The Kronecker product matrix is a very large matrix, compared with $\boldsymbol{B}$ and $\boldsymbol{C}$ themselves. Hence, one may try it in reverse: if a matrix $\boldsymbol{A}$ is very large and has its indices structured as combined indices, it may be represented by relatively small parameter matrices. The Kronecker principal component (KPC) model takes the form:

$$\boldsymbol{A} = \boldsymbol{B} \otimes \boldsymbol{C} + \boldsymbol{E}. \tag{6.3}$$

$\boldsymbol{B}$ and $\boldsymbol{C}$ are called the Kronecker principal component matrices of $\boldsymbol{A}$, if they are estimated under the least squares criterion

$$\text{LS} = \text{trace}(\boldsymbol{E}'\boldsymbol{E}) \stackrel{!}{=} \min. \tag{6.4}$$

An obvious goodness of fit index for Kronecker PCs is

$$\text{Kpcfit}(\boldsymbol{A}) = 1 - \frac{\text{trace}(\boldsymbol{E}'\boldsymbol{E})}{\text{trace}(\boldsymbol{A}'\boldsymbol{A})}, \tag{6.5}$$

which is the proportion of the sum of squares accounted for by the model.

If $\boldsymbol{A}$ is symmetric, e.g. a multitrait-multimethod correlation matrix, it may be appropriate to exclude the diagonal of $\boldsymbol{E}$ from the residual minimization. We define a residual matrix $\tilde{\boldsymbol{E}}$ where the diagonal elements are deleted,

$$\tilde{\boldsymbol{E}} = \boldsymbol{E} * (\mathbf{1} - \mathbf{I}) \tag{6.6}$$

and call those parameter matrices $\boldsymbol{B}$ and $\boldsymbol{C}$ the Kronecker Minres factors (cf. Harman & Jones 1966) which minimize the LS criterion

$$\tilde{\text{LS}} = \text{trace}(\tilde{\boldsymbol{E}}'\tilde{\boldsymbol{E}}) \stackrel{!}{=} \min. \tag{6.7}$$

The following theorem states that the parameter estimation is surprisingly simple.

Theorem: Kronecker Principal Components: The elements of the Kronecker principal component matrices $\boldsymbol{B} = [b_{jk}]$ and $\boldsymbol{C} = [c_{i\ell}]$ of a matrix $\boldsymbol{A} = [a_{(ji)(k\ell)}]$ are identical to the elements of the principal components:

$$\breve{\boldsymbol{A}} = \breve{\boldsymbol{b}}\breve{\boldsymbol{c}} + \breve{\boldsymbol{E}}, \tag{6.8}$$

where $\breve{\boldsymbol{A}} = [a_{(kj)(i\ell)}]$ is a matrix and $\breve{\boldsymbol{b}} = \text{Vec}\,\boldsymbol{B}' = [b_{(kj)}]$ and $\breve{\boldsymbol{c}} = \text{Vec}\,\boldsymbol{C}' = [c_{(i\ell)}]$ are vectors. Proof: The loss function of Eq.6.8 is:

$$\begin{aligned} \breve{\text{LS}} &= \sum_{ijk\ell}(a_{(kj)(i\ell)} - b_{(kj)}c_{(i\ell)})^2 = \text{trace}(\breve{\boldsymbol{E}}'\breve{\boldsymbol{E}}) \\ &= \sum_{ijk\ell}(a_{(ji)(k\ell)} - b_{kj}c_{i\ell})^2 = \text{trace}(\boldsymbol{E}'\boldsymbol{E}) = \text{LS}. \end{aligned} \tag{6.9}$$

Hence $\breve{\text{LS}}$ is identical to the loss function LS of the KPC model, Eq.6.3. Thus, if $\breve{\boldsymbol{b}}$ and $\breve{\boldsymbol{c}}$ are the principal component vectors of $\breve{\boldsymbol{A}}$, their reordered versions $\boldsymbol{B}$ and $\boldsymbol{C}$ will minimize LS.

Table 6.1: Algorithm for the Kronecker principal component model

Model:	$\boldsymbol{A}$	$=$	$\boldsymbol{B} \otimes \boldsymbol{C} + \boldsymbol{E}, \quad \boldsymbol{B} = [b_{jk}], \quad \boldsymbol{C} = [c_{i\ell}]$
Estimation 1:	LS	$=$	$\operatorname{trace} \boldsymbol{E}\boldsymbol{E}' \overset{!}{=} \min$
		$\Longrightarrow$	Kronecker principal components
Solution 1:	$(\boldsymbol{B}, \boldsymbol{C})$	$:=$	$\mathrm{Kpc}(\boldsymbol{A})$
Algorithm 1:	For arbitrary starting values $c_{i\ell}$ repeat steps #1,...,#7:		
step #1	b_{jk}	$:=$	$\sum_i \sum_\ell a_{(ji)(k\ell)} c_{i\ell}$
step #2	f	$:=$	$\max_{j,k} b_{jk}$
step #3	b_{jk}	$:=$	b_{jk}/f
step #4	g	$:=$	$1/\sum_{jk} b_{jk}^2$
step #5	$c_{i\ell}$	$:=$	$g \sum_j \sum_k a_{(ji)(k\ell)} b_{jk}$
step #7	if no convergence is obtained, go to step #1, else stop.		
Estimation 2:	LS	$=$	$\operatorname{trace}[(\mathbf{1} - \mathbf{I}) * \boldsymbol{E}][(\mathbf{1} - \mathbf{I}) * \boldsymbol{E}]' \overset{!}{=} \min$
		$\Longrightarrow$	Kronecker Minres factors
Solution 2:	$(\boldsymbol{B}, \boldsymbol{C})$	$:=$	$\mathrm{Kpc}(\boldsymbol{A}$ with unknown diagonal)
Algorithm 2:	as algorithm 1, but in each iteration cycle the diagonal elements of $\boldsymbol{A}$ are replaced:		
step #6	$a_{(jk)(jk)}$	$:=$	$b_{jj} c_{kk}$

The KPC algorithm. The model has an indeterminacy regarding the scaling of the parameters, because $(\boldsymbol{B}d) \otimes (\boldsymbol{C}/d)$ for a nonzero constant d gives the same fit to the model. An appropriate restriction for scale unambiguity (RSU) is

$$\max_{j,k} b_{jk} = 1. \tag{6.10}$$

The estimation can be done by means of any conventional eigenvalue routine for rectangular matrices. This requires creating the reordered matrix $\breve{\boldsymbol{A}}$, before calling the eigenvalue routine, and reordering $\breve{\boldsymbol{b}}$ and $\breve{\boldsymbol{c}}$ afterwards. A special algorithm for Kronecker PCs and Kronecker Minres factors is outlined in Table 6.1. For later reference in Tb.6.3 we note that this algorithm is invoked by

$$(\boldsymbol{B}, \boldsymbol{C}, \mathrm{Kpcfit}) := \mathrm{Kpc}(\boldsymbol{A}), \tag{6.11}$$

which shows that $\boldsymbol{A}$ is input to the algorithm and that $\boldsymbol{B}, \boldsymbol{C}, \mathrm{Kpcfit}$ are output.

The Kronecker decomposition can be applied to any type of real matrices with the appropriate index structure, whether it be a data matrix ($\boldsymbol{X} = \boldsymbol{X}_1 \otimes \boldsymbol{X}_2 + \boldsymbol{X}_0$), a covariance matrix ($\boldsymbol{S} = \boldsymbol{S}_1 \otimes \boldsymbol{S}_2 + \boldsymbol{S}_0$), a correlation matrix ($\boldsymbol{R} = \boldsymbol{R}_1 \otimes \boldsymbol{R}_2 + \boldsymbol{R}_0$), a path matrix ($\boldsymbol{B} = \boldsymbol{B}_1 \otimes \boldsymbol{B}_2 + \boldsymbol{B}_0$), a loading matrix ($\boldsymbol{P} = \boldsymbol{P}_1 \otimes \boldsymbol{P}_2 + \boldsymbol{P}_0$), a weight matrix ($\boldsymbol{W} = \boldsymbol{W}_1 \otimes \boldsymbol{W}_2 + \boldsymbol{W}_0$) or a contingency table ($\boldsymbol{M} = \boldsymbol{M}_1 \otimes \boldsymbol{M}_2 + \boldsymbol{M}_0$).

Example. Table 6.2 shows an application to a correlation matrix of three test batteries, each with three subtests (verbal, numerical, spatial abilities).[2] Tb.6.2a

[2]Data from Thurstone & Thurstone 1941, cited from Horst 1965:569.

Table 6.2: Kronecker decomposition of a correlation matrix

Tb.6.2a Original correlation matrix $\boldsymbol{R}$

set j	test i	set $k=1$, test $\ell=1$	2	3	set 2, test 1	2	3	set 3, test 1	2	3
1	1	100								
	2	25	100							
	3	27	40	100						
2	1	64	14	18	100					
	2	18	65	41	9	100				
	3	18	26	61	15	30	100			
3	1	63	19	22	71	10	18	100		
	2	37	53	47	25	54	44	29	100	
	3	28	36	61	19	39	50	24	43	100

Tb.6.2b Reordered correlation matrix $\breve{\boldsymbol{R}}$ and vectors $\breve{\boldsymbol{r}}_1, \breve{\boldsymbol{r}}_2$

set k	set j	test $i=1$, test $\ell=1$	2	3	test 2, test 1	2	3	test 3, test 1	2	3	$\breve{\boldsymbol{r}}_1$
1	1	100	25	27	25	100	40	27	40	100	113
	2	64	18	18	14	65	26	19	53	36	74
	3	63	37	28	19	53	36	22	47	61	76
2	1	64	14	18	18	65	41	18	26	61	74
	2	100	9	15	9	100	30	15	30	100	105
	3	71	25	19	10	54	39	18	44	50	73
3	1	63	19	22	37	53	47	28	36	61	76
	2	71	10	18	25	54	44	19	39	50	73
	3	100	29	24	29	100	43	24	43	100	114

$\breve{\boldsymbol{r}}_2'$ =	100	26	26	26	94	48	26	48	94

Kpcfit(R) = 0.973074

Tb.6.2c Back-transformed Kronecker component matrices $\boldsymbol{R}_1$ and $\boldsymbol{R}_2$

$\boldsymbol{R}_1$ for sets

j	$k=1$	2	3
1	113	74	76
2	74	105	73
3	76	73	114

$\boldsymbol{R}_2$ for tests

i	$\ell=1$	2	3
1	100	26	26
2	26	94	48
3	26	48	94

All table entries are multiplied by 100

shows that the subtests in the same battery are relatively independent and that the corresponding subtests are highly correlated between batteries. The submatrices of $\boldsymbol{R}$ in Tb.6.2a are reordered so as to form the rows of $\breve{\boldsymbol{R}}$ in Tb.6.2b. The principal components of $\breve{\boldsymbol{R}}$, i.e. the vectors $\breve{\boldsymbol{r}}_1$ and $\breve{\boldsymbol{r}}_2$, are shown in the right and the lower margin of Tb.6.2b. The index Kpcfit = 97% reveals that the fit of the model is excellent.

Tb.6.2c gives the reordered Kronecker principal component matrices. The matrix $\boldsymbol{R}_2$ is a weighted average of the correlation of the subtests, averaged within and between batteries. Numerical and spatial ability are a slightly more highly correlated with each other ($r_{23}^{(2)} = 0.48$) than with the verbal ability ($r_{12}^{(2)} = r_{13}^{(2)} = 0.26$). The diagonal elements of $\boldsymbol{R}_2$ show the "validity" of the subtests across batteries, which is $r_{22}^{(2)} = r_{33}^{(2)} = 0.94$ for the numerical and spatial ability, when the coefficient for verbal ability is arbitrarily (see Eq.6.9) set to $r_{11}^{(2)} = 1.00$.

The matrix $\boldsymbol{R}_1$ in Tb.6.2c is a weighted average (with the elements of $\boldsymbol{R}_2$ a weights) of the submatrices of $\boldsymbol{R}$ in Tb.6.2a. The three subtests in the second battery are somewhat more independent of each other than in the first and the third battery; this is mapped into the diagonal of $\boldsymbol{R}_1$, where $r_{22}^{(1)} = 1.05$ is lower than $r_{11}^{(1)} = 1.13$ and $r_{33}^{(1)} = 1.14$.

6.3 The Three-Mode LV Path (LVP3) Model

Let $\boldsymbol{y}_{[K \times 1]}$ denote a vector of manifest variables with covariance matrix $\boldsymbol{S}_{[K \times K]}$, and the variables indexed by a combined index:

$$y_{(k_1 k_2)}, \quad k_1 = 1, \ldots, K_1, \quad k_2 = 1, \ldots, K_2, \quad K = K_1 \times K_2. \tag{6.12}$$

Let $\boldsymbol{Y}_{[J \times 1]}$ denote a vector of latent variables with covariance matrix $\boldsymbol{R}_{[J \times J]}$, and the LVs indexed by a combined index:

$$Y_{(j_1 j_2)}, \quad j_1 = 1, \ldots, J_1, \quad j_2 = 1, \ldots, J_2, \quad J = J_1 \times J_2. \tag{6.13}$$

The LVP3 model is then written as:

$$\boldsymbol{Y} = (\boldsymbol{B}_1 \otimes \boldsymbol{B}_2)\boldsymbol{Y} + \boldsymbol{u} \tag{6.14}$$
$$\boldsymbol{y} = (\boldsymbol{P}_1 \otimes \boldsymbol{P}_2)\boldsymbol{Y} + \boldsymbol{e} \tag{6.15}$$
$$\boldsymbol{Y} = (\boldsymbol{W}_1 \otimes \boldsymbol{W}_2)'\boldsymbol{y} + \boldsymbol{d}, \tag{6.16}$$

where $\boldsymbol{P}_{1[K_1 \times J_1]}$ and $\boldsymbol{P}_{2[K_2 \times J_2]}$ are pseudo loading matrices, $\boldsymbol{W}_{1[K_1 \times J_1]}$ and $\boldsymbol{W}_{2[K_2 \times J_2]}$ pseudo weight matrices, $\boldsymbol{B}_{1[K_1 \times K_1]}$ and $\boldsymbol{B}_{2[K_2 \times K_2]}$ pseudo path matrices and $\boldsymbol{u}, \boldsymbol{e}, \boldsymbol{d}$ residual variables.

Comparing Eqs.6.14,..., 6.16 to the LVP model (Sc.2.2) we see that in the three-mode LV path model the path, loading and weight matrices are decomposed into two matrices:

$$\boldsymbol{B} = \boldsymbol{B}_1 \otimes \boldsymbol{B}_2 + \boldsymbol{B}_0 \tag{6.17}$$
$$\boldsymbol{P} = \boldsymbol{P}_1 \otimes \boldsymbol{P}_2 + \boldsymbol{P}_0 \tag{6.18}$$
$$\boldsymbol{W} = \boldsymbol{W}_1 \otimes \boldsymbol{W}_2 + \boldsymbol{W}_0, \tag{6.19}$$

where B_0, P_0, W_0 are residual matrices which are assumed to be zero in Eqs.6.14,..., 6.16. The covariance may also be decomposed into two pseudo covariance matrices and one residual matrix:

$$S = S_1 \otimes S_2 + S_0 \tag{6.20}$$
$$R = R_1 \otimes R_2 + R_0. \tag{6.21}$$

6.4 Special Cases and Properties

Let us assume that

$$Q \equiv \text{cov}(u) = Q_1 \otimes Q_2 \tag{6.22}$$
$$C \equiv \text{cov}(e) = C_1 \otimes C_2 \tag{6.23}$$
$$\text{cov}(Y; e) = 0. \tag{6.24}$$

The reduced form (Eq.2.19) of the inner model is:

$$\begin{aligned} Y &= (\mathrm{I} - B_1 \otimes B_2)^{-1} u \\ &= (\mathrm{I}_1 \otimes \mathrm{I}_2 - B_1 \otimes B_2)^{-1} u \\ &= ((\mathrm{I}_1 - B_1) \otimes (\mathrm{I}_2 - B_2))^{-1} u \\ &= ((\mathrm{I}_1 - B_1)^{-1} \otimes (\mathrm{I}_2 - B_2)^{-1} u \\ &\equiv (B^* \otimes B^*) u \end{aligned} \tag{6.25}$$

Hence, the LV covariance matrix R and the MV covariance matrices are:

$$\begin{aligned} R &= \text{cov}(B_1^* \otimes B_2^* u) = (B_1^* \otimes B_2^*) Q (B_1^* \otimes B_2^*)' = (B_1^* Q_1 B_1^{*\prime}) \otimes (B_2^* Q_2 B_2^{*\prime}) \\ &\equiv R_1 \otimes R_2 \end{aligned} \tag{6.26}$$

$$\begin{aligned} S &= \text{cov}(P_1 \otimes P_2 Y + e) \\ &= (P_1 \otimes P_2) R (P_1 \otimes P_2)' + (P_1 \otimes P_2) \,\text{cov}(Y; e) + \text{cov}(e; Y)(P_1 \otimes P_2) + C \\ &= (P_1 \otimes P_2)(R_1 \otimes R_2)(P_1 \otimes P_2)' + 0 + 0 + C_1 \otimes C_2 \\ &= (P_1 R_1 P_1' + C_1) \otimes (P_2 R_2 P_2' + C_2) \\ &\equiv S_1 \otimes S_2. \end{aligned} \tag{6.27}$$

If it is assumed that Q and C have Kronecker structure, it follows that R and S have a Kronecker structure, too (cf. Browne 1984, his equation 4.1). An algorithm based on assumptions Eqs.6.22,..., 6.24 proceeds by the following steps:

$$\begin{aligned} (S_1, S_2) &:= \text{Kpc}(S) \\ (B_1, P_1, W_1) &:= \text{LVPath}(S_1) \\ (B_2, P_2, W_2) &:= \text{LVPath}(S_2), \end{aligned} \tag{6.28}$$

where LVPath(·) denotes the algorithm described in Tb.2.2. However, Eqs.6.22,..., 6.24 are assumptions on the structure of the residual covariance matrices which are not typical for PLS modeling.

Table 6.3: The PLS3 algorithm for LVP3 models

Step	Comment, Operation
#1	Initialization
#11	read $\boldsymbol{S}$
#12	read design matrices of $\boldsymbol{B}_1, \boldsymbol{B}_2, \boldsymbol{P}_1, \boldsymbol{P}_2, \boldsymbol{W}_1, \boldsymbol{W}_2, \boldsymbol{R}_1, \boldsymbol{R}_2$
#13	read starting values of $\boldsymbol{W}_1, \boldsymbol{W}_2$
#2	Iterative estimation of the LV weights Starting at step #26, repeat steps #21 to #26 until convergence
#21	Patterned orthogonalization of $\boldsymbol{Y}$: $\boldsymbol{Y} := \boldsymbol{O}\boldsymbol{Y}$
#211	$(\boldsymbol{R}_1, \boldsymbol{R}_2) := \text{Kpc}(\boldsymbol{R})$
#212	$\boldsymbol{O} := \text{Pattor}(\boldsymbol{R}_1, \text{Design}(\boldsymbol{R}_1)) \otimes \text{Pattor}(\boldsymbol{R}_2, \text{Design}(\boldsymbol{R}_2))$
#213	$\boldsymbol{R} := \boldsymbol{O}\boldsymbol{R}\boldsymbol{O}'$
#214	$\boldsymbol{T} := \boldsymbol{T}\boldsymbol{O}'$
#22	Inner weights $\boldsymbol{V}$
#221	$\boldsymbol{B} := \text{Path}(\boldsymbol{R}, \text{Design}(\boldsymbol{B}_1) \otimes \text{Design}(\boldsymbol{B}_2))$
#222	$(\boldsymbol{B}_1, \boldsymbol{B}_2) := \text{Kpc}(\boldsymbol{B})$
#223	$(\boldsymbol{R}_1, \boldsymbol{R}_2) := \text{Kpc}(\boldsymbol{R})$
#224	$\boldsymbol{V}_1 := V(\boldsymbol{R}_1, \boldsymbol{B}_1, \text{Design}(B1), \text{weighting scheme})$
#225	$\boldsymbol{V}_2 := V(\boldsymbol{R}_2, \boldsymbol{B}_2, \text{Design}(B2), \text{weighting scheme})$
#226	$\boldsymbol{V} := \boldsymbol{V}_1 \otimes \boldsymbol{V}_2$
#227	$\boldsymbol{V} := \boldsymbol{V}[\text{I} * (\boldsymbol{V}\boldsymbol{R}\boldsymbol{V}')]^{-1/2}$
#23	Inside approximation $\tilde{\boldsymbol{Y}}$: $\tilde{\boldsymbol{Y}} := \boldsymbol{V}\boldsymbol{Y}$
#231	$\boldsymbol{R} := \boldsymbol{V}\boldsymbol{R}\boldsymbol{V}'$
#232	$\boldsymbol{T} := \boldsymbol{T}\boldsymbol{V}'$
#24	Patterned orthogonalization of $\tilde{\boldsymbol{Y}}$: $\tilde{\boldsymbol{Y}} := \boldsymbol{O}\tilde{\boldsymbol{Y}}$
#241	$(\boldsymbol{R}_1, \boldsymbol{R}_2) := \text{Kpc}(\boldsymbol{R})$
#242	$\boldsymbol{O} := \text{Pattor}(\boldsymbol{R}_1, \text{Design}(\boldsymbol{R}_1)) \otimes \text{Pattor}(\boldsymbol{R}_2, \text{Design}(\boldsymbol{R}_2))$
#243	$\boldsymbol{R} := \boldsymbol{O}\boldsymbol{R}\boldsymbol{O}'$
#244	$\boldsymbol{T} := \boldsymbol{T}\boldsymbol{O}'$
#25	Outer weights $\boldsymbol{W}$
#251	$(\boldsymbol{P}_1, \boldsymbol{P}_2) := \text{Kpc}(\boldsymbol{T} * (\text{Design}(\boldsymbol{P}_1) \otimes \text{Design}(\boldsymbol{P}_2)))$
#252	$\boldsymbol{W} := \boldsymbol{P}_1 \otimes \boldsymbol{P}_2$
#253	$(\boldsymbol{W}_1, \boldsymbol{W}_2) := \text{Kpc}(\boldsymbol{S}\boldsymbol{W}')$
#254	$\boldsymbol{W} := \boldsymbol{W}_1 \otimes \boldsymbol{W}_2$
#255	$\boldsymbol{W} := \boldsymbol{W}[\text{I} * (\boldsymbol{W}\boldsymbol{S}\boldsymbol{W}')]^{-1/2}$
#26	Outside approximation, $\boldsymbol{Y} := \boldsymbol{W}'\boldsymbol{y}$
#261	$\boldsymbol{R} := \boldsymbol{W}\boldsymbol{S}\boldsymbol{W}'$
#262	$\boldsymbol{T} := \boldsymbol{S}\boldsymbol{W}'$
#3	Termination
#31	Print results
#32	Compute and print residuals
#33	Compute and print summary table
#34	Stop

6.5 The PLS Estimation of LVP3 Models

The algorithm for the estimation of LVP3 model, as outlined in Table 6.3, follows the steps of the extended method of LVP analysis (see Tb.2.2). However, the LVP3 algorithm includes additional computational steps to satisfy the structural implications of the Kronecker structure of path, loading and weight matrix. The program PLS3 has several options which are not shown in Tb.6.3. It has four options that change the computation of the inner weights in step #22 and four options that change the computation of the outer weights in step #25. This introduces considerable flexibility into the algorithm — and, hence, uncertainty which combination of options is the best for a given model and data set. Here follow some comments on the single steps:

Step #21,#24. The user can specify certain LVs as orthogonal. The zero LV correlations can be specified for $\boldsymbol{R}_1, \boldsymbol{R}_2$ or for the total correlation matrix $\boldsymbol{R}$. It has been found that a more cautious orthogonalization than step #212 often gives better results, such as:

$$
\begin{aligned}
(\boldsymbol{R}_1, \boldsymbol{R}_2) &:= \mathrm{Kpc}(\boldsymbol{R}) \\
\boldsymbol{O}_1 &:= \mathrm{Pattor}(\boldsymbol{R}_1, \mathrm{Design}(\boldsymbol{R}_1)) \\
\boldsymbol{O} &:= \boldsymbol{O}_1 \otimes \mathrm{I}_1 \\
\boldsymbol{R} &:= \boldsymbol{O}\boldsymbol{R}\boldsymbol{O}' \\
(\boldsymbol{R}_1, \boldsymbol{R}_2) &:= \mathrm{Kpc}(\boldsymbol{R}) \\
\boldsymbol{O}_2 &:= \mathrm{Pattor}(\boldsymbol{R}_1, \mathrm{Design}(\boldsymbol{R}_2)) \\
\boldsymbol{O} &:= \mathrm{I}_1 \otimes \boldsymbol{O}_2 \\
\boldsymbol{R} &:= \boldsymbol{O}\boldsymbol{R}\boldsymbol{O}' \\
\boldsymbol{O} &:= \mathrm{Pattor}(\boldsymbol{R}, \mathrm{Design}(\boldsymbol{R}_1) \otimes \mathrm{Design}(\boldsymbol{R}_2)) \\
\boldsymbol{R} &:= \boldsymbol{O}\boldsymbol{R}\boldsymbol{O}'.
\end{aligned}
$$

Step #22. The computation of the inner model can be performed as outlined in step #22. Starting with a path analysis of the total matrix $\boldsymbol{R}$, the path matrix $\boldsymbol{B}$ is obtained, which is then decomposed into the pseudo-matrices $\boldsymbol{B}_1$ and $\boldsymbol{B}_2$. Alternatively, one can start with the decomposition of $\boldsymbol{R}$ in the pseudo-correlation matrices $\boldsymbol{R}_1$ and $\boldsymbol{R}_2$ which are then subjected to separate path analyses to obtain $\boldsymbol{B}_1$ and $\boldsymbol{B}_2$. Furthermore, the inner weights $\boldsymbol{V}_1$ and $\boldsymbol{V}_2$ can be formed separately for the two modes and than multiplied, as shown in steps #224,...,#226. Alternatively, one can compute the inner weights by $\boldsymbol{V} := V(\boldsymbol{B}, \boldsymbol{R}, \text{weighting scheme})$.

Step #25. Step #25 of Tb.6.3 shows the ModeA computation of the outer weight matrix $\boldsymbol{W}$. The computation of the loadings is incorrect and must by refined by the methods shown in Tb.2.4, when more than one LV is attached to an MV. As an alternatively to Step #253, one can compute the pseudo weight matrices separately by $(\boldsymbol{S}_1, \boldsymbol{S}_2) := \mathrm{Kpc}(\boldsymbol{S})$, $\boldsymbol{W}_1 := \boldsymbol{S}_1\boldsymbol{P}_1$, $\boldsymbol{W}_2 := \boldsymbol{S}_2\boldsymbol{P}_2$ for ModeA and by $\boldsymbol{W}_1 := \boldsymbol{S}_1^{-1}\boldsymbol{P}_1$, $\boldsymbol{W}_2 := \boldsymbol{S}_2^{-1}\boldsymbol{P}_2$ for ModeB weights. Step #255 rescales the weights and standardizes the LVs. Notice that the scaling diagonal matrix is not assumed to have a Kronecker structure.

The program PLS3 is in the course of being developed and explored. It has been found that the different alternatives of the algorithm have paramount influence on convergence and numerical results. Although the first decimal seems to be stable under variations of the algorithm which do converge, the second decimal is unreliable. PLS3 has default values for the eight options which control the iteration algorithm. As long as the underlying assumptions and the consequences of these options have not been well explored, it is not recommended to use other specifications than the default values.

6.6 Application: Longitudinal Data

The data used in this section are taken from the Augsburg Longitudinal Study (Hanke, Lohmöller & Mandl 1980) and have in part been used already in Sc.4.3; see also Lohmöller and Wold (1980, 1982). We will now take the half yearly marks in 8 subjects, measured over a period of three years, the first school year having been omitted. $K_2 = 8$ school marks, measured on $K_1 = 6$ occasions, give a total of $K = 48$ manifest variables. The model for these data, as presented in Table 6.4, introduces 12 LVs. The eight subjects are reduced on $J_2 = 2$ orthogonal subject factors; the time points are not reduced, $J_1 = K_1 = 6$.[3] The path model for the time points assumes a second-order autoregression process, i.e. each time point t is regressed on the one directly preceding, $t - 1$, and the time point before that, $t - 2$.[4]

Outer model. The loading matrix $\boldsymbol{P}_2$ for the subjects discloses the first factor as a general factor, loading mostly on the academic achievement subjects German, Arithmetic, Elementary Science, Religion, Penmanship and Arts; the second factor is concerned with the nonacademic subjects, Music and Sport. The (diagonal) loading matrix $\boldsymbol{P}_1$ shows how well the pattern of $\boldsymbol{P}_2$ fits for each time point, when the coefficient for the marks at the end of the fourth grade are arbitrarily set to $p_{66}^{(1)} = 1$: The lowest value of $\boldsymbol{P}_1$ is found for the mid-grade marks in the second year, $p_{22}^{(1)} = 0.97$, indicating that the pseudo loading matrix $\boldsymbol{P}_2$ fits equally well to all time points and that the loading matrix remains the same for all time points (stationary weights, Meredith & Tisak 1982, Bieber & Meredith 1985).

Inner model. The path diagram in Tb.6.4 shows the LVs ordered in two chains, one for the Achievement factors and one for the Nonacademic factors. If the averaged stability of the Achievement factors is arbitrarily set at $b_{11}^{(2)} = 1.00$, the relative stability of the Nonacademic stability is $b_{22}^{(2)} = 0.85$. The pseudo path matrix $\boldsymbol{B}_1$ shows the stabilities over time gaps. The half-year stabilities are higher

[3]Kroonenberg (1983b) would call it a TUCKER2 model, because the third mode is not reduced.

[4]The time points are labeled 2m = 2nd grade - mid-grade, 2e = 2nd grade - end-grade, 3e = 3rd grade - mid-grade, etc.

Table 6.4: LVP3 model for school marks in elementary school, Augsburg Longitudinal Study

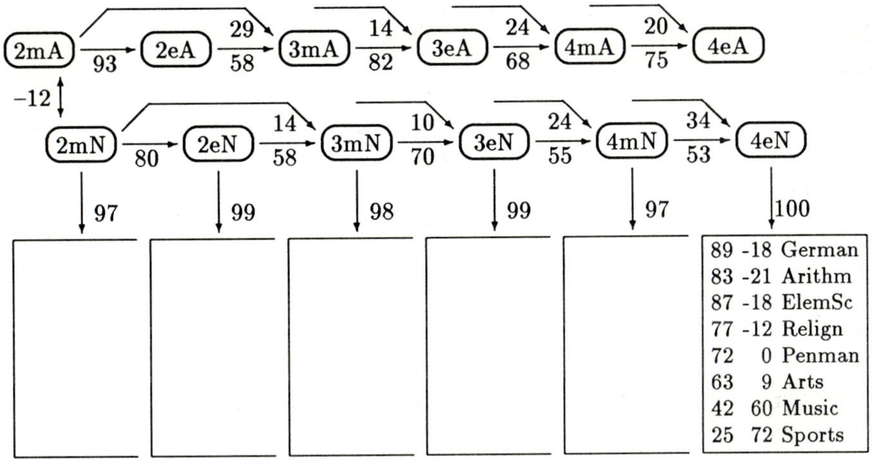

Path matrix decomposition:

$B = B_1 \otimes B_2 + B_0$

B												
2mA	.	.	.	.	.	.	.	.	.	.	.	.
2mN	.	.	.	.	.	.	.	.	.	.	.	.
2eA	93	.	.	.	.	.	.	.	.	.	.	.
2eN	.	80	.	.	.	.	.	.	.	.	.	.
3mA	29	.	58	.	.	.	.	.	.	.	.	.
3mN	.	14	.	58	.	.	.	.	.	.	.	.
3eA	.	.	14	.	82	.	.	.	.	.	.	.
3eN	.	.	.	10	.	70	.	.	.	.	.	.
4mA	.	.	.	.	24	.	68	.	.	.	.	.
4mN	.	.	.	.	.	24	.	55	.	.	.	.
4eA	.	.	.	.	.	.	20	.	75	.	.	.
4eN	.	.	.	.	.	.	.	34	.	53	.	.

=

B_1						
2m	.	.	.	.	.	.
2e	94	.	.	.	.	.
3m	24	62	.	.	.	.
3e	.	13	82	.	.	.
4m	.	.	26	66	.	.
4e	.	.	.	29	70	.

$\otimes$

B_2		
Achievment	100	.
Nonacademic	.	85

All coefficients are multiplied by 100. The loading matrix has to be understood six times repeated, each time multiplied by the proportionality constant written at the vertical arrow.

from mid-grade to end-grade than from end-grade to mid-grade of the next year. This is especially true for the transition from the end of the second grade to the middle of the third grade; through transition, all students had new teachers.

Table 6.5 shows three different models on the same data. Model A is identical to the model in Tb.6.4. In Model B the outer model is restricted and 8 weight coefficients are set to zero, giving no substantial change in the other parameters of the model. In Model C a cross-lag influence of the Achievement factor at time t on the Nonacademic factor at time $t+1$ is admitted. This cross-lag coefficient $b_{21}^{(2)} = 0.10$ is of nearly the same size as found by Lohmöller (1978a, 1978b) from the

Table 6.5: Three versions of the model in Tb.6.4

Tb.6.5a $\boldsymbol{P}_2$, pseudo loading matrix of school marks	Model A	Model B	Model C
German	89 18	88 .	88 .
Arithmetic	83 21	82 .	82 .
Elementary Science	87 18	86 .	86 .
Religion	77 12	77 .	72 .
Penmanship	72 0	72 .	77 .
Arts and Handicraft	63 -9	64 .	64 .
Music	42 -60	. 85	. 86
Sports	25 -72	. 71	. 70

Tb.6.5b $\boldsymbol{B}_2$, pseudo path matrix of school mark LVs	Model A	Model B	Model C
A = Achievement	100 .	100 .	100 .
N = Nonacademic Achvm.	. 85	. 90	10 86

Tb.6.5c $\boldsymbol{P}_1$, pseudo loading matrix of time points	Model A	Model B	Model C
2nd mid-grade	97	99	98 . . . . .
2nd end	. 99	. 99	. 99 . . . .
3rd mid-grade	. . 98 . . .	. . 98 . . .	. . 98 . . .
3rd end	. . . 99 . .	. . . 99 . .	. . . 99 . .
4th mid-grade	. . . . 97 .	. . . . 97 .	. . . . 98 .
4th end	 100	 100	 100

Tb.6.5d $\boldsymbol{B}_1$, pseudo path matrix of time points	Model A	Model B	Model C
2nd mid-grade	. . . . . .	. . . . . .	. . . . . .
2nd end	94	93	93 . . . . .
3rd mid-grade	24 62	22 63	23 62 . . . .
3rd end	. 13 82 . . .	. 13 82 . . .	. 13 82 . . .
4th mid-grade	. . 26 66 . .	. . 25 67 . .	. . 25 67 . .
4th end	. . . 29 70 .	. . . 27 71 .	. . . 27 72 .

Model A: The two LVs "A" and "N" of each time point are urged to be orthogonal and come out with average correlation $\text{cor}(A; N) = -.09$.

Model B: The factor loadings and weights are specified as disjoint; the LVs come out as correlated with $\text{cor}(A; N) = 0.45$.

Model C: Cross-lag path from "A" at $t-1$ to "N" at time point t is specified, $b = 0.10$

All coefficients are multiplied by 100. Dot indicates fixed zero.

same data, though with unsuitable and obsolete methods. The positive cross-lag influence must be interpreted as a trait integration.

6.7 Concluding Remarks

Nearly everybody who has published a three-mode analysis has also written his own three-mode program. To understand the method it seems necessary to understand the algorithm exactly. However, the potentials of three-mode data are often understood, and analyses below the appropriate level of statistical sophistication do occur.

After Tucker had developed his three-mode component model (Tucker 1963, 1964a, 1964b) and his estimation method (Tucker 1966, 1972, cf. Lohmöller 1979a), consisting of three separate PC analyses, the model was reformulated by Bloxom (1968, 1984) as a stochastic factor model. Bentler and Lee (1978, 1979) give an ML estimation method for the Tucker-Bloxom model.

Most of the three-mode estimation methods are based on Wold's PLS estimation method. The very first estimation method for a three-mode model was developed by Svante Wold in 1965 and published by Herman Wold (1966a). The model is a unidimensional multiplicative model:

$$y_{ktn} = b_k g_t Y_n + e_{ktn}, \tag{6.29}$$

$$\sum_k b_k^2 = 1, \quad \sum_t g_t^2 = 1, \quad \text{var}(e) = \sum_{ktn} e_{ktn} \stackrel{!}{=} \text{min}. \tag{6.30}$$

The algorithm is extended by Carroll and Chang (1970) to several dimensions and their algorithm CANDECOMP handles even seven-mode data. Lastovicka (1981) discusses the extension of Tuckers model on four-mode data, Kroonenberg and de Leeuw (1977, 1980) and Sands and Young (1980) the problem of MVs with different measurement levels. The CANDELINC method by Carroll, Pruzansky and Kruskal (1980) adds linear constraints on parameters.

An overall view of three-mode modeling is given in an annotated bibliography by Kroonenberg (1983a) and in his dissertation (Kroonenberg 1983b, 1984). A wealth of contributions to three-mode analysis is assembled in the "Tucker Festschrift" edited by Law, Snyder, Hattie and McDonald (1984).

Multiplicative vs. additive models. The three-mode path and factor models treated in Sc.6 are multiplicative: the parameter of the one mode (b_k as in Eq.6.29, say) is multiplied by the parameter of the other mode (g_t, say), which is expressed by the Kronecker product. The additive model for three-mode data is:

$$y_{ktn} = b_k Y_n + g_t Z_n + e_{ktn}. \tag{6.31}$$

Let y_{ktn} denote multitrait-multioccasion intelligence data, with index k denoting the subtest, t the time point of measurement, and n the observation, the latent variable Y denotes the (stabile) General Intelligence, and Z the "Time factor", i.e. something that is related to time and added to the intelligence. It is unclear what time factors are, as opposed to trait factors, when the data are multitrait-multioccasion data, or what method factors are, when multitrait-multimethod data are analysed. In the past, the omission of a subject-related choice between an additive and a multiplicative model was only justified by the lack of an appropriate general model for multiplicative models. For a discussion of additive vs. multiplicative models see Campbell and O'Connell (1967); for an additive model Jöreskog (1969); for multiplicative models Lohmöller (1976, 1978a, 1978b, 1979f), also Hanke, Lohmöller and Mandl (1980); for longitudinal models McArdle (1987a, 1987b), McArdle, Anderson and Aber (1987), McArdle, Connell and Goldsmith (1980), McArdle, Horn and Aber (1985), Short, Horn and McArdle (1984), Millsap (1986), and Millsap and Meredith (1988).

Chapter 7

PLS Programs and Applications

7.1 PLS Programs

Except for special cases and simple models, PLS models require special PLS programs. Principal component and canonical correlation analysis can be performed by nearly all standard statistical computer packages (like SPSS, BMDP, OSIRIS, SAS). How simple models can be estimated by SPSS is shown in Table 7.1. For every iteration cycle the program has to be restarted, with new weight coefficients for the LVs each time.

Special PLS programs. Already several special-purpose PLS programs have been written.

Areskoug (1982) mentions his own program TWOBLOCK for two-block PLS models (cf. Sc.3.4).

Apel (1977a, 1977b, 1977c) has written a Fortran program, which uses the raw data as input and performs the estimation in several dimensions. The Stone-Geisser test can be applied to one block of endogenous variables. Apel (1982) mentions another PLS program written in Basic and which runs on a microcomputer.

Hui (1978) published his APL program which performs a Fix-Point estimation for non-recursive LV path models; the classical statistical tests for the parameters are performed by this program.

Noonan (cf. Noonan & Wold 1977, 1980, 1983, 1982, 1986) has written a program that copes with very big models, also allowing for hierarchic LV structures.

The program SIMCA, written by Svante Wold, performs the Stone-Geisser test.

Other PLS programs have been written or used by Kowalski, Gerlach and Wold (1982), S.Wold (1982), Scholing (1983) and Schieber (1983).

PLS 1.8. The program used throughout here is now called PLS 1.8 (Lohmöller 1986, 1988b). It grew out of three separate programs, the first being the raw data program PLSX (Lohmöller 1979f), based on Apel's program, the second the covariance program PLSC (Lohmöller 1979c), the third the three-mode LV path analysis program PLS3 (Lohmöller & Wold 1980, 1982). In 1981 and 1982 the programs were combined into one program system, then called LVPLS 1.6 with main programs LVPLSX, LVPLSC, LVPLS3. The program conforms to a very

Table 7.1: SPSS commands for one PLS iteration cycle of a three-block model

```
VARIABLE LIST  X1 X2 X3 Y4 Y5 Y6 Z7 Z8 Z9
COMPUTE        X = w1*X1 + w2*X2 + w3*X3
COMPUTE        Y = w4*Y4 + w5*Y5 + w6*Y6
COMPUTE        Z = w7*Z7 + w8*Z8 + w9*Z9
COMPUTE        XSTAR = Y + Z
COMPUTE        YSTAR = X + Z
COMPUTE        ZSTAR = X + Y
REGRESSION     VARIABLES=ALL/ DESCRIPTIVES=CORR/
               DEPENDENT=Z/ ENTER X Y/
               DEPENDENT=XSTAR/ ENTER X1 X2 X3/
               DEPENDENT=YSTAR/ ENTER Y4 Y5 Y6/
               DEPENDENT=ZSTAR/ ENTER Z7 Z8 Z9/
```

Note: The weight coefficients $w_1, w_2, ..$ have to be changed for each iteration cycle. They are taken from the regressions. The latent variables are denoted by **X, Y, Z**, the inside approximation by **XSTAR, YSTAR, ZSTAR**.

restricted Fortran IV / Fortran 77 standard and has been tested on a wide range of mainframe computers. A MS-DOS version for PCs is also available. Documentation of PLS comprises the user manual (Lohmöller 1981a), the implementation manual (Lohmöller 1982b) and the instant manual (Lohmöller 1982c). Up to now PLS has been installed in nearly 100 university computer centers. Comparisons of PLS, LISREL and other covariance structure programs are found with Lohmöller (1980, 1981b, 1988a). A primer on PLS 1.8 is written by Falk (1988).

7.2 Applications

PLS modeling has been applied in a wide range of different sciences. The first applications were documented in the group report of a session at the Toronto World Conference of the Econometric Society (Wold 1975d). A second series of applications was presented at the "System Under Indirect Observations" conference, organized by Karl G. Jöreskog and Herman Wold and held in Cartigny near Geneva in Switzerland in October of 1979; the proceedings of this conference are published under the same title (Jöreskog & Wold 1982a). In his introductory chapter of the Cartigny proceedings, Wold (1982d:36s) refers to 28 applications in 10 different scientific areas, ranging from chemistry to psychology. More applications were presented at the Symposium "PLS (Partial Least Squares) Model Building: Theory and Application" held in Frankfurt am Main, Sept. 23-25, 1987. Some recent applications will be presented here.

7.2.1 Applications to Non-Individual Data

Econometric and politometric data are, typically, on time series (e.g. N=20 years) or cross-sections (e.g. N=100 nations or N=50 universities). These types of data involve certain difficulties as far as classical methods are concerned. Time series data are not independent observations. Cross-sectional data of nations or organizations are often full-survey data where the sample is identical to the population, leaving no room for inferential statistics beyond the sample. PLS modeling seems to be the cure for both problems, because PLS requires neither independent observations nor exchangeable samples of observations.

Time series data. Time series models, typically, are affected by the problem of linear trend which often produces all of the correlation between the variables.

Apel (1977c) uses time series on 16 variables, measured in the state Hesse 1962-1973; his analysis is reproduced by PLSX (Lohmöller 1981a:chapter 5). In his 1982 paper Apel uses 47 time series, from 1965 to 1978, in a path model with 7 LVs (Apel 1982, Apel & Lohmöller 1988).

Meissner and Uhle-Fassing (1982a, 1982b) use German time series data, from 1960 to 1976, for a model with four LVs, with governmental expenditure for welfare and stabilization policies as target variable, and economic stability, material welfare and popularity of the respective governments of the days as predictors.

Rogowski (1987) models Polish time series data (1963-1984) on shortage, level of technology, investment and efficiency (7 LVs). Ejsmont and Rogowski (1987) report a model on Poland's trade with EEC countries, European socialist countries, and developing economies (33 MVS, 7 LVs).

Cremer and Hujer (1979), Cremer and Knepel (1980) and Knepel (1980, 1981) use time series data from the Federal Republic of Germany for a model on the labor market.

A reader on structural models on longitudinal data and time series is edited by Möbus and Schneider (1986).

A special development with nonlinear relations in the inner model is noted. A linear equation like $\eta = \beta\xi + \epsilon$ predicts for a given ξ only one η-value. The equation $\eta^3 = \xi\eta + \zeta + \epsilon$, however, predicts (for certain values of ξ and ζ) two η-values which give the same optimum and represent two equally good points of balance and equilibrium. This equation is used by Wold and Kaasch (1985) and Wold and Mensch (1983, 1985) to explain why in the course of long swings, given identical conditions ξ= rationalizing and ζ= innovating investments, the economic performance η may differ.

Cross-national data.

The Adelman and Morris (1967, 1973) data bank on 74 developing non-communist countries and Adelman's four-block model for these data have served as "an experimental rabbit for the development of PLS", as Herman Wold likes to recall (Adelman, Abrahamsson & Areskoug 1975; Adelman, Abrahamsson, Areskoug, Lorentzson & Wallmyr 1975; Apel & Wold 1982; Bookstein 1980, 1982b; Adelman, Lohmöller & Morris 1987, 1988).

Rossa (1982) and Hopple (1979) use data on 56 countries, measured on characteristics related to interstate behavior.

Scholing and Timmermann (1987) report a model on growth rates in underdeveloped countries (113 MVs, 18 LVs).

Cross-level inference. Nowadays, questionnaire surveys are conducted to find out, for which party unemployed people are inclined to cast their vote. Individual-level data for this type of question are not generally available for prewar elections.

Election results and unemployment figures may be available on an aggregated level, i.e. of counties or districts. The statistical problem is that of inference from the aggregated level to the individual level. For manifest variables the problem is discussed under the heading of ecological inference and ecological regression (Bernstein 1932, Robinson 1950, Goodman 1953, 1959, Lohmöller & Falter 1986). To put it simply, when the regression within each aggregate can be assumed to be not very different from the between-aggregate regression, the between-aggregate regression is an estimator of the total regression, i.e. of the individual level regression.

The same argument may be carried over from the regression between two manifest variables to the regression between LV and MV and also between two LVs. Hence, an LV path analysis of aggregated data requires no assumptions beyond the ones implied by LVP modeling on the one hand and ecological inference on the other. This argument forms the basis for the LVP model for election data published by Falter and his coworkers. ML estimation methods are clearly inappropriate, because, on one hand, the variables are categorical on the individual level which is contradictory to the normality assumption, and, on the other hand, the data are full-survey data.

Falter (1981), Falter, Link, Lohmöller, de Rijke and Schumann (1982, 1983), Falter, Lohmöller, Link and de Rijke (1982, 1985), Lohmöller, Falter, Link and de Rijke 1985 use census and election data on some 800 counties, mainly to predict the rise of the NSDAP (Nazi party) at the polls.

Wellhofer (1985) predicts the outcome of the voting in late Victorian Britain (Elections 1885-1910), when religious and territorial cleavages were gradually replaced by social class cleavages generated by industrialization (N=115 surrogate constituencies, 19 LVs, 49 MVs).

Wellhofer (1982) reports additionally a similar model on the British elections in the 1918-1950 period (21 LVs, 80 MVs, N=161 constituencies)

Engel (1987) reports four models on the relations of party representatives to their local constituencies (75 MVs, 27 LVs).

van Donkelaar and van der Knaap (1982) explain the varying residential mobility in Rotterdam by the demographic characteristics and the stock of housing. (8 MVs, 3 LVs, N = 57 city quarters); cf. Blommestein (1985).

Institutional data.

Grüner, Jädicke and Ruhland (1988) use PLS to model the determinants of house construction policy in German cities (11 MVs, 7 LVs, N=54 cities).

Roeder, Baumert, Naumann and Trommer (1987a, 1987b) present models for explaining the determinants of scientific productivity (11 MVs, 4 LVs, N=50) and reputation (18 MVs, 7 LVs, N=52) of West German universities.

Historical events. Wold (1986) analyses a data set comprising 28 attributes of 10 historical events which are unequally spaced over several decades. The events are economic sanctions, the first against the Soviet Union in 1933, the last against Rhodesia in 1965. The target variable of the model is success/failure of the sanction. Typically for historical data, there are less events than predictors and a multiple regression with 27 predictors for 10 cases would break down. (In fact, with only two predictors, $R^2 = 1$ is reached.) As Wold demonstrates by the Stone-Geisser test, an LVP model with two blocks of predictors actually has predictive relevance. This encourages applied PLS modeling of small-sized data sets, as, for example, in quantitative historical research.

Chemistry and sciences. The PLS technique has taken a separate development in chemistry and food research. Most of the models have only one or two blocks, so the stress is more on the PLS algorithm (on components, aggregation of information, calibration of measurement procedures) than on latent variable path modeling (on theoretical constructs, validity, complex networks of causal relations).

M.Martens (1987) edited a report on statistical and computational aspects of PLS, with contributions by

Geladi (1987) on the history of PLS,

Esbensen (1987) on the philosophy of PLS regression,

Helland (1987) with theoretical considerations,

Næs (1987) on PLS versus other statistical prediction methods,

Manne (1987) on PLS as a generalized inverse (for this topic cf. S.Wold, Ruhe, H.Wold and Dunn 1984)

Schweder (1987) on canonical regression versus PLS,

Spjøtvoll (1987) with critical comments on PLS and

H.Martens (1987) telling why users like PLS regression.

More on data analysis and exploration by Lorber, Wangen and Kowalski (1987), S.Wold *et al.* (1982, 1984a, 1984b, 1985), S.Wold, Geladi, Esbensen and Öhman (1986), S.Wold, Hellberg, Lundstedt, Sjöström and H.Wold (1987).

Applications in chemistry:

M.Martens, Fjeldsenden, Russwurm and H.Martens (1982) report a two-block model predicting sensory qualities (taste etc.) of carrots from chemical properties, a structure vs. performance approach.

Jensen, Munck and H.Martens (1982) treat the botanical constituents of wheat by a PLS model.

Geladi, Esbensen and S.Wold (1987) apply PLS for the analysis of multivariate images.

7.2.2 Applications in Psychological and Educational Research

From the beginning, latent variables have been more latent in psychology than in other sciences. (Psychologists often resist even to assume that the subject, or object, as the case may me, of their research, the "soul", exists.) No wonder that latent variables were introduced by a psychologist (Charles Spearman 1904) and that the most convincing applications of LVP analysis are made in psychological, sociological, and educational research. Take, for example, the dissertations by Gattringer (1980), Schulte (1981), Engfer (1982), Uehlinger (1986), and the research reports by Lohmöller (1979e), Bartl, Unverdorben and Lohmöller (1981), Engfer and Schneewind (1981), Lohmöller and Schulte-Cloos (1986), Schumann (1982), Lohmöller and Wold (1982), and Winteler (1982, 1984a, 1984b), which all used the program PLS. See also other publications by Hui (1978, 1982), Boardman, Hui and Wold (1981), Noonan (1981, 1987), the joint work of Noonan and Wold (1977, 1980, 1982, 1983), Noonan, Abrahamsson, Areskoug, Lorentzson and Wallmyr (1975), Fornell and Bookstein (1982).

Schneewind, Beckmann and Engfer (1983) report in their monograph on "Parents and Children" six PLS models (LV:MV = 14:56, 8:25, 8:31, 8:27, 8:33, 9:38) which explain child behavior (personality, social activity, TV consumption, creativity, participation) by social structure, personality and educational behavior of the parents.

Engfer and Gavranidou (1984, 1987)

Hansson and Berglund (1987) model the factor influencing treatment outcome and patient satisfaction in a short-term psychiatric ward (15 MVs, 9 LVs).

Sampson, Streissguth, Vega-Gonzalez, Barr and Bookstein (1987) and Sampson, Streissguth, Barr and Bookstein (1987) apply PLS (and Breiman and Friedman's (1985) nonlinear scaling of categorical variables) for predicting classroom behavior ratings by prenatal alcohol exposure (61 MVs, 10 LVs).

Hasselhorn and Körkel (1985), Schneider (1980, 1985) and Schneider, Borkowski, Kurtz and Kerwin (1985) report models on the effect of metamemory and motivation.

Schneider and Bös (1985) explore components of school success by PLS models.

Stary (1983) models the conditions of political attitudes of youths.

Survey data. Typically, survey data are representative samples of defined populations, often with large arrays of variables.

Huber, Gabriel and Ehrmann (1988) surveyed workers laid off from closing factories in Berlin and Bremen; both samples are not representative for any population, and different models for the explanation of their chances on the local labor market had to be developed (16 MVs, 6 LVs; 13 LVs, 17 MVs which include categorical variables).

Voges and Lohmöller (1988) predict early retirement of workers by structural (industrial branch of chemistry or paper production, size of factory, information policy of unions) and individual characteristics (education, social status, income, age) (23 MVs, 8 LVs).

Jain, Pinson and Malhotra (1987) conduct a causal investigation of customers' shopping predispositions (store loyalty, deal proneness) which they model as caused by individual traits (cognitive differentiation, information processing confidence), patronage related behavior (extent of shopping, product information search, confidence in shopping) and store and product related perceptions (13 LVs, 23 MVs).

Balderjahn (1984, 1985) and, on the same data set, Lohmöller (in press a) model the relationship between attitudes and intended behavior regarding environmental issues (10 categorical MVs, 5 LVs).

Balderjahn (1987) presents a model of ecologically concerned consumption behavior (51 MVs, 17 LVs).

Löfgren (1986) reports a model on bilingual instruction of immigrant children (10 LVs).

Voges and Hoschka (1985, 1986) report a similar model of the school success of Italian migrant children in Germany (35 MVs, 14 LVs).

Winteler (1984a, 1984b) develops a big PLS model (83 MVs, 23 LVs) to explain study satisfaction and drop out tendency.

Treinies and Einsiedler (1986) report a causal model on developmental conditions and leisure time play behavior of kindergarten children (47MVs, 13 LVs).

Slavenburg (1986) reports a model on the effects of compensatory education and family involvement programs on first grade achievement (14 MVs, 5 LVs).

Jönsson (1984) reports a longitudinal model on TV consumption in preschool age and schoolmarks from grade 1 to 6 (17 LVs).

Löfgren (1984) develops a causal explanation of working milieu in school classes (14 LVs, 20 MVs).

Schumann's model on nondemocratic attitudes. Schumann (1982, 1985, 1986) uses questionnaire data from students of the Bundeswehr University in Munich. Apart from background variables concerning family, education and military career, the questionnaire contains personality and attitude scales. Schumann postulates two types of non-democratic attitudes which are related to two personality types. The "first chain" starts with positive family relationships, proceeds via conservatism, Super-Ego strength and intolerance of ambiguity, and results in non-democratic attitude Type A, which is that of favouring efficiency over democracy. The "second chain" starts with negative family relations, is mediated by neuroticism, anomia, agressivity, and results in non-democratic attitude Type B, which is indicated by political alienation and the perception of any confrontation with policy as disturbing.

Schulte's model on therapy success. Schulte (1981) has collected measurements on 96 variables of patients during a cure. During the process of data gathering (before, during, at the end and 6 weeks after the end of the cure) the sample has decreased from N=77 at the beginning to N=30 at the end. Schulte's LVP model (cf. Lohmöller & Schulte-Cloos 1986) has 18 blocks with 26 LVs, three blocks with three orthogonal LVs each and two blocks with two orthogonal LVs. Thus it is an extended PLS model (cf. Sc.2.2). The aim of the modeling was to describe stability and/or sources of change in personality traits and in states of subjective well-being.

Winteler's cross-validation. Winteler (1982) presents a very convincing cross-validatory assessment for the stability of PLS weights. He has questionnaire data from the Bundeswehr Universities (universities of the Federal Armed Forces) in Munich and Hamburg. The quality of the Munich sample (N=500, low return rate) is inferior to the Hamburg sample (N=1100, higher return). The Munich sample was chosen as an exploration sample, and the LVP model with 80 MVs and 15 LVs being elaborated until a satisfying balance between systematic (path coefficients) and unsystematic effects (residual covariances) was found.

The weights of the Munich sample were, then, used as starting values for the Hamburg confirmatory sample. The results after the zero-th iteration of PLSC represent what Bentler (1978) calls a weak cross-validation of a model, i.e. the outer model is strictly cross-validated and kept constant across samples, the inner model, though, is re-estimated for the new sample and, thus, weakly cross-validated. Hence the communalities (H^2) are strictly cross-validated, while all other parameters are weakly cross-validated. It must be expected that all parameters are lower in the new sample than in the exploration sample.

Surprisingly enough, the communalities in the confirmation sample are higher than in the exploration sample: $H^2_{\text{Hamburg}} = 44\% > H^2_{\text{Munich}} = 43\%$ and $r_{\text{BB(Hamburg)}} = 65\% > r_{\text{BB(Munich)}} = 56\%$. The only explanation for this difference is found in the different qualities of the samples. The insufficient return rate and the smaller number of cases make the correlation matrix of the Munich sample less reliable, sufficiently reliable, however, to estimate one reliable weight coefficient for each MV. When this weight coefficient is applied to the Hamburg sample data, its quality becomes apparent. (For more on the crossvalidation of structural models cf. Balderjahn 1988 and Opwis, Gold and Schneider 1987).

Bibliography

Abrahamsen,Y. (1987) *Comparative study of Tukes's jackknife standard error and Stone-Geiser's test on predictive relevance versus other methods of model evaluation.* (Paper, symposium PLS Model Building: Theory and Application, Frankfurt am Main, Sept.23-25, 1987). St. Gall, Switzerland: University.

Adelman,I., Abrahamsson,A., & Areskoug,B. (1975) Perturbation analysis of NIPALS parameter estimates. In H. Wold (Ed.), *Modelling in complex situations with soft information.* (Paper, Third World Congress on Econometrics; Toronto, Canada; August 21-26, 1975). (Research Report 1975:5). Göteborg, Sweden: University, Institute of Statistics.

Adelman,I., Abrahamsson,A., Areskoug,B., Lorentzson,L.O., & Wallmyr,J. (1975) Applications of methods I-II to Adelman-Morriss's data. In H. Wold (Ed.), *Modelling in complex situations with soft information.* (Paper, Third World Congress on Econometrics; Toronto, Canada; August 21-26, 1975). (Research Report 1975:5). Göteborg, Sweden: University, Institute of Statistics.

Adelman,I., Lohmöller,J.B., & Morris,C.T. (1987) *A latent variable regression model of nineteenth century economic development.* (Paper, symposium PLS Model Building: Theory and Application, Frankfurt am Main, Sept.23-25, 1987). (Working Paper No. 439, Department of Agricultural and Resource Economics). Berkeley: University of California.

Adelman,I., Lohmöller,J.B., & Morris,C.T. (1988) *Economic development then and now.* (Working Paper No. 470, Department of Agricultural and Resource Economics). Berkeley: University of California.

Adelman,I. & Morris,C.T. (1967) *Society, politics, and ecomic development: A quantitative approach.* Baltimore, MD: John Hopkins Press.

Adelman,I. & Morris,C.T. (1973) *Economic growth and social equity in developing countries.* Stanford, CA: Standford University Press.

Allen,D.M. (1971) *The prediction sum of squares as a criterion for selecting predictor variables.* (Technical Report No. 23). University of Kentuckey: Department of Statistics.

Andrews,F.M. & Messenger,R.C. (1975) *Multivariate nominal scale analysis: A report on a new technique and a computer program.* Ann Arbor,Michigan: The University of Michigan, Institute for Social Research.

Apel,H. (1977a) *Estimation of soft models – A guidance to a raw data input computerprogram.* Philadelphia: University.

Apel,H. (1977b) *NIPALS – Generalprogramm* (Programmdokumentation). Frankfurt: Johann Wolfgang von Goethe-Universität.

Apel,H. (1977c) *Simulation sozio-ökonomischer Zusammenhänge: Kritik und Modifikation of Systems Analysis.* Darmstadt: Toeche Mittler.

Apel,H. (1978a) *Consistency at large; first investigation.* Frankfurt: University Institute of Economics.

Apel,H. (1978b) *Soft models of Herman Wold's type: models of environmental problems and underdevelopment.* Frankfurt: Johann Wolfgang von Goethe-Universität.

Apel,H. (1979) *Soft modeling in socio-economic applications.* Frankfurt am Main: University Institute of Econometrics.

Apel,H. (1982) *Umweltqualität aus ökonomischer Sicht: Eine theoretische und empirische Annäherung.* Frankfurt am Main: University Institute of Econometrics. (unpublished paper).

Apel,H. & Lohmöller,J.B. (1988) Ökonomie und Umweltqualität: Quantitative Bestimmung umweltökonomischer Zusammenhänge mit PLS-Modellen. In L. Hildebrandt, G. Rudinger, & P. Schmidt (Eds.), *Kausalanalyse in der Umweltforschung* (pp.__-__). Stuttgart: Gustaf Fischer. (im Druck).

Apel,H. & Wold,H. (1982) Soft modeling with latent variables in two or more dimension: PLS estimation and testing for predictive relevance. In K.G. Jöreskog & H. Wold (Eds.), *Systems under indirect observation: Causality, structure, prediction* (Vol. 2, pp. 209-248). Amsterdam: North Holland.

Areskoug,B. (1975) Speed of convergence for NIPALS path models with one or two latent variables. In H. Wold (Ed.), *Modelling in complex situations with soft information.* (Paper, Third World Congress on Econometrics; Toronto, Canada; August 21-26, 1975). (Research Report 1975:5). Göteborg, Sweden: University, Institute of Statistics.

Areskoug,B. (1982) The first canonical correlation: Theoretical PLS analysis and simulation experiments. In K.G. Jöreskog & H. Wold (Eds.), *Systems under indirect observation: Causality, structure, prediction* (Vol. 2, pp. 95-118). Amsterdam: North Holland.

Asher,H.B. (1983) *Causal Modeling* (Second edition). (Sage Unversity Paper series on Quantitative Applications in the Social Sciences, series no. 07-003). Beverly Hills: Sage.

Balderjahn,I. (1984) *Das Umwelt- und Konsumbewußtsein von Studenten: Multivariate Statistische Analysen.* Berlin: Technische Universität Berlin, Fachbereich 20 – Informatik, Institut für Quantitative Methoden.

Balderjahn,I. (1985) Strukturen sozialen Konsumbewußtseins: Reanalyse und Versuch einer Bestimmung. *Marketing Zeitschrift für Forschung und Praxis,* __: 253-262.

Balderjahn,I. (1987) Prognose umweltbewußter Komsumgewohnheiten mit dem Verfahren der partiellen kleinsten Quadrate (PLS-Verfahren). In *Operations Research Proceedings 1986* (pp. 432-439).

Balderjahn,I. (1988) Die Kreuzvalidierung von Kausalmodellen. *Marketing Zeitschrift für Forschung und Praxis,* __: 61-73.

Bartl,P., Unverdorben,K., & Lohmöller,J.B. (1981) *Soziale Probleme im Grundwehrdienst: Eine Pfadanalyse zu Alkohol- und Suicidproblemen.* (Forschungsbericht 81.01 Fachbereich Pädagogik). München: Hochschule der Bundeswehr. (114p).

Bentler,P.M. (1976) Multistructure statistical model applied to factor analysis. *Multivariate Behavioral Research, 11*: 3-25.

Bentler,P.M. (1978) The interdependence of theory, methodology, and empirical data: Causal modeling as an approach to construct validation. In D.B. Kandel (Ed.), *Longitudinal research on drug use: Empirical findings and methodological issues* (pp. 267-302). New York: Wiley.

Bentler,P.M. (1980) Multivariate analysis with latent variables: Causal modeling. *Annual Review of Psychology, 31*: 419-456.

Bentler,P.M. (1982) Linear systems with multiple levels and types of latent variables. In K.G. Jöreskog & H. Wold (Eds.), *Systems under indirect observation: Causality, structure, prediction* (Vol. 1, pp. 101-130). Amsterdam: North Holland.

Bentler,P.M. (1983) Some contributions to efficient statistics in structural models: Specification and estimation of moment structures. *Psychometrika, 48*: 493-517.

Bentler,P.M. (1985) *Theory and implementation of EQS, a structural equations program.* Los Angeles: BMDP Statistical Software, Inc.

Bentler,P.M. (1986) Structural modeling and Psychometrika: A historical perspective on growth and achievement. *Psychometrika, 51*: 35-51.

Bentler,P.M. & Bonett,D.G. (1980) Significance tests and goodness of fit in the analysis of covariance structures. *Psychological Bulletin, 88*: 588-606.

Bentler,P.M. & Huba,G.J. (1979) *Symmetric and asymmetric rotations in canonical correlation analysis: New methods with drug variable examples.* Los Angeles: University of California.

Bentler,P.M. & Lee,S.Y. (1978) Statistical aspects of a three-mode factor analysis model. *Psychometrika, 43*: 343-352.

Bentler,P.M. & Lee,S.Y. (1979) A statistical development of three-mode factor analysis. *British Journal of Mathematical and Statistical Psychology, 32*: 97-104.

Bentler,P.M. & Lee,S.Y. (1983) Covariance structures under polynomial constraints: Applications to correlation and alpha-type structural models. *Journal of Educational Statistics*, *8*: 207-222.

Bentler,P.M. & Weeks,D.G. (1979) Interrelations among models for the analysis of moment structures. *Multivariate Behavioral Research*, *14*: 169-186.

Bentler,P.M. & Weeks,D.G. (1980) Linear structural equations with latent variables. *Psychometrika*, *45*: 289-308.

Bentler,P.M. & Weeks,D.G. (1982) Multivariate analysis with latent variables. In P.R. Krishnaiah & L.N. Kanal (Eds.), *Classification, pattern recognition, and reduction of dimensionality* (pp. 747-771). (Handbook of statistics, Vol. 2). Amsterdam: North Holland.

Benzécri,J.P. (1976) *L'Analyse des données; tome 1: La taxinomie; tome 2: L'analyse des correspondances.* Paris: Dunod (2-e ed.).

Bergström,R. (1981) *FIXP – a computer program.* Uppsala, Sweden: University, Department of Statistics.

Bergström,R. (1986) *Jackknifing of some econometric estimators.* (Paper, symposium PLS Model Building: Theory and Application, Frankfurt am Main, Sept.23-25, 1987). (Research Report 86-4). Uppsala, Sweden: University, Department of Statistics.

Bergström,R. & Wold,H. (1980) *On the location of Fix-Point estimates for Generalized and Classical Interdependent systems.* (Paper, Fourth World Congress of the Econometric Society; Aix-en-Provence, France; 1980). Uppsala, Sweden: University, Department of Statistics.

Bergström,R. & Wold,H. (1982) *Fix-Point estimation in theory and praxis.* Göttingen: Vandenhoeck und Rupprecht.

Bernert,C. (1983) The career of causal analysis in American sociology. *British Journal of Sociology*, *34*: 230-254.

Bernstein,F. (1932) Über eine Methode, die soziologische und bevölkerungsstatistische Gliederung von Abstimmungen bei geheimen Wahlverfahren statistisch zu ermitteln. *Allgemeines Statistisches Archiv*, *22*: 253-256.

Berry,W.D. (1984) *Nonrecursive causal models.* (Sage Unversity Paper series on Quantitative Applications in the Social Sciences, series no. 07-037). Beverly Hills: Sage.

Bertholet,J.L. (1987) *PLS and the structure analysis of contingency tables.* (Paper, symposium PLS Model Building: Theory and Application, Frankfurt am Main, Sept.23-25, 1987). Geneva, Switzerland.

Bertholet,J.L. & Wold,H. (1984) *Recent developments on categorical data analysis by PLS modeling.* (Paper, ICUS XIII; J. D. Marriott Hotel, Washington, D.C.; 1984 September 3 - 5). Geneva, Switzerland: University, Department of Econometrics.

Bertholet,J.L. & Wold,H. (1985) Recent developments in categorical data analysis by PLS. In P. Nijkamp, H. Leitner, & N. Wrigley (Eds.), *Measuring the unmeasurable* (pp. 253-286). (NATO ASI Series D No. 22). Dordrecht, Boston, Lancaster: Martinus Nijhoff.

Bieber,S.L. & Meredith,W. (1985) Transformation to achieve a longitudinally stationary factor pattern matrix. *Psychometrika*, *50*: 535-547.

Bishop,Y.M., Fienberg,S.E., & Holland,P. (1975) *Discrete multivariate analysis: Theory and practice.* Cambridge, Mass.: The MIT Press.

Blalock,H.M. (1961) *Causal inferences in nonexperimental research.* Chapel Hill: The University of North Carolina Press.

Blalock,H.M. (1963) Making causal inferences for unmeasured variables from correlations among indicators. *The American Journal of Sociology*, *69*: 53-62.

Blalock,H.M. (1969) *Theory construction: From verbal to mathematical formulation.* Englewood Cliffs, NJ.: Prentice-Hall.

Blalock,H.M. (Ed.) (1972) *Causal models in the social sciences.* Chicago: Aldine.

Blommestein,H.J. (1985) Soft modelling and spatial econometrics: Towards an integreated approach. In P. Nijkamp, H. Leitner, & N. Wrigley (Eds.), *Measuring the unmeasurable* (pp. 287-300). (NATO ASI Series D No. 22). Dordrecht: Martinus Nijhoff.

Bloxom,B. (1968) A note on invariance in three-mode factor analysis. *Psychometrika*, *33*: 347-350.

Bloxom,B. (1984) Tucker's three-mode factor analysis model. In H.G. Law, C.W. Snyder, J.A. Hattie, & R.P. McDonald (Eds.), *Research methods for multimode data analysis* (pp. 104-121). New York: Praeger.

Boardman,A.E., Hui,B.S., & Wold,H. (1981) The Partial Least Squares-Fix Point method of estimating interdependent systems with latent variables. *Communications in Statistics, A - Theory and Methods, 10*: 613-639.

Bock,R.D. (1975) *Multivariate statistical methods in behavioral research.* New York: McGraw-Hill.

Bodewig,E. (1956) *Matrix calculus.* Amsterdam: North Holland.

Bodin,L. (1970) Recursive Fix-Point (RFP) estimation. In E.J. Mosbaek & H. Wold (Eds.), *Interdependent systems: Structure and estimation.* Amsterdam: North Holland.

Bodin,L. (1974) *Recursive Fix-Point estimation: Theory and applications.* (Doctoral thesis at Uppsala University 1974). (Selected Publications, Vol. 32). Uppsala, Sweden: University, Department of Statistics.

Bookstein,F.L. (1980) Data analysis by partial least squares. In J. Kmenta & J.B. Ramsey (Eds.), *Evaluation of econometric methods* (pp.75-90). New York: Academic.

Bookstein,F.L. (1982a) *Convergence of the regression cycles in Wold's soft modeling.* University of Michigan: Department of Radiology and Center for Human Growth.

Bookstein,F.L. (1982b) Panel discussion – modeling and method. In K.G. Jöreskog & H. Wold (Eds.), *Systems under indirect observation: Causality, structure, prediction* (Vol. 2, pp. 317-322). Amsterdam: North Holland.

Bookstein,F.L. (1982c) The geometric meaning of soft modeling, with some generalizations. In K.G. Jöreskog & H. Wold (Eds.), *Systems under indirect observation: Causality, structure, prediction* (Vol. 2, pp. 55-74). Amsterdam: North Holland.

Boomsma,A. (1982) The robustness of LISREL against small sample sizes in factor analysis models. In K.G. Jöreskog & H. Wold (Eds.), *Systems under indirect observation: Causality, structure, prediction* (Vol. 1, pp. 149-174). Amsterdam: North Holland.

Brandstätter,J. (1976) Soziale Schicht, Umwelt und Intelligenz: Eine Pfadanalyse der Korrelationsbefunde von Marjoribanks. *Psychologische Beiträge, 18*: 35-53.

Breiman,L. & Friedman,J.H. (1985) Estimating optimal transformations for multiple regression and correlation (with comments and rejoinder). *Journal of the American Statistical Association, 80*: 580-619.

Browne,M.W. (1979) The maximum-likelihood solution in inter-battery factor analysis. *British Journal of Mathematical and Statistical Psychology, 32*: 75-86.

Browne,M.W. (1980) Factor analysis of multiple batteries by maximum likelihood. *British Journal of Mathematical and Statistical Psychology, 33*: 184-199.

Browne,M.W. (1984) The decomposition of multitrait-multimethod matrices. *British Journal of Mathematical and Statistical Psychology, 37*: 1-21.

Burt,C. (1950) The factorial analysis of qualitative data. *British Journal of Statistical Psychology, 2*: 166-185.

Campbell,D.T. & O'Connell,E.J. (1967) Method factors in multimethod-multitrait research: Multiplicative rather than additive? *Multivariate Behavioral Research, 2*: 409-426.

Carroll,J.D. (1963) Generalization of canonical correlation analysis to three or more sets of variables. *Proceedings 76th Annual Convention, APA, 76*: 227-228.

Carroll,J.D. & Chang,J.J. (1970) Analysis of individual differences in multidimensional scaling via an *N*-way generalisation of "Eckart-Young" decomposition. *Psychometrika, 35*: 283-319.

Carroll,J.D., Pruzansky,S., & Kruskal,J.B. (1980) CANDELINC – A general approach to multidimensional analysis of many-way arrays with linear constraints on parameters. *Psychometrika, 45*: 2-24.

Cattell,R.B. (1966) The data box: Its ordering of total resources in terms of possible relational systems. In R.B. Cattell (Ed.), *Handbook of multivariate experimental psychology* (pp. 67-128). Chicago: Rand McNally.

Cohen,J. (1982) Set correlation as a general multivariate data-analytic method. *Multivariate Behavioral Research, 17*: 301-341.

Cohen,P. & Cohen,J. (1977) Inefficient redundancy. *Multivariate Behavioral Research, 12*: 167-169.

Coleman,J.S. (1981) Response to Page and Keith. *Educational Researcher, 10*, No. 8: 18-20.

Conger,A.J. (1971) Evaluation of multimethod factor analysis. *Psychological Bulletin, 75*: 416-420.

Coxhead,P. (1974) Measuring the relationship between two sets of variables. *British Journal of Mathematical and Statistical Psychology, 27*: 205-212.

Cramér,H. (1946) *Mathematical methods of statistics.* Princeton NJ: Princeton University Press.

Cramer,E.M. (1974) A generalization of vector correlation and its relation to canonical correlation. *Multivariate Behavioral Research, 9*: 347-351.

Cremer,R. & Hujer,R. (1979) "Weiche" Modelle für Arbeitsmarkt und Beschäftigung: Ein Beitrag zur Ergänzung ökonometrischer Ansätze? *Konjunkturpolitik, 25*: 277-296.

Cremer,R. & Knepel,H. (1980) Ein Indikatormodell für sozioökonomische Problembereiche des Arbeitsmarktes: Spezifikationsprobleme, methodische Grundlagen, erste Ergebnisse. *Mitteilungen aus der Arbeitsmarkt- und Berufsforschung, 13*: 125-136.

Cronbach,L.J. & Meehl,P.E. (1955) Construct validity in psychological testing. *Psychological Bulletin, 52*: 281-302.

Darlington,R.B., Weinberg,S.L., & Walberg,H.J. (1973) Canonical variate analysis and related techniques. *Review of Educational Research, 43*: 433-454.

Davis,J.A. (1985) *The logic of causal order.* (Sage Unversity Paper series on Quantitative Applications in the Social Sciences, series no. 07-055). Beverly Hills: Sage.

Dawson-Saunders,B.K. & Tatsuoka,M.M. (1983) The effect of affine transformation on redundancy analysis. *Psychometrika, 48*: 299-302.

de Leeuw,J. (1984) *Least squares and maximum likelihood for causal models with discrete variables.* Leiden, NL: University of Leiden.

de Pijper,W.M. & Saris,W.E. (1982) The effect of identification restrictions on the test statistics in latent variable models. In K.G. Jöreskog & H. Wold (Eds.), *Systems under indirect observation: Causality, structure, prediction* (Vol. 1, pp. 175-182). Amsterdam: North Holland.

DeSarbo,W.S. (1981) Canonical-redundancy factoring analysis. *Psychometrika, 46*: 307-329.

Dijkstra,T.K. (1981) *Latent variables in linear stochastic models: Reflections on Maximum Likelihood and Partial Least Squares methods.* (Dissertation). Groningen, NL: Rijksuniversiteit te Groningen, Economische Wetenschappen.

Dijkstra,T.K. (1982) Appendix to the Whittle and Adelman chapter. In K.G. Jöreskog & H. Wold (Eds.), *Systems under indirect observation: Causality, structure, prediction* (Vol. 2, pp. 290-292). Amsterdam: North Holland.

Dijkstra,T.K. (1983) Some comments on maximum likelihood and partial least squares methods. *Journal of Econometrics, 22*: 67-90.

Draper,N.R. & Smith,S. (1981) *Applied regression analysis.* New York: Wiley.

Duncan,O.D., Haller,A.O., & Portes,A. (1968) Peer influence on aspirations: A reinterpretation. *American Journal of Sociology, 74*: 119-137.

Dupačová,J. & Wold,H. (1975) NIPALS vs. Maximum Likelihood modelling. In H. Wold (Ed.), *Modelling in complex situations with soft information.* (Paper, Third World Congress on Econometrics; Toronto, Canada; August 21-26, 1975). (Research Report 1975:5). Göteborg, Sweden: University, Institute of Statistics.

Dupačová,J. & Wold,H. (1982) On some identification problems in ML modeling of systems with indirect observations. In K.G. Jöreskog & H. Wold (Eds.), *Systems under indirect observation: Causality, structure, prediction* (Vol. 2, pp. 293-316). Amsterdam: North Holland.

Ejsmont,Z. & Rogowski,J. (1987) *The foreign trade structure of economic development – The PLS soft-model.* (Paper, symposium PLS Model Building: Theory and Application, Frankfurt am Main, Sept.23-25, 1987). Białystok, Poland: Warsaw University.

Engel,A. (1987) *Wahlen und Parteien im lokalen Kontext: Ein vergleichende Untersuchung des Basisbezugs lokaler Parteiakteure in 24 nordhessischen Kreisparteiorganisationen von CDU, FDP und SPD.* Frankfurt am Main: Peter Lang.

Engfer,A. (1982) *Bedingungen und Auswirkungen harten elterlichen Strafens.* (Dissertation). Trier: Universität.

Engfer,A. & Gavranidou,M. (1984) *Scources of stress and support – their impact on the mother-child relationship.* (Paper presented an the Inaugural European Conference on Development

Psychology: Individual Development and Human Welfare, Groningen, Netherland, August 28-31, 1984.). Munich, Germany: University.

Engfer,A. & Gavranidou,M. (1987) Antecedents and consequences of maternal sensivity: A longitudinal study. In H. Rauh & H.C. Steinhausen (Eds.), *Psychobiology and early development* (pp. 71-99). Amsterdam: Elsevier.

Engfer,A. & Schneewind,K.A. (1981) *Causes and consequences of harsh parental punishment: An empirical investigation in a representative sample of 570 German families.* (Third International Congress on Child Abuse and Neglect; Amsterdam, NL; April 21-25, 1981). München: Institut für Psychologie, Universität München.

Ennis,R.H. (1982) Abandon Causality? *Educational Researcher, 11*, No. 8: 25-27.

Ericson,D.P. (1982) Letter to the Editor on noncausal inquiry in education. *Educational Researcher, 11*, No. 2: 25.

Esbensen,K. (1987) Aspects of the philosophy of PLS regression and the strategy of PLS data analysis. In M. Martens (Ed.), *Data-approximation by PLS methods* (pp. 14-24). (Report no. 800). Olso, Norway: Norwegian Computing Center.

Falk,R.F. (1988) *A primer for soft modeling.* Berkeley, CA: University of California, Institute of Human Development.

Falter,J.W. (1977) Zur Validierung theoretischer Konzepte: Wissenschaftstheoretische Aspekte des Validierungskonzepts. *Zeitschrift für Soziologie, 6*: 349-385.

Falter,J.W. (1981) Kontinuität und Neubeginn: Die Bundestagswahl 1949 zwischen Weimar und Bonn. *Politische Vierteljahresschrift, 22*: 236-263.

Falter,J.W., Link,A., Lohmöller,J.B., de Rijke,J., & Schumann,S. (1982) *Unemployment and the rise of National Socialism: Some new findings.* (Paper, 7th annual conference of the Western Association for German Studies; University of Texas, El Paso; Oct. 1982). München: Hochschule der Bundeswehr, Fachbereich Pädagogik.

Falter,J.W., Link,A., Lohmöller,J.B., de Rijke,J., & Schumann,S. (1983) Arbeitslosigkeit und Nationalsozialismus: Eine empirische Analyse des Beitrags der Massenerwerbslosigkeit der NSDAP 1932 und 1933. *Kölner Zeitschrift für Soziologie und Sozialpsychologie, 35*: 525-554.

Falter,J.W. & Lohmöller,J.B. (1982) Manifeste Schwächen im Umgang mit latenten Variablen: Ein Kommentar zu Hans-Hermann Hoppes Theologie der LV-Pfadmodell in ZfS Juli 1981. *Zeitschrift für Soziologie, 11*: 69-77.

Falter,J.W., Lohmöller,J.B., Link,A., & de Rijke,J. (1982) Die Arbeiter machten den Großteil der Wählerschaft Hitlers aus. *Frankfurter Rundschau, 1982-12-21*: 12.

Falter,J.W., Lohmöller,J.B., Link,A., & de Rijke,J. (1985) Hat Arbeitslosigkeit tatsächlich den Aufstieg des Nationalsozialismus bewirkt? *Jahrbücher für Nationalökonomie und Statistik, 200*: 121-136.

Falter,J.W. & Ulbricht,K. (1981) *Zur Kausalanalyse von Kontingenztafeln.* Frankfurt: Peter Lang.

Fearing,J. (1981) Letter to the Editor on noncausal inquiry in education. *Educational Researcher, 10*, No. 8: 25.

Fienberg,S.E. (1977) *The analysis of cross-classified categorical data.* Cambridge, Mass: The MIT Press.

Fisher,R.A. (1940) The precision of discriminant functions. *Annals of Eugenics, 10*: 422-429.

Flaig,G. (1983) *Die Analyse von qualitativen Variablen mit Latenten-Struktur-Modellen.* (Dissertation, Universität Mannheim, Wirtschaftwissenschaften). Mannheim: Universität.

Fletcher,R. & Powell,M.J.D. (1963) A rapidly convergent descent method for minimization. *Computer Journal, 6*: 163-168.

Fletcher,R. & Reeves,C.M. (1964) Function minimization by conjugate gradients. *Computer Journal, 7*: 149-154.

Fornell,C. (1978) Three approaches to canonical analysis. *Journal of the Market Research Society, 20*: 166-181.

Fornell,C. (Ed.) (1982) *A second generation of multivariate analysis, Vol. 1 - measurement and evaluation.* New York: Praeger.

Fornell,C. & Bookstein,F.L. (1981) *A comparative analysis of two structural equation models: LISREL and PLS applied to market data.* Ann Arbor: University of Michigan, Graduate School of Business Administration.

Fornell,C. & Bookstein,F.L. (1982) Two structural equation models: LISREL and PLS applied to consumer exit-voice theory. *Journal of Marketing Research*, *19*: 440-452.

Fortier,J.J. (1966) Simultaneous linear prediction. *Psychometrika*, *31*: 369-381.

Fraser,C. (1980 ca.) *COSAN user's guide*. Ontario, Canada: The Ontario Institute for Studies in Education.

Frederiksen,C.H. (1974) Canonical analysis in several populations with predictor weights constrained across populations. *British Journal of Mathematical and Statistical Psychology*, *27*: 213-228.

Gattringer,H. (1980) *Die Entwicklung des Leistungsmotivs und dessen Bedeutung für die Schulleistungsprognose*. (Dissertation). Universität Salzburg: Philosophische Fakultät.

Geisser,S. (1974) A predictive approach to the random effect model. *Biometrika*, *61*: 101-107.

Geladi,P. (1987) History and uses of partial least squares (PLS). In M. Martens (Ed.), *Data-approximation by PLS methods* (pp. 5-13). (Report no. 800). Olso, Norway: Norwegian Computing Center.

Geladi,P., Esbensen,K., & Wold,S. (1987) *Soft modeling on multivariate images*. (Paper, symposium PLS Model Building: Theory and Application, Frankfurt am Main, Sept.23-25, 1987). Umeå, Sweden: University, Institute of Chemistry.

Gerbner,G. (1984) *Political functions of television viewing: A cultivation analysis*. In G. Melischek, K.E. Rosengren, & J. Stappers (Eds.), *Cultural indicators: An international symposium* (pp.329-343). (Philosophisch-Historische Klasse, Sitzungsberichte, 416. Band). Vienna, Austria: Verlag der Österreichischen Akademie der Wissenschaften.

Gerbner,G., Gross,L., Morgan,M., & Signorielli,N. (1980) The Mainstreaming of America: Violence Profile No.11. *Journal of Communication*, *30*: 10-29.

Gerbner,G., Gross,L., Morgan,M., & Signorielli,N. (1982) *Violence Profile No.11: Trends in network television drama and viewer conception of social reality 1967-1979*. (Technical Report). Philadelphia: University of Pennsylvania, The Annenberg School of Communication.

Giegler,H. (1983) *LISREL - kritisch gesehen aus soziologischer und politologischer Anwenderperspektive*. (Arbeitspapier, Methodensektion der Deutschen Gesellschaft für Soziologie; München; 30.9. - 1.10. 1983). Gießen: Universität.

Gittins,R. (1985) *Canonical analysis: A review with applications in ecology*. Berlin: Springer.

Gilula,Z., Krieger,A.M., & Ritov,Y. (1988) Ordinal association in contingency tables: Some interpretative aspects. *Journal of the American Statistical Association*, *83*: 540-545.

Gleason,T.C. (1976) On redundancy in canonical analysis. *Psychological Bulletin*, *83*: 1004-1006.

Glymour,C., Scheines,R., Spirtes,P., & Kelly,K. (1987) *Discovering causal structure: Artificial intelligence, philosophy of science, and statistical modeling*. Orlando: Academic Press.

Goldberger,A.S. & Duncan,O.D. (Eds.) (1973) *Structural equation models in the social sciences*. New York: Seminar Press.

Golding,S.L. & Seidman,E. (1974) Analysis of multitrait-multimethod matrices: A two-step principal components procedure. *Multivariate Behavioral Research*, *9*: 479-496.

Golub,G.H. & Van Loan,C.F. (1983) *Matrix Computations*. Baltimore MD: The John Hopkins University Press.

Goodman,L.A. (1953) Ecological regression and behavior of individuals. *American Journal of Sociology*, *18*: 663-664.

Goodman,L.A. (1959) Some alternatives to ecological correlation. *Journal of Sociology*, *64*: 610-625.

Goodman,L.A. (1978) *Analyzing qualitative-categorical data: Log-linear models and latent-structure analysis*. Cambridge, Mass.: Abt Books.

Gorsuch,R.L. (1974) *Factors analysis*. Philadelphia: W.B.Saunders.

Graff,J. & Schmidt,P. (1982) A general model for decomposition of effects. In K.G. Jöreskog & H. Wold (Eds.), *Systems under indirect observation: Causality, structure, prediction* (Vol. 1, pp. 131-148). Amsterdam: North Holland.

Greene,V.L. (1978) Simultaneous optimization of factor assessibility and representativeness: An old solution to a new problem. *Psychometrika*, *43*: 273-275.

Grizzle,J.E., Starmer,C.F., & Koch,G.G. (1969) Analysis of categorical data by linear models. *Biometrics*, *25*: 489-504.

Gross,L. (1984) *The cultivation of intolerance: Television, blacks and gays.* In G. Melischek, K.E. Rosengren, & J. Stappers (Eds.), *Cultural indicators: An international symposium* (pp.345-363). (Philosophisch-Historische Klasse, Sitzungsberichte, 416. Band). Vienna, Austria: Verlag der Österreichischen Akademie der Wissenschaften.

Grüner,H., Jädicke,W., & Ruhland,K. (1988) Rote Politik im schwarzen Rathaus? Bestimmungsfaktoren der wohnungspolitischen Ausgaben bundesdeutscher Großstädte. *Politische Vierteljahresschrift, 29*: 42-57.

Guttman,L. (1941b) An outline of the statistical theory of prediction. In P. Horst (Ed.), *The prediction of personal adjustment; supplementary study B-I* (pp. 253-311). (SSRC bulletin No. 48). New York: Social Science Research Council.

Guttman,L. (1941c) The quantification of a class attributes: A theory and method of scale construction. In P. Horst (Ed.), *The prediction of personal adjustment; supplementary study B-III* (pp. 319-348). (SSRC bulletin No. 48). New York: Social Science Research Council.

Guttman,L. (1950a) Relation of scalogram analysis to other techniques. In S.A. Stouffer & al (Eds.), *Studies in social psychology in World War II; volume IV, Measurement and prediction* (pp. 172-212). Gloucester,Mass.: Peter Smith (reprint).

Guttman,L. (1950b) The principal components of scale analysis. In S.A. Stouffer & al (Eds.), *Studies in social psychology in World War II; volume IV, Measurement and prediction* (pp. 312-361). Gloucester,Mass.: Peter Smith (reprint).

Hakstian,A.R. (1973) Procedures for the factor analytic treatment of measures obtained on different occasions. *British Journal of Mathematical and Statistical Psychology, 26*: 219-239.

Hakstian,A.R. (1974) The development of a class of oblique factor solutions. *British Journal of Mathematical and Statistical Psychology, 27*: 100-114.

Hakstian,A.R. (1976) Two-matrix orthogonal rotation procedure. *Psychometrika, 41*: 267-272.

Hakstian,A.R. (1977) Transformation of axes in interbattery factor analysis. *Multivariate Behavioral Research, 12*: 159-165.

Hakstian,A.R., Zidek,J.V., & McDonald,R.P. (1977) Best univocal estimates of orthogonal common factors. *Psychometrika, 42*: 627-629.

Hanke,B., Lohmöller,J.B., & Mandl,H. (1980) *Schülerbeurteilung in der Grundschule: Ergebnisse aus der Augsburger Längsschnittuntersuchung.* München: Oldenbourg. (302p).

Hansson,L. & Berglund,M. (1987) Factors influencing treatment outcome and patient satisfaction in a short-term psychiatric ward. *European Archives of Psychiatry and Neurolocigal Sciences, 236*: 269-275.

Harman,H.H. (1967) *Modern factor analysis.* Chicago: The University of Chicago Press.

Harman,H.H. & Jones,W.H. (1966) Factor analysis by minimizing residuals (Minres). *Psychometrika, 31*: 351-368.

Harris,C.W. (1967) On factors and factor scores. *Psychometrika, 32*: 363-379.

Harris,C.W. & McArthur,D.L. (1974) Another view of the relation of environment to mental abilities. *Journal of Educational Psychology, 66*: 457-459.

Hasselhorn,M. & Körkel,J. (1985) *Metacognitive versus traditional reading instructions: The mediating role of domain-specific knowledge of children's text-processing.* München: Max Planck Institut für Psychologische Forschung.

Hauser,R.M. & Goldberger,A.S. (1971) The treatment of unobservable variables in path analysis. In H.L. Costner (Ed.), *Sociological Methodology 1971* (pp. 81-117). San Francisco: Jossey-Bass.

Hayduk,L.A. (1987) *Structural equation modeling with LISREL: Essentials and advances* Baltimore: The Johns Hopkins University Press.

Heise,D.A. (1975) *Causal Analysis.* New York: Wiley.

Helland,I. (1987) Some theoretical considerations around PLS. In M. Martens (Ed.), *Data-approximation by PLS methods* (pp. 25-34). (Report no. 800). Olso, Norway: Norwegian Computing Center.

Hildebrand,D.K., Laing,J.D., & Rosenthal,H. (1977) *Prediction analysis of cross classifications.* New York: Wiley.

Hildebrandt,L., Rudinger,G., & Schmidt,P. (Eds.) (in press) *Kausalanalyse in der Umweltforschung.* Stuttgart: Gustav Fischer.

Hirschfeld,H.O. (1935) A connection between correlation and contingency. *Cambridge Philosophical Society Proceedings, 31*: 520-524.

Hoppe,H.H. (1981) Über die Verwendung ungemessener Variablen in Kausalmodellen: Eine epistemologische Kritik. *Zeitschrift für Soziologie, 10*: 307-318.

Hoppe,H.H. (1982) Über ungemessene Variablen: Von einem Fehlschluß und zwei unbeantworteten Fragen (Antwort auf Falter und Lohmöller). *Zeitschrift für Soziologie, 11*: 78-81.

Hopple,G.W. (1979) *Soft modeling in the social sciences: Applications of partial least squares in political sciences.* McLean, VA: IPPRC – International Public Policy Research Corporation.

Horst,P. (1935) Measuring complex attitudes. *Journal of Social Psychology, 6*: 369-374.

Horst,P. (1961) Relations among *m* sets of measures. *Psychometrika, 26*: 129-149.

Horst,P. (1965) *Factor analysis of data matrices.* New York: Holt, Rinehart and Winston.

Hotelling,H. (1933) Analysis of a complex of statistical variables into principal components. *Journal of Educational Psychology, 24*: 417-441,498-520.

Hotelling,H. (1935) The most predictable criterion. *The Journal of Educational Psychology, 1*: 139-142.

Huba,G.J., Newcomb,M.D., & Bentler,P.M. (1981) Comparison of canonical correlation and interbattery factor analysis on sensation seeking and drug use domains. *Applied Psychological Measurement, 5*: 291-306.

Huber,M., Gabriel,J., & Ehrmann,A. (1988) *Zur Ungleichverteilung von Arbeitsmarktchancen nach einer Betriebsstillegung: Ergebnisse einer multivariaten Analyse.* (Arbeitspapier aus dem Arbeitskreis Sozialwissenschaftliche Arbeitsmarktforschung (SAMF), 1988 - 3). Berlin: Freie Universität Berlin.

Hui,B.S. (1978) *The partial least squares approach to path models of indirectly observed variables with multiple indicators.* (Disseratation, Managerial Science and Applied Economics). Philadelphia, PA: University of Pennsylvania.

Hui,B.S. (1982) On building partial least squares models with interdependent inner relations. In K.G. Jöreskog & H. Wold (Eds.), *Systems under indirect observation: Causality, structure, prediction* (Vol. 2, pp. 249-272). Amsterdam: North Holland.

Hui,B.S. & Wold,H. (1982) Consistency and consistency at large of partial least squares estimates. In K.G. Jöreskog & H. Wold (Eds.), *Systems under indirect observation: Causality, structure, prediction* (Vol. 2, pp. 119-130). Amsterdam: North Holland.

Israels,A.Z. (1984) Redundancy analysis for qualitative variables. *Psychometrika, 49*: 331-346.

Jackson,D.N. (1969) Multimethod factor analysis in the evaluation of convergent and discriminant validity. *Psychological Bulletin, 72*: 30-49.

Jackson,D.C. (1975) Multimethod factor analysis - a reformulation. *Multivariate Behavioral Research, 10*: 259-275.

Jackson,D.N. (1977) Distinguishing trait and method variance in multitrait-multimethodmatrices: A reply to Golding. *The Journal of Multivariate Behavioral Research, 12*: 99-110.

Jain,A.K., Pinson,C., & Malhotra,N.K. (1987) *A Partial Least Squares (PLS) causal investigation of customers' shopping behavior.* Fontainebleau Cedex, France: INSEAD.

Jensen,S.A., Munck,L., & Martens,H. (1982) *The botanical constituents of wheat and wheat milling fractions, I: Quantification by autofluorescence.* (7th World Cereal and Bread Congress; Prague, Czechoslovakia; June 28 - July 2, 1982). DK-2500 Copenhagen-Valby: Carlsberg Research Laboratory, Dept. of Biotechnology, Gamle Carlsbergvej 10.

Johansson,J.K. (1981) An extension of Wollenberg's redundancy analysis. *Psychometrika, 46*: 93-103.

Jönsson,A. (1984) *TV – ett hot eller en resurs för barn? En longitudinell studie av relationen mellan TV och skola.* Lund, Sweden: CWK Gleerup.

Johnson,R.M. (1966) The minimal transformation to orthonormality. *Psychometrika, 31*: 61-66.

Jöreskog,K.G. (1969) *Factoring the multitest-multioccasion correlation matrix.* (Research Bulletin 69-62). Princeton, N.J.: Educational Testing Service.

Jöreskog,K.G. (1970) A general method for analysis of covariance structures. *Biometrika, 57*: 239-251.

Jöreskog,K.G. (1971) Simultaneous factor analysis in several populations. *Psychometrika, 36*: 409-426.

Jöreskog,K.G. (1973a) A general method for estimating a linear structural equation system. In A.S. Goldberger & O.D. Duncan (Eds.), *Structural equation models in the social sciences* (pp. 85-112). New York: Seminar Press.

Jöreskog,K.G. (1973b) Analysis of covariance structures. In P.R. Krishnaiah (Ed.), *Multivariate analysis* (Vol. 3, pp. 263-285). New York: Academic.

Jöreskog,K.G. (1974) Analyzing psychological data by structural analysis of covariance matrices. In D. H.Krantz, R.C. Atkinson, R.D. Luce, & P. Suppes (Eds.), *Contemporary developments in mathematical psychology* (Vol. 2, pp. 1-56). San Francisco: Freeman.

Jöreskog,K.G. (1977a) Statistical models and methods for analysis of longitudinal data. In D.J. Aigner & A.S. Goldberger (Eds.), *Latent variables in socioeconomic models* (pp. 285-325). Amsterdam: North Holland.

Jöreskog,K.G. (1977b) Structural equation models in the social sciences: Specification, estimation and testing. In P.R. Krishnaiah (Ed.), *Applications of statistics* (pp. 265-287). Amsterdam: North Holland.

Jöreskog,K.G. (1978) Structural analysis of covariance and correlation matrices. *Psychometrika, 43*: 443-477.

Jöreskog,K.G. (1979a) Basic ideas of factor analysis and component analysis. In K.G. Jöreskog & D. Sörbom (Eds.), *Advances in factor analysis and structural equation models* (pp. 5-20). Cambridge, Massachusetts: Abt Books.

Jöreskog,K.G. (1979b) Statistical estimation of structural models in longitudinal-developmental investigations. In J.R. Nesselroade & P.B. Baltes (Eds.), *Longitudinal research in the study of behavior and development* (pp. 303-351). New York: Academic Press.

Jöreskog,K.G. (1981) Basic issues in the application of LISREL. *Data, communications computer data analysis, 1*: 1-6.

Jöreskog,K.G. (1982) The LISREL approach to causal model-building in the social sciences. In K.G. Jöreskog & H. Wold (Eds.), *Systems under indirect observation: Causality, structure, prediction* (Vol. 1, pp. 81-100). Amsterdam: North Holland.

Jöreskog,K.G. & Goldberger,A.S. (1975) Estimation of a model with multiple indicators and multiple causes of a single latent variable. *Journal of the American Statistical Association, 70*: 631-639.

Jöreskog,K.G. & Sörbom,D. (1978) *LISREL IV – A general computer program for estimation of linear structural equation systems by maximum likelihood methods.* Uppsala, Sweden: University, Department of Statistics.

Jöreskog,K.G. & Sörbom,D. (1979) *Advances in factor analysis and structural equation models.* Cambridge, Massachusetts: Abt Books.

Jöreskog,K.G. & Sörbom,D. (1981) *LISREL V – Analysis of linear structural relationships by maximum likelihood and least squares methods.* (Research Report 81-8). Uppsala, Sweden: University, Department of Statistics.

Jöreskog,K.G. & Sörbom,D. (1987) *LISREL VII program manual.* Chicago: International Educational Services.

Jöreskog,K.G. & van Thillo,M. (1973) *LISREL – a general computer program for estimating a linear structural equation system involving multiple indicators of unmeasured variables.* (Research Report 1973-5). Uppsala, Sweden: University, Department of Statistics.

Jöreskog,K.G. & Wold,H. (Eds.) (1982a) *Systems under indirect observation: Causality - structure - prediction* (2 vols.). Amsterdam: North Holland.

Jöreskog,K.G. & Wold,H. (1982b) The ML and PLS techniques for modeling with latent variables: Historical and comparative aspects. In K.G. Jöreskog & H. Wold (Eds.), *Systems under indirect observation: Causality, structure, prediction* (Vol. 1, pp. 263-270). Amsterdam: North Holland.

Kaiser,H.F. (1967) Uncorrelated linear composites maximally related to a complex of correlated observations. *Educational and Psychological Measurement, 27*: 3-6.

Kenny,D.A. (1979) *Correlation and causation.* New York: Wiley.

Kettenring,J.R. (1971) Canonical analysis of several sets of variables. *Biometrika, 58*: 433-451.

Knepel,H. (1980) Modelle mit unbeobachteten Variablen: der PLS-Ansatz. *Statistische Hefte, 22*: 248-279.

Knepel,H. (1981) *Sozioökonomische Indikatormodelle zur Arbeitsmarktanalyse.* (SFB 3, Mikroanalytische Grundlagen der Gesellschaftpolitik, Band 7). Frankfurt, New York: Campus.

Kowalski,B.R., Gerlach,R.W., & Wold,H. (1982) Chemical systems under indirect observation. In K.G. Jöreskog & H. Wold (Eds.), *Systems under indirect observation: Causality, structure, prediction* (Vol. 2, pp. 191-208). Amsterdam: North Holland.

Kristof,W. (1967) Orthogonal inter-battery factor analysis. *Psychometrika, 32*: 199-227.

Kristof,W. & Wingersky,B. (1971) Generalization of the orthogonal procrustes rotation procedure to more than two matrices. *Proceedings 79th Annual Convention, APA, 79*: 89-90.

Kroonenberg,P.M. (1983a) Annotated bibliography of three-mode factor and principal component analysis. *British Journal of Mathematical and Statistical Psychology, 36*: 81-113.

Kroonenberg,P.M. (1983b) *Three-mode principal component analysis: Theory and applications.* (Dissertation, Rijksuniversiteit te Leiden). Leiden NL: DSWO Press.

Kroonenberg,P.M. (1984) Three-mode principal component analysis: Illustrated with an example from attachment theory. In H.G. Law, C.W. Snyder, J.A. Hattie, & R.P. McDonald (Eds.), *Research methods for multimode data analysis* (pp. 64-103). New York: Praeger.

Kroonenberg,P.M. & de Leeuw,J. (1977) *TUCKALS2 – a principal component analysis of three mode data.* (RB 001-77). Leiden, NL: Rijksuniversiteit te Leiden, Faculty of Social Sciences, Department of Datatheory.

Kroonenberg,P.M. & de Leeuw,J. (1980) Principal component analysis of three-mode data by means of alternating least squares algorithms. *Psychometrika, 45*: 69-97.

Krus,D.J., Reynolds,T.J., & Krus,P.H. (1976) Rotation in canonical variate analysis. *Educational and Psychological Measurement, 36*: 725-730.

Lancaster,H.O. (1969) *The chi-squared distribution.* New York: Wiley.

Lastovicka,J. (1981) The extension of component analysis to four-mode matrices. *Psychometrika, 46*: 47-57.

Law,H.G., Snyder,C.W., Hattie,J.A., & McDonald,R.P. (Eds.) (1984) *Research methods for multimode data analysis.* New York: Praeger.

Lebart,L., Morineau,A., & Tabard,N. (1978) *Techniques de la description statistique.* Paris: Editions Dunod.

Lee,S.Y. & Tsui,K.L. (1982) Covariance structure analysis in several populations. *Psychometrika, 47*: 297-308.

Levine,M.S. (1977) *Canonical analysis and factor comparison.* (Sage Unversity Paper series on Quantitative Applications in the Social Sciences, series no. 07-006). Beverly Hills, CA: Sage.

Loehlin,J.C. (1987) *Latent variable models: An introduction to factor, path and structural analysis.* Hillsdale, NJ: Erlbaum.

Löfgren,H. (1984) Arbtsmiljön i skolan: en studie av elever i åskurserna 7 och 9 i Malmö kommun. In P. Sennerfeldt (Ed.), *Hälsa är helhet – Rapport från en konferens med temat forskning om hälsa.*

Löfgren,H. (1986) *Bilingual instruction of immigrant children: A theoretical overview and results from empirical research.* (Didakometry No. 70). Malmö, Sweden: School of Education.

Lohmöller,J.B. (1976) *Multivariate Veränderungsmessung mit dem Modell der trimodalen Faktorenanalyse.* (Vortrag, Tagung Entwicklungspsychologie, Bochum, 13.Apr.1976). Augsburg: Universität, Institut für Psychologie.

Lohmöller,J.B. (1977) *Varimax rotation in multiple set factor analysis.* (Paper, conference of the Society for Multivariate Experimental Psychology (SMEP), European Branch; Schliersee, Bavaria; Dec. 12, 1977). München: Hochschule der Bundeswehr, Fachbereich Pädagogik.

Lohmöller,J.B. (1978a) *How longitudinal factor stability, continuity, differentiation and integration are portrayed into the core matrix of three-mode factor analysis.* (Paper, European Meeting on Psychometrics and Mathematical Psychology; Uppsala, Sweden; June 16, 1978). München: Hochschule der Bundeswehr, Fachbereich Pädagogik.

Lohmöller,J.B. (1978b) Stabilität und Kontinuität in Längsschnittdaten, analysiert durch T- und trimodale Faktorenanalyse. In G. Rudinger (Ed.), *Methoden der Entwicklungspsychologie* (in preparation since 1978).

Lohmöller,J.B. (1979a) Die trimodale Faktorenanalyse von Tucker: Skalierungen, Rotationen, andere Modelle. *Archiv für Psychologie, 131*: 137-166.

Lohmöller,J.B. (1979b) *Estimating parameters of linear structural relation models under partial least squares criteria.* (Paper presented at the annual meeting of the American Educational Research Association, San Francisco, April 1979). (Forschungsbericht 79.01 Fachbereich Pädagogik). München: Hochschule der Bundeswehr.

Lohmöller,J.B. (1979c) *Pfadanalyse mit latenten Variablen: Das Programm PLSC.* München: Hochschule der Bundeswehr, Fachbereich Pädagogik. (Mimeo).

Lohmöller,J.B. (1979d) *Residuals in PLS modelling.* (Paper, Workshop on Soft Modelling Applications; Department of Political Sciences, University of Geneva, Switzerland; Oct 22-23, 1979). München: Hochschule der Bundeswehr, Fachbereich Pädagogik. (Mimeo).

Lohmöller,J.B. (1979e) *Stabilität, Kontinuität und Transformation von Schulleistungen.* (Vortrag, Tagung Entwicklungspsychologie; Berlin; September 1979). München: Hochschule der Bundeswehr, Fachbereich Pädagogik.

Lohmöller,J.B. (1979f) *Strukturgleichungsmodelle unter Partialkleinstquadratkriterien: Darstellung und Beschreibung des Programms PLSX.* München: Hochschule der Bundeswehr, Fachbereich Pädagogik. (Mimeo).

Lohmöller,J.B. (1980) *LV-Pfadanalyse und Probleme mit LISREL.* München: Hochschule der Bundeswehr, Fachbereich Pädagogik. (unpubl. manuscript).

Lohmöller,J.B. (1981a) *LVPLS 1.6 program manual: Latent variables path analysis with partial least squares estimation.* (Forschungsbericht 81.04 Fachbereich Pädagogik). München: Hochschule der Bundeswehr.

Lohmöller,J.B. (1981b) *Pfadmodelle mit latenten Variablen: LVPLSC ist eine leistungsfähige Alternative zu LISREL.* (Vortrag, 23. Tagung experimentell arbeitender Psychologen; Berlin; 12.-16.Apr.1981). (Forschungsbericht 81.02 Fachbereich Pädagogik). München: Hochschule der Bundeswehr.

Lohmöller,J.B. (1982a) *An Overview of latent variables path analysis.* (Paper, AERA (American Educational Research Association) Annual Meeting; New York; March 19-23 1982). München: Hochschule der Bundeswehr, Fachbereich Pädagogik. (Mimeo).

Lohmöller,J.B. (1982b) *LVPLS 1.6 implementation notes.* München: Hochschule der Bundeswehr, Fachbereich Pädagogik. (28p, tape file).

Lohmöller,J.B. (1982c) *LVPLS 1.6 instant manual.* München: Hochschule der Bundeswehr, Fachbereich Pädagogik. (17p, tape file).

Lohmöller,J.B. (1983) *Path models with latent variables and Partial Least Squares (PLS) estimation.* (Dissertation,). München: Hochschule der Bundeswehr, Fachbereich Pädagogik.

Lohmöller,J.B. (1986) Das Programmsystem LVPLS für Pfadmodelle mit latenten Variablen und Partialkleinstquadratschätzung. In W. Lehmacher & A. Hörmann (Eds.), *Statististik-Software: 3. Konferenz über die wissenschaftliche Anwendung von Statistik-Software, 1985* (pp.179-188). Stuttgart: Gustav Fischer Verlag.

Lohmöller,J.B. (1988a) Die LV-Pfadprogramme PLS, LISREL, EQS, COSAN und RAM im Vergleich. In F. Faulbaum & H.M. Uehlinger (Eds.), *Fortschritte der Statistik-Software 1: 4. Konferenz über die wissenschaftliche Anwendung von Statistik-Software, Heidelberg, 1987* (pp.54-64). Stuttgart: Gustav Fischer.

Lohmöller,J.B. (1988b) The PLS program system: Latent variables path analysis with partial least squares estimation. *Multivariate Behavioral Research, 23*:125-127.

Lohmöller,J.B. (in press a) Die Partialkleinstquadratmethode für Pfadmodelle mit latenten Variablen und das Programmsystem LVPLS. In L. Hildebrandt, G. Rudinger, & P. Schmidt (Eds.), *Kausalanalyse in der Umweltforschung* (pp._-_). Stuttgart: Gustav Fischer.

Lohmöller,J.B. (in press b) Basic principles of scientific model building: Specification, estimation, evaluation. In H. Wold (Ed.), *Theoretical Empiricism: A general rationale for scientific model building* (pp._-_). New York: Paragon.

Lohmöller,J.B. & Falter,J.W. (1986) Some further aspects of ecological regression analysis. *Quality and Quantity, 20*: 109-125.

Lohmöller,J.B., Falter,J.W., Link,A., & Rijke,J.d. (1985) Unemployment and the rise of National Socialism: Contradicting results from different regional aggregatations. In P. Nijkamp, H. Leitner, & N. Wrigley (Eds.), *Measuring the unmeasurable* (pp.357-370). (NATO ASI Series D No. 22). Dordrecht, Boston, Lancaster: Martinus Nijhoff.

Lohmöller,J.B. & Oerter,R. (Eds.) (1979) *Medien in der Erzieherausbildung: Erprobung des Medienverbundes "Vorschulische Erziehung im Ausland"*. München: Oldenbourg. (280p).

Lohmöller,J.B. & Schulte-Cloos,C. (1986) Therapieeffekte: Ein Pfadmodell mit latenten Variablen unter Partialkleinstquadratschätzung (PLS) im Rahmen einer Evaluationsstudie klinisch-therapeutischer Interventionen. In K. Möbus & W. Schneider (Eds.), *Strukturmodelle für Längsschnittdaten und Zeitreihen: LISREL, Pfad- und Varianzanalysen* (pp.187-197). Bern, Switzerland: Huber.

Lohmöller,J.B. & Wold,H. (1980) *Three-mode path models with latent variables and Partial Least Squares (PLS) parameter estimation.* (Paper presented at the European Meeting of the Psychometric Society; University of Groningen, The Netherlands; June 18-21 1980). (Forschungsbericht 80.03 Fachbereich Pädagogik). München: Hochschule der Bundeswehr.

Lohmöller,J.B. & Wold,H. (1981) *Path models with latent variables: Comparative aspects of PLS and ML parameter estimation.* (Paper, conference of the Society of Multivariate Experimental Psychology (SMEP), European Branch; Freudenberg, FRG; Dec. 2-4, 1981). München: Hochschule der Bundeswehr, Fachbereich Pädagogik.

Lohmöller,J.B. & Wold,H. (1982) Pfad- und faktorenanalytische Ansätze zur differentiellen Entwicklungsbeschreibung: Die trimodale Pfadanalyse mit latenten Variablen (Band 1, pp.36-43). In R. Oerter (Ed.), *Bericht über die 5. Tagung Entwicklungspsychologie in Augsburg, 21-23.Sept.1981.* Augsburg: Universität.

Lohmöller,J.B. & Wold,H. (1984) Introduction to PLS estimation of path models with latent variables, including some recent developments on mixed scales variables. In G. Melischek, K.E. Rosengren, & J. Stappers (Eds.), *Cultural indicators: An international symposium* (pp.501-520). (Philosophisch-Historische Klasse, Sitzungsberichte, 416. Band). Vienna, Austria: Verlag der Österreichischen Akademie der Wissenschaften.

Long,J.S. (1983) *Covariance structure models: An introduction to LISREL.* (Sage Unversity Paper series on Quantitative Applications in the Social Sciences, series no. 07-034). Beverly Hills, CA: Sage.

Lorber,A., Wangen,L.E., & Kowalski,R.L. (1987) A theoretical foundation for the PLS algorithm. *Journal of Chemometrics, 1*: 19-31.

Lunneborg,C.E. (1967) Developing prediction weights be matching battery factorings. *Psychometrika, 32*: 133-141.

Lyttkens,E. (1966) On the Fix-Point property of Wold's iterative estimation method for principal components. In P.R. Krishnaiah (Ed.), *Multivariate analysis* (pp.335-350). New York: Academic.

Lyttkens,E. (1973) The Fixed-Point method for estimating interdependent systems with the underlying model specification. *Journal of the Royal Statistical Society, Series A, 136*: 353-394.

Lyttkens,E. (1982) Instrumental variables estimation of interdependent systems. In K.G. Jöreskog & H. Wold (Eds.), *Systems under indirect observation: Causality, structure, prediction* (Vol. 1, pp. 125-262). Amsterdam: North Holland.

Lyttkens,E., Areskoug,B., & Wold,H. (1975) *The convergence of NIPALS estimation procedures for six path models with one or two latent variables.* (Research report 75:3). Göteborg, Sweden: University, Institute of Statistics.

MacCorquodale,K. & Meehl,P.E. (1948) On a distinction between hypothetical constructs and intervening variables. *Psychological Review, 55*: 95-107.

MacDuffee,C.C. (1933) *The theory of matrices.* Berlin: Verlag von Julius Springer.

Manne,R. (1987) PLS as a generalized inverse. In M. Martens (Ed.), *Data-approximation by PLS methods* (pp. 50-53). (Report no. 800). Olso, Norway: Norwegian Computing Center.

Mardia,K.V., Kent,J.T., & Bibby,J.M. (1979) *Multivariate analysis.* London: Academic Press.

Marjoribanks,K. (1972) Environment, social class, and mental abilities. *Journal of Educational Psychology, 63*: 103-109.

Marjoribanks,K. (Ed.) (1974) *Environment for Learning.* London: National Foundation for Educational Research Publications.

Martens,M. (Ed.) (1987) *Data-approximation by PLS methods.* (Report no. 800). Olso, Norway: Norwegian Computing Center.

Martens,H. (1987) Why the users like PLS regression. In M. Martens (Ed.), *Data-approximation by PLS methods* (pp. 68-87). (Report no. 800). Olso, Norway: Norwegian Computing Center.

Martens,M., Fjeldsenden,B., Russwurm,H., & Martens,H. (1982) *Relationships between sensory and chemical quality criteria for carrots studied by multivariate data analysis.* (Proceedings, "Sensory quality in foods and beverages; its definition, measurement and control", a joint society of chemical industry (sensory panel); Long Ashton Research Station, Bristol; April 4-8, 1982). N-1432 Aas-NLH: Norwegian Food Research Institute, P.O.Box 50.

McArdle,J.J. (1979) *Reticular analysis modeling (RAM) theory: The simplicity and generality of structural equations.* Denver, CO: University of Denver, Department of Psychology.

McArdle,J.J. (1980) Causal modeling applied to psychonomic systems simulation. *Behavior Research Methods and Instrumentation, 12*: 193-209.

McArdle,J.J. (1981) *The Reticular Action Meta-model: A simplified representation for structural equation modeling.* Denver: University of Denver, Department of Psychology.

McArdle,J.J. (1982) *Modeling hierarchical latent variable systems: Comparative results from LV-PLS and LISREL.* (Paper presented at the American Educational Research Association Annual Meeting, New York City, March 19, 1982). Denver, Colorado: University of Denver, Department of Psychology.

McArdle,J.J. (1987a) Dynamic and structural equation modeling with repreated measures data. In J.R. Nesselroade & R.B. Cattell (Eds.), *Handbook of multivariate experimental psychology* (Vol. 2). New York: Plenum.

McArdle,J.J. (1987b) *The age variable in structural equation modeling.* Charlottesville, VA: University of Virginia, Department of Psychology.

McArdle,J.J., Anderson,E., & Aber,M.S. (1987) *Convergence hypotheses modeled and tested with linear structural equations.* (Proceedings of the 1987 Public Health Conference on Records and Statistics, Juli 13-15, 1987, Mayflower Hotel, Washington, DC.). Charlottesville, VA: University of Virginia, Department of Psychology.

McArdle,J.J., Connell,J.P., & Goldsmith,H.H. (1980) *Latent variable approaches to measurement structure, longitudinal stability, and genetic influences.* Denver, Colorado: University of Denver, Department of Psychology.

McArdle,J.J., Horn,J.L., & Aber,M.S. (1985) *Aging and abilities: A meta-analysis of repeated-measures studies on the WAIS.* Charlottesville, VA: University of Virginia, Department of Psychology.

McArdle,J.J. & McDonald,R.P. (1984) Some algebraic properties of the Reticular Action Model for moment structures. *British Journal of Mathematical and Statistical Psychology, 37*: 234-251.

McCullagh,P. & Nelder,J.A. (1983) *Generalized linear models.* London: Chapman and Hall.

McDonald,R.P. (1968) A unified treatment of the weighting problem. *Psychometrika, 33*: 351-381.

McDonald,R.P. (1970) Three common factor models for groups of variables. *Psychometrika, 35*: 111-128.

McDonald,R.P. & Burr,E.J. (1967) A comparison of four methods of constructing factor scores. *Psychometrika, 32*: 381-401.

McDonald,R.P., Torii,Y., & Nishisato,S. (1979) Some results on proper eigenvalues and eigenvectors with applications to scaling. *Psychometrika, 44*: 211-227.

McKeon,J.J. (1966) Canonical analysis - some relations between canonical correlations, factor analysis, discriminant function analysis, and scaling theory. *Psychometric Monograph, 13*: 1-43.

Meissner,W. & Uhle-Fassing,M. (1982a) PLS-modeling and estimation of politometric models. In K.G. Jöreskog & H. Wold (Eds.), *Systems under indirect observation: Causality, structure, prediction* (Vol. 2, pp. 161-176). Amsterdam: North Holland.

Meissner,W. & Uhle-Fassing,M. (1982b) *Weiche Modelle und iterative Schätzung: Eine Anwendung auf Probleme der Neuen Politischen Ökonomie.* (Campus Forschung Band 293). Frankfurt: Campus.

Meredith,W. (1964) Canonical correlations with fallible data. *Psychometrika, 29*: 55-65.

Meredith,W. & Tisak,J. (1982) Canonical analysis of longitudinal and repeated measures data with stationary regression weights. *Psychometrika, 47*: 47-67.

Miller,R.G. (1974) The jackknife – a review. *Biometrika, 61*: 1-15.

Millsap,R.E. (1986) *Component vs. factor analytic approaches to longitudinal data.* (Paper presented at the annual meeting of the Psychometric Society, June 21-24, 1986, Toronto, Ontario, Canada). New York, NY: City University, Baruch College.

Millsap,R.E. & Meredith,W. (1988) Component analysis in cross-sectional and longitudinal data. *Psychometrika, 53*: 123-134.

Möbus,C. & Schneider,W. (Eds.) (1986) *Strukturmodelle für Längsschnittdaten und Zeitreihen: LISREL, Pfad- und Varianzanalyse.* Bern: Huber.

Mooijaart,A. (1982) Latent structure analysis for categorical variables. In K.G. Jöreskog & H. Wold (Eds.), *Systems under indirect observation: Causality, structure, prediction* (Vol. 1, pp. 1-18). Amsterdam: North Holland.

Morgan,M. (1984) *Symbolic victimization and real world fear* In G. Melischek, K.E. Rosengren, & J. Stappers (Eds.), *Cultural indicators: An international symposium* (pp.365-376). (Philosophisch-Historische Klasse, Sitzungsberichte, 416. Band). Vienna, Austria: Verlag der Österreichischen Akademie der Wissenschaften.

Mosbaek,E.J. & Wold,H.O. (1970) *Interdependent systems: Structure and estimation.* Amsterdam: North Holland.

Mulaik,S.A. (1972) *The foundations of factor analysis.* New York: McGraw-Hill.

Mulaik,S.A. (1981) A note on sufficient conditions that a common factor will be determinate in an infinite domain of variables. *Psychometrika, 46*: 105-107.

Mulaik,S.A. & McDonald,R.P. (1978) The effect of additional variables on factor indeterminacy in models with a single common factor. *Psychometrika, 43*: 177-192.

Muller,K.E. (1981) Relationships between redundancy analysis, canonical correlation, and multivariate regression. *Psychometrika, 46*: 139-142.

Muthén,B. (1978) Contributions to factor analysis of dichotomous variables. *Psychometrika, 43*: 551-560.

Muthén,B. (1982) Some categorical response models with continuous latent variables. In K.G. Jöreskog & H. Wold (Eds.), *Systems under indirect observation: Causality, structure, prediction* (Vol. 1, pp. 65-80). Amsterdam: North Holland.

Muthén,B. (1984) A general structural equation model with dichotomous, ordered categorical, and continous latent variables indicators. *Psychometrika, 49*: 115-132.

Muthén,B. (1985) *TOBIT factor analysis.* (Paper presented at the fourth European meeting of the Psychometric Society, July 2-5, Cambridge, England). Los Angeles, CA: University of California, Graduate School of Education.

Næs,T. (1987) PLS versus some other statistical prediction methods. In M. Martens (Ed.), *Data-approximation by PLS methods* (pp. 35-49). (Report no. 800). Olso, Norway: Norwegian Computing Center.

Nishisato,S. & Sheu,W.J. (1980) Piecewise method of reciprocal averages for dual scaling of multiple-choice data. *Psychometrika, 45*: 467-478.

Noonan,R. (1981) School environments and school outcomes: An empirical comparative study using the IEA data. In M. Nissen & J. Peschar (Eds.), *Comparative research on education: Overview, strategy and applications in eastern and western Europe.* Oxford: Pergamon Press.

Noonan,R. (1987) *Evaluation of school systems using Partial Least Squares (PLS): An application in the analysis of open systems.* (Paper, symposium PLS Model Building: Theory and Application, Frankfurt am Main, Sept.23-25, 1987). Stockholm: Inter Sciences Research Corp.

Noonan,R., Abrahamsson,A., Areskoug,B., Lorentzson,L.O., & Wallmyr,J. (1975) Applications of methods I-II to the I.E.A. data bank. In H. Wold (Ed.), *Modelling in complex situations with soft information.* (Paper, Third World Congress on Econometrics; Toronto, Canada; August 21-26, 1975). (Research Report 1975:5). Göteborg, Sweden: University, Institute of Statistics.

Noonan,R. & Wold,H. (1977) NIPALS path modeling with latent variables: Analysing school survey data using Nonlinear Iterative Partial Least Squares. *Scandinavian Journal of Educational Research, 21*: 33-61.

Noonan,R. & Wold,H. (1980) PLS path modelling with latent variables: Analysing school survey data using Partial Least Squares - Part II. *Scandinavian Journal of Educational Research, 4*: 1-24.

Noonan,R. & Wold,H. (1982) PLS path modelling with indirectly observed variables: A comparison of alternative estimates for the latent variable. In K.G. Jöreskog & H. Wold (Eds.), *Systems under indirect observation: Causality, structure, prediction* (Vol. 2, pp. 75-94). Amsterdam: North Holland.

Noonan,R. & Wold,H. (1983) Evaluating school systems using Partial Least Squares. *Evaluation in Education - An International Review Series, 7*: 219-364.

Noonan,R. & Wold,H. (1986) Partial Least Squares path analysis. In T. Husén & T.N. Postlethwaite (Eds.), *International Encyclopedia of Education* (pp. 3769-3775). Oxford: Pergamon.

Opwis,K., Gold,A., & Schneider,W. (1987) Möglichkeiten der Kreuzvalidierung von Strukturgleichungsmodellen. *Psychologische Beträge, 29*: 60-77.

Pfeifer,A. & Schmidt,P. (1987) *LISREL - Die Analyse komplexer Strukturgleichungsmodelle.* Stuttgart: Gustav Fischer.

Pruzek,R.M., Lehrer,R., & Kim,K.L. (1979) *Generalized canonical correlation of relationships among qualitative variables.* (Paper presented at the annual meeting of the American Educational Research Association, San Francisco, April 1979). Albany, NY: State University of New York.

Rao,C.R. (1964) The use and interpretation of principal components analysis in applied research. *Sankhyā, A, 26*: 329-358.

Rao,C.R. (1966) Generalized inverse for matrices and its applications in mathematical statistics. In F.N. David (Ed.), *Festschrift for J. Neyman: Research papers in statistics* (pp. 263-279). New York: Wiley.

Rao,B.R. (1969) Partial canonical correlation. *Trabasjo de Estadistica y de Investigacion operative, 20*: 211-219.

Rindskopf,D. (1983) Parameterizing inequality constraints on unique variances in linear structural models. *Psychometrika, 48*: 73-83.

Rindskopf,D. (1984) Using phantom and imaginary latent variables to parameterize constraints in linear structural models. *Psychometrika, 49*: 37-47.

Rindskopf,D. & Rose,T. (1985) *Some Theory and applications of confirmatory second-order factor analysis.* (Paper presented at the fourth European meeting of the Psychometric Society, July 2-5, Cambridge, England). New York, NY 10036: City University of New York, Graduate Center, Educational Psychology Dept.

Robinson,W.S. (1950) Ecological correlations and the behavior of individuals. *American Sociological Review, 15*: 351-357.

Roeder,P.M., Baumert,J., Naumann,J., & Trommer,L. (1987) *Über institutionelle Bedingungen wissenschaftlicher Produktivität.* (Nr.17/SuU). Berlin: Max-Planck-Institut für Bildungsforschung.

Roeder,P.M., Baumert,J., Naumann,J., & Trommer,L. (1987) *Zur institutionellen Stratifizierung im Hochschulsystem der Bundesrepublik Deutschland.* (Nr.16/SuU). Berlin: Max-Planck-Institut für Bildungsforschung.

Rogowski,J. (1987) *Application of soft modeling to Kornai's shortage theory.* (Paper, symposium PLS Model Building: Theory and Application, Frankfurt am Main, Sept.23-25, 1987). Białystok, Poland: Warsaw University.

Rossa,P.J. (1982) Explaining international political behavior and conflict through partial least squares modeling. In K.G. Jöreskog & H. Wold (Eds.), *Systems under indirect observation: Causality, structure, prediction* (Vol. 2, pp. 130-160). Amsterdam: North Holland.

Rozeboom,W.W. (1965) Linear correlation between sets of variables. *Psychometrika, 30*: 57-71.

Rozeboom,W.W. (1982) The determinacy of common factors in large item domains. *Psychometrika, 47*: 281-295.

Sampson,P.D., Streissguth,A.P., Barr,H.M., & Bookstein,F.L. (1987) *Application of Partial Least Squares methodology to the behavioral Teratology of alcohol.* (Paper, symposium PLS Model Building: Theory and Application, Frankfurt am Main, Sept.23-25, 1987). Seattle: University of Washington.

Sampson,P.D., Streissguth,A.P., Vega-Gonzalez,S.C., Barr,H.M., & Bookstein,F.L. (1987) *Predicting classromm behavior ratings by prenatal alcohol exposure: Latent variable modeling and nonlinear scaling.* (Paper, symposium PLS Model Building: Theory and Application, Frankfurt am Main, Sept.23-25, 1987). Seattle: University of Washington.

Sands,R. & Young,F.W. (1980) Component models for three-way data: An alternating least squares algorithm with optimal scaling features. *Psychometrika, 45*: 39-67.

Saporta,G. (1978) *About some remarkable properties of generalized canonical analysis.* F-75016 Paris: Institut Universitaire de Technologie.

Saris,W.E. & Stronkhorst,L.H. (1984) *Causal modelling in nonexperimental research: An introduction to the LISREL approach.* Amsterdam: Sociometric Research Foundation.

Schieber,N. (1983) *PLSPATH Version A – program manual.* Hamburg: University, Department of Education.

Schneewind,K.A., Beckmann,M., & Engfer,A. (1983) *Eltern und Kinder: Umwelteinflüsse auf das familiäre Verhalten.* Stuttgart: Kohlhammer.

Schneider,W. (1980) Kausalmodelle in der Pädagogischen Psychologie: Ergänzende Analysen zu zwei einschlägigen Untersuchungen. *Zeitschrift für Entwicklungspsychologie und Pädagogische Psychologie, 12*: 88-95.

Schneider,W. (1985) Metagedächtnis, gedächtnisbezogene Verhalten und Gedächtnisleistung: Eine Analyse der empirischen Zusammenhänge bei Grundschülern der dritten Klasse. *Zeitschrift für Entwicklungspsychologie und Pädagogische Psychologie, 17*: 1-16.

Schneider,W., Borkowski,J.G., Kurtz,B.E., & Kerwin,K. (1985) Metamemory and Motivation: A comparison of strategy use and performance in German and American children. (Paper presented at the annual meeting of the American Educational Research Association, Chicago, 1985.). *Journal of Cross-Cultural Psychology,* __: __-__.

Schneider,W. & Bös,K. (1985) Exploratorische Analysen zu Komponenten des Schulerfolgs. *Zeitschrift für Entwicklungspsychologie und Pädagogische Psychologie, 17*: 325-340.

Scholing,E. & Timmermann,V. (1987) *Why LCD-Growth rates differ – Measuring "unmeasurable" influences.* (Paper, symposium PLS Model Building: Theory and Application, Frankfurt am Main, Sept.23-25, 1987). Hamburg: University.

Scholing,E. (1983) *ILVPLS Programm, Programmdokumentation* Hamburg: Universität, Sozialökonomisches Seminar.

Schönemann,P.H. (1981) *On the use of intelligence factors for the prediction of easter sunday.* Lafayette: Purdue University.

Schönfeld,P. (1969) *Methoden der Ökonometrie* (2 Vols). Berlin: Verlag Franz Vahlen.

Schulte,C. (1981) *Überlegungen zur Genese und Aufrechterhaltung psychosomatischer Störungen: Das verhaltensmedizinische Gruppentraining (VMGT), seine Entwicklung und Evaluation in der klinischen Rehabilitation.* (Dissertation Universität Tübingen). (VM-Texte zur Verhaltensmedizin). Garmisch-Partenkirchen: Kroll Verlag.

Schumann,S. (1982) *Persönlichkeit und undemokratische Einstellungen, untersucht an 1822 Studenten der HSw-München, sowie Analysen zur Sozialisation der Studenen an der HSBw-München.* (Forschungsbericht 82.04 Fachbereich Pädagogik). München: Hochschule der Bundeswehr München.

Schumann,S. (1985) *Wahlverhalten und Persönlichkeit: Einige empirische Forschungsergebnisse.* Berlin: Freie Universität Berlin, ZISOWIFO.

Schumann,S. (1986) *Politische Einstellungen und Persönlichkeit: Ein Bericht über empirische Forschungsergebnisse.* Frankfurt am Main: Peter Lang.

Schweder,T. (1987) Canonical regression versus PLS. In M. Martens (Ed.), *Data-approximation by PLS methods* (pp. 54-61). (Report no. 800). Olso, Norway: Norwegian Computing Center.

Seidel,G. (1987) *Identification problems in LISREL models and their algebraic clarification by LISRAN.* Frankfurt am Main: Deutsches Institut für Internationale Pädagogische Forschung.

Short,R., Horn,J.L., & McArdle,J.J. (1984) Mathematical-statistical model building in analysis of developmental data. In R.N. Emde & R.J. Harmon (Eds.), *Continuities and discontinuities in development* (pp. 371-401). New York: Plenum.

Skinner,H.A. (1977) Exploring relationships among multiple data sets. *The Journal of Multivariate Behavioral Research, 12*: 199-220.

Skinner,H.A. (1978) The art of exploring predictor-criterion relationships. *Psychological Bulletin, 85*: 327-337.

Slavenburg,J.H. (1986) *Onderwijsstimulering en gezinsactivering: Effecten van Programma's van het project Onderwijs en Sociaal Milieu op de schoolpresaties von eersteklassers lager onderwijs.* (Doctoral dissertation, University of Groningen). Den Haag: Stichting voor Onderzoek van her Onderwijs. (Summary in English and French).

Smith,N.L. (1981) Noncausal inquiry in education. *Educational Researcher, 10*, No. 3: 23.

Sockloff,A.L. (1976) The analysis of nonlinearity via linear regression with polynomial and product variables: An examination. *Review of Educational Research, 46*: 267-291.

Sörbom,D. (1974) A general method for studying differences in factor means and factor structure between groups. *British Journal of Mathematical and Statistical Psychology, 27*: 229-239.

Sörbom,D. (1978) An alternative to the methodology for analysis of covariance. *Psychometrika, 43*: 381-396.

Sörbom,D. (1982) Structural equation models with structured means. In K.G. Jöreskog & H. Wold (Eds.), *Systems under indirect observation: Causality, structure, prediction* (Vol. 1, pp. 182-196). Amsterdam: North Holland.

Spearman,C. (1904) General intelligence objectively determined and measured. *American Journal of Psychology, 15*: 210-293.

Spjøtvoll,E. (1987) Critical comment on PLS. In M. Martens (Ed.), *Data-approximation by PLS methods* (pp. 62-67). (Report no. 800). Olso, Norway: Norwegian Computing Center.

Stary,J. (1983) *Politische Werthaltungen Jugendlicher und deren Einflußbedingungen.* (Referat beim 5. Workshop Politische Psychologie, Jugend und Werte; Berlin; 17. - 19. Nov. 1983). Wien: Universität, Institut für pädagogische Psychologie.

Steel,R.D. (1951) Minimum generalized variance for a set of linear functions. *Annals of Mathematical Statistics, 22*: 456-460.

Steiger,J.H. (1979a) Factor indeterminacy in the 1930's and the 1970's: Some interesting parallels. *Psychometrika, 44*: 157-167.

Steiger,J.H. (1979b) The relationship between external variables and common factors. *Psychometrika, 44*: 93-97.

Stelzl,I. (1986) Changing a causal hypothesis without changing the fit: Some rules for generating equivalent path models. *Multivariate Behavioral Research, 21*: 309-331.

Stevens,S.S. (1951) Mathematics, Measurement and Psychophysics. In S.S. Stevens (Ed.), *Handbook of experimental psychology* (pp. 1-49). New York: Wiley.

Stewart,D. & Love,W. (1968) A general canonical correlation index. *Psychological Bulletin, 70*: 160-163.

Steyer,R. (1983) Modelle zur kausalen Erklärung statistischer Zusammenhänge. In J. Bredenkamp & H. Feger (Hrsg.), *Strukturierung und Reduzierung von Daten* (pp. 59-153). (Enzyklopädie der Psychologie, Serie Forschungsmethoden der Psychologie, Band 4). Göttingen: Verlag für Psychologie – Hogrefe.

Steyer,R. (1985) *The randomized experiment and the concept of weak causal regressive dependence.* Trier, Germany: Universität, Psychologisches Institut.

Stone,M. (1974) Cross-validatory choice and assessment of statistical predictions. *Journal of the Royal Statistical Society, series B, 36*: 111-147.

Strotz,R.H. & Wold,H.O.A. (1960) Recursive vs. nonrecursive systems: An attempt at synthesis (Part I of a triptych on causal chain systems). *Econometrica, 28*: 417-427.

Suppes,P. (1973) New foundations of objective probability: Axioms for propensities. In P. Suppes, L. Henkin, A. Joja, & G.C. Moisil (Eds.), *Logic, methodology, and philosophy of science IV: Proceedings of the Fourth International Congress for Logic, Methodology, and Philosophy of Science, Bucharest, 1971* (pp. 515-529). Amsterdam: North Holland.

ten Berge,J.F. (1984) A joint treatment of Varimax rotation and the problem of diagonalizing symmetric matrices simultaneously in the least-squares sense. *Psychometrika, 49*: 347-358.

ten Berge,J.F. & Knol,D.L. (1984) Orthogonal rotations to maximal agreement for two or more matrices of different column orders. *Psychometrika, 49*: 49-55.

Thompson,B. (1982) CANBAK – A program which performs stepwise canonical correlation analysis. *Educational and Psychological Measurement, 42*: 849-851.

Thorndike,R.M. (1977) Canonical analysis and predictor selection. *Multivariate Behavioral Research, 12*: 75-87.

Thurstone,L.L. & Thurstone,T.G. (1941) Factorial studies of intelligence. *Psychometric Monograph, 2.*

Travers,R.M.W. (1969) *An introduction to educational research* (3rd ed.). New York: Macmillan.

Travers,R.W. (1981) Letter to the Editor on Noncausal inquiry in education. *Educational Researcher, 10,* No. 6: 32.

Treinies,G. & Einsiedler,W. (1986) *Analyse explorativer Kausalmodelle zu Zusammenhängen zwischen häuslicher Entwicklungsbedingungen und Freizeitspielverhalten von Kindergartenkindern.* Nürnberg: Universität Erlangen-Nürnberg, Erziehungswissenschaftliche Fakultät.

Tucker,L.R. (1958) An inter-battery method of factor analysis. *Psychometrika, 23*: 111-136.

Tucker,L.R. (1963) Implications of factor analysis of three-way matrices for measurement of change. In C.W. Harris (Ed.), *Problems in measuring change.* Madison, Wisconsin: The University of Wisconsin Press.

Tucker,L.R. (1964a) Experiments in multi-mode factor analysis. In *Proceedings of the 1964 Invitational Conference on Testing Problems* (pp. 46-57). Princeton, N.J.: Educational Testing Service.

Tucker,L.R. (1964b) The extension of factor analysis to three-dimensional matrices. In N. Frederiksen & H. Gulliksen (Eds.), *Contributions to mathematical psychology* (pp. 110-127). New York: Holt, Rinehart & Winston.

Tucker,L.R. (1966) Some mathematical notes on three-mode factor analysis. *Psychometrika, 31*: 279-311.

Tucker,L.R. (1971) Relations of factor score estimates to their use. *Psychometrika, 36*: 427-436.

Tucker,L.R. (1972) Relations between multidimensional scaling and three-mode factor analysis. *Psychometrika, 37*: 3-27.

Tucker,L.R. & Lewis,C. (1973) A reliability coefficient for maximum likelihood factor analysis. *Psychometrika, 38*: 1-10.

Tyler,D.E. (1982) On the optimality of the simultaneous redundancy transformation. *Psychometrika, 47*: 77-86.

Uehlinger,H.M. (1986) *Strukturen und Erklärungsmodelle politischer Partizipation in der Bundesrepublik Deutschland.* Zürich: Universität.

Valette-Florence,P. (1986) *A second generation of multivariate analysis: Contribution an applications for marketing research.* Grenoble, France: Université, Ecole superieur des affaires de Grenoble.

van de Geer,J.P. (1971) *Introduction to multivariate analysis for the social sciences.* San Francisco: Freeman.

van de Geer,J.P. (1984) Linear relations among *k* sets of variables. *Psychometrika, 49*: 79-94.

van de Geer,J.P. (1986) *Introduction to linear multivariate data analysis* (2 vols.). Leiden: DSWO Press.

van den Wollenberg,A.L. (1977) Redundancy analysis: An alternative for canonical correlation analysis. *Psychometrika, 42*: 207-219.

van der Burg,E. & de Leeuw,J. (1980) *Non-linear canonical correlation.* Leiden NL: University, Datatheory.

van Donkelaar,J.H. & van der Knaap,G.A. (1982) *Kausale relaties in multivariate analyse: De PLS-methode.* (Working Papers Series A, nr.82-2). Rotterdam, NL: Erasmus Universiteit Rotterdam, Economisch Geographisch Instituut.

van Driel,O.P. (1978) On various causes of improper solutions in maximum likelihood factor analysis. *Psychometrika, 43*: 225-243.

van Rijckevorsel,J.L.A., Bettonvil,B., & de Leeuw,J. (1980) *Recovery and stability in nonlinear principal components analysis.* (European Meeting, Psychometric Society; Groningen NL; June 18-21, 1980). Leiden NL: University, Faculty of Scocial Sciences, Department of Datatheory.

Vinograde,B. (1950) Canonical positive matrices under internal linear transformations. *Proceedings of the American Mathematical Society, 1*: 159-161.

Voges,W. & Hoschka,A. (1985) Schulerfolg und außerschulische Betreuung: Analyse von Wirkungszusammenhängen schulbezogener Sozialisationshilfen für italienische Migrantenkinder. *Angewandte Sozialforschung, 13*: 387-398.

van Driel,O.P. (1978) On various causes of improper solutions in maximum likelihood factor analysis. *Psychometrika, 43*: 225-243.

van Rijckevorsel,J.L.A., Bettonvil,B., & de Leeuw,J. (1980) *Recovery and stability in nonlinear principal components analysis.* (European Meeting, Psychometric Society; Groningen NL; June 18-21, 1980). Leiden NL: University, Faculty of Scocial Sciences, Department of Datatheory.

Vinograde,B. (1950) Canonical positive matrices under internal linear transformations. *Proceedings of the American Mathematical Society, 1*: 159-161.

Voges,W. & Hoschka,A. (1985) Schulerfolg und außerschulische Betreuung: Analyse von Wirkungszusammenhängen schulbezogener Sozialisationshilfen für italienische Migrantenkinder. *Angewandte Sozialforschung, 13*: 387-398.

Voges,W. & Hoschka,A. (1986) *Die Wirksamkeit der Doposcuola: Eine empirische Untersuchung über schulbegleitende Hilfen für italienische Schüler der ersten bis fünften Klasse.* München: DJI Verlag Deutsches Jugendinstitut.

Voges,W. & Lohmöller,J.B. (1988) *Bedingungen vorzeitiger Beendigung der Erwerbsphase: Ein PLS-Modell zur Erklärung der Kausalzusammenhänge am Beispiel des Vorruhestandes.* (SFB 186 – Arbeitspapier Nr. 1). Bremen: Universität, Sonderforschungsbereich Statuspassagen und Risikolagen im Lebensverlauf.

von Frenckell,R. (1980) *PRINQUAL – a method of optimal scaling to test the ordinality of items in rating scales.* (European Meeting of the Psychometric Society; Groningen NL; June 8-21, 1980). Liege, B-4000: Clinique Psychiatrique Universitaire.

von Mises,R. & Pollaczek-Geiringer,H. (1929) Praktische Verfahren der Gleichungsauflösung. *Zeitschrift für angewandte Mathematik und Mechanik, 9*: 58-77,152-164.

Walberg,H.J. & Marjoribanks,K. (1973) Differential mental abilities and home environment: A canonical analysis. *Developmental Psychology, 9*: 363-368.

Walberg,H.J. & Marjoribanks,K. (1974) Social environment and cognitive development: Toward a generalized causal analysis. In K. Marjoribanks (Ed.), *Environment for Learning* (pp.259-273). London: National Foundation for Educational Research Publications.

Walberg,H.J. & Marjoribanks,K. (1976) Family environment and cognitive development: Twelve analytic models. *Review of Educational Research, 46*: 527-551.

Weber,R.P. (1982) *Content-analytic cultural indicators.* (Paper, Symposium on Cultural Indicators for the Comparative Study of Change, Vienna, February 16-19, 1982). Havard: University.

Weede,E. (1977) *Hypothesen, Gleichungen und Daten: Spezifikations- und Meßprobleme für Daten aus einer oder mehreren Beobachtungsperioden.* (ZUMA Monographien Sozialwissenschaftlicher Methoden, Band 1.). Kronberg,Ts.: Athenaeum.

Wegge,L.F. (1981) *Identification with latent variables.* Bonn: Institut für Ökonometrie und Operations Research, Universität Bonn.

Wellhofer,E.S. (1982) *To educate their volition to dance in their chains: Partisan mobilization and regime stability in Britain 1885-1950.* (Paper, American Political Association meetings; Denver; September 2-5, 1982). University of Denver: Department of Political Science.

Wellhofer,E.S. (1985) "Two Nations": Class and periphery in late Victorian Britain, 1985-1910. *American Political Sciene Review, 79*: 977-993.

Wheaton,B., Muthén,B., Alwin,D. & Summers,G. (1977) Assessing reliability and stability in panel models. In D.R. Heise (Ed.) *Sociological Methodology 1977* (pp. 84-136). San Francisco: Jossey-Bass.

Whittle,P. & Adelman,I. (1982) The fitting of restricted rank regression with prior information. In K.G. Jöreskog & H. Wold (Eds.), *Systems under indirect observation: Causality, structure, prediction* (Vol. 2, pp. 273-289). Amsterdam: North Holland.

Wilks,S.S. (1935) On the independence of *k* sets of normally distributed statistical variables. *Econometrica, 3*: 309-326.

Williams,J.S. (1978) A definition for the common-factor analysis model and the elimination of problems of factor scores indeterminacy. *Psychometrika, 43*: 291-306.

Winer,B. (1971) *Statistical principles in experimental design.* New York: McGraw Hill.

Wold,H. (1966b) Nonlinear estimation by interative least squares procedures. In F.N. David (Ed.), *Festschrift for J. Neyman: Research papers in statistics* (pp. 411-444). London: Wiley.

Wold,H. (1969a) Econometrics as pioneering in nonexperimental model building. (Presidential Address, Economic Society meeting; San Francisco, 1966). *Econometrica, 37*: 369-381.

Wold,H. (1969b) Mergers of economics and philosophy of science: A cruise in shallow waters and deep seas. *Synthese, 20*: 427-482.

Wold,H. (1973) Nonlinear Iterative Partial Least Squares (NIPALS) modelling: Some current developments. In P.R. Krishnaiah (Ed.), *Multivariate analysis* (Vol. 3, pp. 383-407). New York: Academic.

Wold,H. (1975a) From hard to soft modelling. In H. Wold (Ed.), *Modelling in complex situations with soft information.* (Paper, Third World Congress on Econometrics; Toronto, Canada; 1975 August 21-26). (Research Report 1975:5). Göteborg, Sweden: University, Institute of Statistics.

Wold,H. (Ed.) (1975b) *Modelling in complex situations with soft information.* (Paper, Third World Congress of Econometrics; Toronto, Canada; 1975 August 21-26). (Research Report 1975:5). Göteborg, Sweden: University, Institute of Statistics.

Wold,H. (1975c) Path models with latent variables: The NIPALS approach. In H.M. Blalock & al (Eds.), *Quantitative sociology: International perspectives on mathematical and statistical modeling* (pp. 307-357). New York: Academic.

Wold,H. (1975d) Soft Modelling by latent variables: The Non-Linear Iterative Partial Least Squares (NIPALS) approach. In J.Gani (Ed.), *Perspectives in probability and statistics: Papers in honour of M.S.Bartlett on the occasion of his sixty-fifth birthday* (pp. 117-142). London: Applied Probability Trust, Academic.

Wold,H. (1976) *Open path models with latent variables: The NIPALS (Nonlinear Iterative Partial Least Squares) approach.* Uppsala, Sweden: University, Department of Statistics.

Wold,H. (1977a) Modèles de régulation en situations complexes: La méthode itérative non linéare partielle des moindres carrés (dite "NIPALS" en version anglais). In A. Lichnerowicz, F. Perroux, & G. Gadoffre (Eds.), *L'idée des régulation dans les sciences* (pp. 103-115). Paris: Doin.

Wold,H. (1977b) On the transition from pattern cognition to model building. In R. Henn & O. Moeschlin (Eds.), *Mathematical economics and game theory: Essays in honour of Oskar Morgenstern* (pp. 536-549). Heidelberg: Springer.

Wold,H. (1977c) Open path models with latent variables. In H. Albach, E. Helmstädter, & R. Henn (Eds.), *Quantitative Wirtschaftforschung: Wilhelm Krelle zum 60. Geburtstag* (pp. 729-754). Tübingen: Mohr.

Wold,H. (1978a) *Causal-predictive analysis when the problems explored are complex and prior knowledge is scarce.* (Paper, European Meeting on Psychometrics and Mathematical Psychology; Uppsala, Sweden; 1978 June 15-16). Uppsala, Sweden: University, Department of Statistics.

Wold,H. (1978b) *Path models with latent variables: Comparative aspects of the of the PLS (Partial Least Squares) versus the ML (Maximum Likelihood) approach.* (Paper, European Meeting on Psychometrics and Mathematical Psychology; Uppsala, Sweden; 1978 June 15-16). Uppsala, Sweden: University, Department of Statistics.

Wold,H. (1978c) Ways and means of multidisciplinary studies. In *The search for absolute values in a changing world* (pp. 1071-1095). (Sixth International Conference on the Unity of the Sciences; San Francisco; 1977 November 25-27). New York: The International Cultural Foundation.

Wold,H. (1979a) Causal-predictive analysis of problems with high complexity and low informations: Recent developments of soft modeling. In *The re-evaluation of existing values and the search for absolute values* (pp. 1023-1033). (Proceedings of the Seventh International Conference on the Unity of the Sciences; Boston; 1978). New York: The International Cultural Foundation.

Wold,H. (1979b) Mål och medel i kvantitativ systemanalys. In *Matematiska modeller för sociala system* (pp. 33-45). Stockholm: Naturvetenskapliga Forskningsrådet.

Wold,H. (1979c) *Model construction and evaluation when theoretical knowledge is scarce: An example of the use of Partial Least Squares.* (Cahier 79.06 du département d'économétrie, faculté des sciences économiques et sociales). Genève: Université de Genève.

Wold,H. (1980a) Model construction and evaluation when theoretical knowledge is scarce: On the theory and application of Partial Least Squares. In J. Kmenta & J.B. Ramsey (Eds.), *Model evaluation in econometrics* (pp. 47-74). New York: Academic.

Wold,H. (1980b) Soft modelling: Intermediate between traditional model building and data analysis. In *Mathematical Statistics* (Vol. 6, pp. 333-346). Warsaw: Polish Scientific Publishers.

Wold,H. (1981a) *PLS and LISREL, comparative aspects: Latent variables, and other fundamental notions.* (Seminar, 1981 May 6). Uppsala, Sweden: University, Department of Statistics.

Wold,H. (1981b) *Systems under indirect observation using soft modelling.* (Working Paper No. 48). Cleveland,Ohio: Case Western Reserve University, Economics Department.

Wold,H. (Ed.) (1981c) *The fix-point approach to interdependent systems.* Amsterdam: North Holland.

Wold,H. (1981d) *What modes of interaction can be handled by Soft Modeling?.* (Notes for invited lecture, Conference on Modelling of innovation processes and structural change; Science Center Berlin, International Institute of Management; Berlin; 1980 April 2-3).

Wold,H. (1982a) Models for knowledge. In J. Gani (Ed.), *The making of statisticians* (pp. 190-212). London: Applied Probability Trust, Academic.

Wold,H. (1982b) *Predictive modeling of large systems: The FP and PLS approaches.* (Paper, Conference on Large Systems; University of Łódź, Poland; 1982 Dec. 1-3). Uppsala, Sweden: University, Department of Statistics.

Wold,H. (1982c) Soft modeling: The basic design and some extensions. In K.G. Jöreskog & H. Wold (Eds.), *Systems under indirect observation: Causality, structure, prediction* (Vol. 2, pp. 1-54). Amsterdam: North Holland.

Wold,H. (1982d) Systems under indirect observations using PLS. In C. Fornell (Ed.), *A second generation of multivariate analysis* (Vol. 1, pp. 325-347). New York: Praeger.

Wold,H. (1982e) *Utility analysis from the point of view of model building.* (Paper, First International Conference on Foundations of Utility and Risk Theory; Oslo, Norway; 1982 June 21-26). Uppsala, Sweden: University, Department of Statistics.

Wold,H. (1983a) Fix-Point method. In S. Kotz & N.L. Johnson (Eds.), *Encyclopedia of statistical sciences* (3:148-156). New York: Wiley.

Wold,H. (1983b) Quantitative systems analysis: The pedigree and broad scope of PLS (Partial Least Squares) soft modeling. In H. Martens & H. Russwurm (Eds.), *Food Research and data analysis* (pp. 409). (Poster, IUFoST Symposion; Oslo, Norway; 1982 Sept. 20-23). London: Applied Science Publishers.

Wold,H. (1983c) Utility analysis from the point of view of model building. In B.P. Stigum & F. Wenstøp (Eds.), *Foundations of utility and risk theory with Applications* (pp. 87-93). Dordrecht: Reidel.

Wold,H. (1985a) Partial Least Squares. In S. Kotz & N.L. Johnson (Eds.), *Encyclopedia of statistical sciences* (6:581-591). New York: Wiley.

Wold,H. (1985b) Systems analysis by Partial Least Squares. In P. Nijkamp, H. Leitner, & N. Wrigley (Eds.), *Measuring the unmeasurable* (pp. 221-251). (NATO ASI Series D No. 22). Dordrecht, Boston, Lancaster: Martinus Nijhoff.

Wold,H. (1985c) *The come back of Least Squares.* (Contributed paper, poster session, Centenary session of the International Statistical Institute (ISI), Amsterdam, 12th-22nd August, 1985). Uppsala, Sweden: University, Department of Statistics.

Wold,H. (1986) Factors influencing the outcome of economic sanctions: An application of soft modeling. In N. N (Eds.), *Trabajos de estadistica e investigacion operativa: Homenaje al Profesor Sixto Ríos* (pp. __-__). (Paper, University of Haifa International Workshop on Conflict Resolution; Haifa, Israel; 1978 June 19-24). (Paper, Fourth World Congress of Econometric Society; Aix-en-Provence, France; 1980 Aug 28 - Sept 1).

Wold,H. (1987) *Principal theory of my work.* (Paper, symposium PLS Model Building: Theory and Application, Frankfurt am Main, Sept.23-25, 1987). Uppsala, Sweden: University, Department of Statistics.

Wold,H. (1988a) Introduction to the second generation of multivariate analysis. In H. Wold (Ed.), *Theoretical empiricism: A general rationale for scientific model-building* (Chapter I). New York: Paragon. (in press).

Wold,H. (1988b) Specification, predictor. In S. Kotz & N.L. Johnson (Eds.), *Encyclopedia of statistical sciences* (8:587-599). New York: Wiley.

Wold,H. (Ed.) (1988c) *Theoretical empiricism: A general rationale for scientific model-building.* New York: Paragon. (in press).

Wold,H. & Areskoug,B. (1975) Path models with latent variables as proxies for blocks of manifest variables. In H. Wold (Ed.), *Modelling in complex situations with soft information.* (Paper, Third World Congress on Econometrics; Toronto, Canada; 1975 August 21-26). (Research Report 1975:5). Göteborg, Sweden: University, Institute of Statistics.

Wold,H. & Bertholet,J.L. (1982a) *From quantitative to qualitative data analysis in the PLS (partial least squares) approach.* Geneva, Switzerland: University, Department of Economics.

Wold,H. & Bertholet,J.L. (1982b) The PLS (Partial Least Squares) approach to multidimensional contingency tables (Part I). *Metron, 40*: 303-326.

Wold,H. & Bertholet,J.L. (1982c) *The PLS (Partial Least Squares) approach to multidimensional contingency tables* (Part II). (Paper, Annual Meeting of the American Educational Reserach Association; New York; 1982 March 19-23). Geneva, Switzerland: University, Department of Economics.

Wold,H. & Kaasch,K. (1985) Transfer between industrial branches in the course of Schumpeter-Mensch long swings. In T. Vasko (Ed.), *The long-wave debate* (pp. 405-419). Berlin: Springer.

Wold,H. & Lyttkens,E. (Eds.) (1969) *Nonlinear Iterative Partial Least Squares (PLS) estimation procedures.* (Group report, session of International Statistical Institute; London; 1969 Sept. 3-11).

Wold,H.O. & Mensch,G.O. (1983) *Nonlinear extensions in PLS Soft Modeling.* (WSOM WP # 83-011). Cleveland, Ohio: Case Western Reserve University, Weatherhead School of Management.

Wold,H.O. & Mensch,G.O. (1985) *The BIEQ-PLS model of a Schumpeterian Wave.* (WSOM WP # 85-017). Cleveland, Ohio: Case Western Reserve University, Weatherhead School of Management.

Wold,S. (1978) Cross-validatory estimation of the number of components in factor and principal component models. *Technometrics, 20*: 397-405.

Wold,S. & al. (1981) Pattern recognition by means of disjoint principal components models (SIMCA): Philosophy and Methods. In A.Höskuldson, K. Esbensen, & al (Eds.), *Symposium on applied statistics.* Copenhagen: University computer center, and Danish Society of Theoretical Statistics.

Wold,S. & al. (1982) The indirect observation of molecular chemical systems. In K.G. Jöreskog & H. Wold (Eds.), *Systems under indirect observation: Causality, structure, prediction* (Vol. 2, pp. 177-190). Amsterdam: North Holland.

Wold,S. & al. (1984a) Modelling data tables by principal components and PLS: class patterns and qualitatitive predictive relations. *Analusis, 12*: 447-485.

Wold,S. & al. (1984b) Multivariate data analysis in chemistry. In B.R. Kowalski (Ed.), *Proceedings, NATO Advanced Study Institute on Chemometrics, Cosenza, Italy, September 1983* (pp. 17-95). Dordrecht: Reidel.

Wold,S. & al. (1985) *Multivariate data analysis: Converting chemical data tables to plots.* (7th International Conference on Computers in Chemical Research and Education, Garmisch-Partenkirchen, FRG, 1985, June 10-14). (#166). Umeå, Sweden: University, Research Group for Chemometrics.

Wold,S., Geladi,P., Esbensen,K., & Öhman,J. (1986) *Principal components- and PLS-analyses generalized to multi-way (multi-order) data arrays.* (Copenhagen Symposion on Applied Statistics, Jan 1986). (#175). Umeå Sweden: University, Research Group for Chemometrics.

Wold,S., Hellberg,S., Lundstedt,T., Sjöström,M., & Wold,H. (1987) *PLS Modeling with latent variables in two or more dimensions.* (Paper, symposium PLS Model Building: Theory and

Application, Frankfurt am Main, Sept.23-25, 1987). Umeå, Sweden: University, Institute of Chemistry.

Wold,S., Lorentzson,L.O., & Wallmyr,J. (1975) Methods I, Pattern recognition by disjoint principal component models. In H. Wold (Ed.), *Modelling in complex situations with soft information.* (Paper, Third World Congress on Econometrics; Toronto, Canada; August 21-26, 1975). (Research Report 1975:5). Göteborg, Sweden: University, Institute of Statistics.

Wold,S., Ruhe,A., Wold,H., & Dunn,W.J. (1984) The collinearity problem in linear regression: The partial least squares (PLS) approach to generalized inverses. *SIAM J. Sci. Stat. Comput.*, *5*: 735-743.

Wold,S., Wold,H., & al (1981) Pattern recognition by means of disjoint principal component models (SIMCA): Philosophy and methods. In *Symposium i Anvendt Statistik 1981* (pp. 183-218). Denmark: Danmarks tekniske Højskole, Edb-center.

Wonnacott,T.H. & Wonnacott,R.J. (1981) *Regression: A second course in statistics.* (Wiley Series in Probability and Mathematical Statistics). New York: Wiley.

Woodward,J.A. & Bentler,P.M. (1979) Application of optimal sign-vectors to reliability and cluster analysis. *Psychometrika*, *44*: 337-341.

Young,F.W. (1981) Quantitative analysis of qualitative data. *Psychometrika*, *46*: 357-388.

Young,F.W., de Leeuw,J., & Takane,Y. (1976) Regression with qualitative and quantitative variables: An alternating least squares method with optimal scaling features. *Psychometrika*, *41*: 505-529.

Zurmühl,R. (1964) *Matrizen und ihre technische Anwendung* (4.ed.). Berlin: Springer Verlag.

Author Index

Subject Index

The PLS 1.8 Program System

for Latent Variable Path Analysis with Partial Least Squares Estimation

by Jan-Bernd Lohmöller
Free University Berlin · Department of Political Science
Schwendenerstr. 53 · D–1000 Berlin (West) 33

PLS 1.8 is a program system for latent variable path analysis following the Partial Least Squares methods developed by Herman Wold and programmed by Jan-Bernd Lohmöller. PLS 1.8 consists of three main programs:

PLSX for raw data analysis: PLSX reads empirical data from a file or uses simulated data generated by the utility PLSGEN. PLSX handles incomplete data, and each variable can have a different missing data code. PLSX 1.8 offers new techniques to form latent variable estimates: One LV can be a function (not of MVs but) of another LV: $Y_{jn} = Y_{in}^2$, $Y_{jn} = \text{lag}\, Y_{in}$. PLSX uses Wold's (1983) Basic Method for modeling and estimation and delivers jackknife standard errors of parameters. The Stone-Geisser test on predictive validity of the total model is applied.

PLSC for covariance matrix analysis: PLSC reads empirical data from file or uses simulated data. PLSC uses the Extended Method (Lohmöller 1984): Several LVs per block can be estimated simultaneously. PLSC generates a complete input deck for the structural equation modeling programs LISREL, COSAN, EQS and RAMIT.

PLS3 for Three-Mode Path Analysis with LVs: The outer model is Tucker's three-mode component model, the inner model is the three-mode path model.

PLS 1.8 is written in Fortran 77 and and can be customized in single, mixed, and double precision. PLS 1.8 has been tested and is available on a wide variety of mainframe operating systems. The "Special version" is customized for a special operating system (7.000 lines of Fortran, no comments; academic price DM 500); the "General version" can be converted to any operating system (13.000 lines, with comments; DM 1.900).

Special attention deserves PLS-PC 1.8, the MS-DOS version for PC/XT/AT /386 microcomputer. PLS-PC comes ready to run on two 360 KB diskettes, together with a collection of examples and test data (DM 200). It will run with or without a numerical coprocessor (e.g., 8087). The PLS 1.8 Program System may be obtained from:

- Zentralarchiv für Empirische Sozialforschung der Universität zu Köln, att. Harald Rohlinger, Bachemer Str. 40, D-5000 Köln 41 / West Germany
- VGSPS – Vertriebsgesellschaft für statistische Programmsysteme mbH, Goetheallee 19, D-5300 Bonn 3 / West Germany
- Jack McArdle, 1929 Lewis Mountain Road, Charlottesville, VA 22901 / U.S.A.

May 10, 1989